COMMERCIAL OPPORTUNITIES IN SPACE

Edited by
F. Shahrokhi
The University of Tennessee Space Institute
Tullahoma, Tennessee

C. C. Chao
Institute of Aeronautics and Astronautics
National Cheng Kung University
Tainan, Taiwan, R.O.C.

K. E. Harwell
The University of Tennessee Space Institute
Tullahoma, Tennessee

Volume 110
PROGRESS IN
ASTRONAUTICS AND AERONAUTICS

Martin Summerfield, Series Editor-in-Chief
Princeton Combustion Research Laboratories, Inc.
Monmouth Junction, New Jersey

Technical papers selected from the Symposium on Commercial Opportunities in Space: Roles of Developing Countries, Taipei, Taiwan, Republic of China, April 1987, and subsequently revised for this volume.

Published by the American Institute of Aeronautics and Astronautics, Inc.
370 L'Enfant Promenade, SW, Washington, DC 20024

American Institute of Aeronautics and Astronautics, Inc.
Washington, DC

Library of Congress Cataloging in Publication Data
Main entry under title:

Commercial opportunities in space / edited by F. Shahrokhi,
C.C. Chao, K.E. Harwell.

    p. cm. – (Progress in astronautics and aeronautics; v. 110) Papers
from a Symposium on Commercial Opportunities in Space, held in
Taipei, Taiwan, Apr. 19–24, 1987; and sponsored by The University of
Tennessee Space Institute et al.
    Bibliography: p.
    Includes index.
    1. Space industrialization – Congresses.　I. Shahrokhi, F.
II. Chao, C.C. (Chi-Chang), 1923-　III. Harwell, K.E.　IV. Symposium
on Commercial Opportunities in Space (1987: Taipei, Taiwan)
V. University of Tennessee (System). Space Institute.　VI. Series
TL507.P75 vol. 110　　　　　　629.1 s – dc19　　　　　88-12280 CIP
[TL797]　　　　　　　　　　　[620'.419]
ISBN 0-930403-39-8

Progress in Astronautics and Aeronautics

Series Editor-in-Chief
Martin Summerfield
Princeton Combustion Research Laboratories, Inc.

Series Editors

Burton I. Edelson
Johns Hopkins University—SAIS

Allen E. Fuhs
Carmel, California

A. Richard Seebass
University of Colorado

Assistant Series Editor

Ruth F. Bryans
Ocala, Florida

Norma J. Brennan
Director, Editorial Department
AIAA

Jeanne Godette
Series Managing Editor
AIAA

Table of Contents

Preface

The Symposium on Commercial Opportunities in Space: Roles of Developing Countries was held in Taipei, Taiwan, Republic of China, April 19-24, 1987. The Symposium was attended by representatives from twenty seven countries, eleven specialized agencies, and four international organizations. The principal sponsors were The University of Tennessee Space Institute and the National Cheng Kung University—Institute of Aeronautics and Astronautics, with the co-sponsorship of the American Institute of Aeronautics and Astronautics and the International Academy of Astronautics.

The objectives of the Symposium were: a) to examine the practical benefits to be derived from technical and scientific achievements of space research and exploration and to determine the extent to which nonspace nations, especially the developing countries, may enjoy the benefits of participation in space activities, and b) to examine the opportunities available to nonspace powers for international cooperation in space activities.

His Excellency Premier K. H. Yu and His Excellency Minister K. T. Lee opened the Symposium. These two leaders stated in part: "The exploration and the use of space has been marked by outstanding scientific and technological achievements. It is not unreasonable to assume, therefore, that the future exploration and use of space might now be marked by an increased emphasis on practical applications that will directly benefit all countries of the world, contribute to solutions of their economic and social problems, and help to bridge the gap between the industrially advanced countries and the developing nations."

In the preparation of the Symposium's summary and presentation of the selected papers for this publication, the Editors felt overwhelmed because vast canvases were covered and so much of deep significance was discussed that it seemed difficult to present an adequate summary of the Symposium in such a brief volume. The ideas that were presented on subjects related to reality, and some that still appear to be in the realm of fantasy, enriched the experience at the Symposium. The question often asked by developing countries: "Can we afford to undertake space research and activities?" was contrasted by many who asked, "Can anyone afford to ignore the applications of space research?" The applications of space research touch every facet of life so that, in adopting them with imagination and understanding, nations have an opportunity to dedicate themselves to meaningful tasks of direct relevance to their development.

The early space experiments and the new commitments from several nations have provided new opportunities for potential economic returns from space. The early success of communications satellites makes attractive the prospect of new commercial Earth observation services. Additional economic opportunities likely will arise from the utilization of the space environment for understanding materials processing, product development, and the development of space facilities.

The activities reported in this volume have been concentrated in the areas of: "generic" materials processing, effective commercialization of remote sensing, real-time satellite mapping, macromolecular crystallography, space processing of engineering materials, crystal growth techniques, molecular beam epitaxy development, and space automation robotics. These activities have proven successful in attracting new commercial coalitions by stimulating innovation in arrangements among commercial and noncommercial entities. The current facilities conduct research, development, analysis, and other necessary activities to understand and develop processes, products, services and other results having commercial space-related potential. The authors have identified other research and development areas that may benefit from the unique attributes of the space environment such as: fluid behavior and management, glasses and ceramics, combustion science, space power, space propulsion, space materials and structures, guidance and control, Earth and ocean observation, communication and data systems, automation and robotics, life sciences and human factors.

Since substantial portions of the technological base and motivation reside in the private sector, several governments have entered into transactions and have taken necessary and proper actions to achieve objectives of their national technological interests through joint action with their domestic industries. These transactions and actions include, but are not limited to, engaging in joint research programs having as an end objective enhancement of long-lead-time commercial space leadership. The benefits from the commercial use of space literally dazzle the imagination. The production of rare medicines possessing the potential of saving thousands of lives and millions of dollars, the manufacturing of superchips for the computer market, the establishment of space observatories enabling scientists to see out to the edge of the universe, the production of special alloys and biological materials, all benefit greatly from the space environment. Today, just a quarter-century after the first human being ventured into space, hundreds of corporations are directly involved in the space business. Twenty-five years from now the scale of industrial activities in space may rival that of today's computer industry. The next-century after that will bring even greater triumphs. Visionaries tell us to expect miles-wide solar arrays in space, beaming power down to Earth by microwave, mining operations on asteroids and the moon, and even perhaps a human colony in space. Space technology in the twenty-first century probably will be what aviation, electronics, and computers were in this century. It is the next evolutionary step for mankind. How fast these new economic frontiers will be opened depends in large part on the action of developed countries. Like the opening of the frontiers with railroad and the national highway and airway systems that propelled the countries into an era of rapid mobility, space ventures require investments beyond the capacity of the private sector alone.

Cooperation between nations on matters concerning space has a long and rather fruitful record, though recent trends that can lead to the extension of the arms race into outer space present a cause of great concern to the international community. It is hoped that space does not become a new arena for rivalry between nations, not only for the sake of continuing peace, but also as a stimulus to mutually beneficial development. In fact, the recognition of this necessity and benefit has led to very successful examples of cooperation such as the operational international communication and meteorological data systems. Space activities have shown a way for different countries with widely varying political systems, levels of development, and culture to work for mutual benefit. The use of communication satellites provides an example of multilateral intergovernmental cooperation. Bilateral cooperation between countries in the field of space is widespread and has had very successful results. A partial list of activities fostered through such cooperation includes: provision for launch, loan of group equipment, permitting the reception of data, exchange of scientists, rendezvous of space vehicles, integrating payloads, etc. These activities indicate the very wide range of bilateral cooperation that has and continues to take place in the field of space. An overall assessment of multilateral and bilateral cooperation in space indicates a rather positive picture and many concrete achievements. Nevertheless, it seems clear that the full potential of the existing possibilities opened up by space technology remains to be tapped. Greater benefits from space can be derived by increasing the level of international cooperation. The technologically advanced nations have special responsibilities to assist less developed countries for the benefit of mankind. The developing countries deposit their widely varying levels of economic, scientific, technological, and industrial development—recognize the similarity of their problems and the complimentarity of their needs and resources. Varying levels of scientific, technological, and industrial development can, in fact, provide the basis for mutually beneficial cooperation in space applications technology and science. It is therefore highly desirable that developing nations collectively make the most of their existing knowledge. Since the technical and economic advantages of jointly-owned systems appear obvious in many situations, the developing countries should take steps to implement cooperative programs. While short-term and immediate benefits from space technology can be expected to be reaped by outright purchase of systems and turn-key contracts, including operation of facilities, in the long run it maybe desirable—even essential—for each country to have its own pool of knowledgeable experts. The size of this pool, the depth of knowledge, and breath of disciplines required would vary from country to country.

Governments must invest substantially in order to create the infrastructure needed to exploit space. Consistent with the free-market philosophy, most of this infrastructure must be turned over to the private corporations in order to encourage business enterprise in space. The following paragraphs survey the new technologies and industries that lie over the horizon.

The first industry in space, satellite communications, is already mature. This industry generates some $5 billion a year from the transmission of radio and television broadcasts, telephone conversations, electronic mail, and business data. Future satellite communication from geosynchronous orbit presents opportunities for advances in communication technology through the development of large communication platforms, higher frequency systems, and in dynamic and directed techniques such as beam switching. All of these new technologies give rise to potential opportunities for commercial exploitation. In addition, satellite communications opportunities exist in the development and marketing of new and innovative services not currently available to the public. Some examples include: direct audio broadcasting; the provision of satellite communications services to automobiles, trucks, emergency service units, and aircraft; the use of low-orbiting satellites to provide "electronic mailman" services to remote areas; and to provide for the location and tracking of very simple, low-cost transmitters mounted on commercial rail, maritime, and trucking fleets. In addition, servicing, transportation to and retrieval from geosynchronous orbit, refueling, etc. all offer opportunities for future economic development.

The second industry to appear on the scene, remote sensing, is at least a decade behind the first. The satellite-relayed weather maps and information presented on local news programs, which have dramatically improved forecasts and reduced deaths from hurricanes and the like, are an obvious example. Remote-sensing satellites can also detect air pollution (such as sources of acid rain) and measure ozone and other critical atmospheric elements. An important direction of experimental and future operational meteorology will be the monitoring of slowly changing atmospheric variables and implications of these changes for long-term weather patterns. Satellite sensors are being developed to recognize global albedo and the amount and distribution of carbon dioxide, chlorofluorocarbons, ozone and dust. These factors are affected by man's activities, and long-term changes may give rise to changes in weather patterns with harmful environmental and economic effects. Satellite sensors provide the least-cost means of accurately monitoring such global phenomena, and the future development of such sensors should be continued.

From a commercial standpoint the most promising use of remote-sensing satellites is in providing images of the Earth's surface. The information gathered from the five Landsat satellites launched since the early 1970's and the recently launched SPOT have proved invaluable in the mapping of remote areas. They have revealed unknown lakes, islands, and underwater shoals and reefs. They have been used to map routes for railroads, pipelines, and electric-power lines; to provide population estimates; and to guide ships through iceberg-infested waters. The most extensive commercial uses of Landsat and SPOT data are in mineral exploration and agriculture. By studying satellite images of known oil, gas, and mineral deposits, scientists have learned to identify features that might point to new deposits. Oil companies look for folds and domes capable of trapping oil or gas. Mineral experts search for deposits. Landsat's thematic mapper records the reflection of light off the Earth with a sensitivity to color far surpassing that of the human eye; by enhancing colors digitally, explorers can detect features they would have missed if they had walked the land themselves. In agriculture satellite images are used chiefly for crop forecasting to estimate the international supply and demand.

Position determination in real time for navigation and with an acceptable time delay for geosciences is an increasingly important requirement in a wide range of human activities. Position measurements by satellites provide geocentric coordination of surface points and thus data on physical movements of the Earth's surface without assumption and hypothesis and internal structure of the Earth. Beside this, satellite technology has demonstrated its unique capability for high-precision geodetic measurements and navigation on a world-wide scale, covering also with the same accuracy the vast ocean areas. Other important tasks of geodynamics, in particular direct measurement of distances of plate-tectonic layer movements with sufficient accuracy, monitoring variations of polar motion with very high accuracy, the study of terrestrial tides and monitoring the Earth's rotational period with an accuracy down to 100 microseconds, can only be solved by means of modern satellite and space technology. Navigation and geodesy are important applications of space technology. Navigation has always been of great importance to man, and while navigation errors have sometimes led to unexpected discoveries, they have more frequently led to disaster. In the modern world, accurate navigation is becoming increasingly important not only for safety, but also for optimizing steadily growing traffic flows and minimizing fuel consumption through appropriate routing. Reliable position determination is also needed for a variety of other economic activities—e.g., off-shore oil drilling and sea-bed mining—or scientific studies related to geodesy, plate-tectonics, geodynamics, etc. Although some of the scientific studies require very accurate geodetic measurements using satellite laser radar or very long baseline interferometry, almost all safety and economic activities can effectively be carried out by using satellite navigation based on Doppler measurements. While much of the work in this area is currently undertaken by the governments, there is an increasing need to move proven remote-sensing technology to the commercial sector and to involve the private sector in research and development in such a manner to facilitate future commercial ventures in space. Research in sensor technology to define configurations optimized for commercial payloads represents a major area that might be addressed.

The economic potential of communication and remote-sensing satellites not withstanding, there is a feeling among space buffs that the real revolution will come only when we are actually manufacturing in space. Space-based "materials processing," as the production of materials as diverse as drugs, alloys, and crystals is called, may be more important commercially than genetic engineering. On Earth, gravity influences every physical process, however minutely. When two metals of different densities are mixed to form an alloy, gravity causes nonuniformity so that by-products are not as strong or durable as expected from the theory. In a microgravity environment, this effect is avoided so that superior alloys can be produced. Corporations can also exploit the environment of space to learn how to refine the manufacture on Earth of such heavy products as steel and lead alloys, and complex three-dimensional structures. Given the cost of space transport, probably only products of extremely high value per pound, such as pharmaceuticals and crystals, will actually be manufactured there for some time. Later we can expect to see manufacture of complete assemblies in space, including some with novel structures and interconnections. Some such as mokainase (an enzyme that dissolves blood clots), can be readily produced in microgravity but can be produced on Earth only in minute quantities and at great expense.

Space provides an environment that cannot easily be reproduced on Earth: a virtual absence of gravity (microgravity), access to that part of the cosmic spectrum of radiations which cannot so easily be sampled and studied on Earth and a virtually infinite source of near-vacuum. There are many well-known effects of gravity in fluid systems including buoyancy, convection, sedimentation and segregation. In a microgravity environment, the relative importance of convection, diffusion and surface forces may be changed drastically, leading to experiments of broad significance. In space, crucible-free forms—melting of single crystallive materials or metals is possible without restriction by grivity effects. Foam metals become possible.

The search for high-temperature superconductivity with a novel superconducting mechanism is one of the most challenging tasks of condensed-matter physics and material sciences. To obtain a commercially-usable superconducting state above the technological temperature barrier of $77°K$, (the liquid-nitrogen boiling point), will be one of the greatest triumphs of scientific endeavors.

An obvious advantage of manufacturing in space is that molds and container walls are not necessary. Crystals can be grown without touching a surface and absorbing alien molecules. The first commercial products manufactured in space, in fact, were tiny spheres made for the National

Bureau of Standards. The bureau sells "standard reference materials": used in a device for measuring microscopic electronic components, for instance, or for calibrating instruments like filters and porous membranes, which measure tiny spheres. The spheres—droplets of polystyrene—are 1/2500th of an inch in diameter, or about the size of a red blood cell. Businesses will use the spheres in counting blood cells, measuring particulate pollution, producing finely ground products such as paint pigments and chemicals. This was the first product manufactured in space for private enterprise. Numerous genetic engineering, culturing, fermentation and separation technologies are being used in ground-based biological processing. However, the combination of separation, resolution and throughput is significantly hampered by thermal convection, sedimentation, gravity-driven flows, and ensuing turbulence. Processing of biological materials to obtain specific cell types, all components, hormones, antigens, proteins and other organic substances to obtain greater purity and throughput in the microgravity environment shows significant promise. Continuous flow electrophoresis and isoelectric focusing are but two separation methods that benefit greatly from the low-gravity environment.

Space-processing also seems very promising for the production of pharmaceuticals such as erythropoietin to stimulate red-cell production in fighting kidney and blood diseases, antihemophiliac factors to clot blood in hemophiliacs, mokainase to reduce clotting in stroke and phlebitis victims, and beta cells which might provide a single-injection cure for diabetes. Electrophorectic chambers for the experimental processing of such biological products will be flown as established space facilities. Important experimental facilities already exist or are being developed in many countries for research in: basic physiological phenomena with respect to the three main sensing systems (vestibular, somatosensory, and visual) related to spatial orientation, control posture and locomotion; cardiovascular phenomena and conditioning of the cardiovascular system during prolonged exposure to weightlessness; effect of microgravity on cell-prolification kinetics; and cell and molecular biology. Established space facilities are important to the scientists in developing countries that do not now have the capability of setting up their own experimental facilities for such work in a cooperative and coordinated manner. The result thus obtained would be useful for clinical medicine and public health care. Development of this area of applications has enormous implications in the health care field. The shopping list in this category may include interferon, shin-growth agents, and a potential treatment for emphysema.

Crystal growth is one of the promising commercial opportunities in space. One of the first products will be crystals made from gallium and arsenic, two soft metals. Gallium arsenide crystals that conduct electrons ten times faster than silicon can be used in computer chips, lasers, switching devices in fiber-optic systems, high-frequency antennas, and solar-power arrays. One of the particular characteristics of these materials is they are quite resistant to radiation and heat. Crystals are normally grown in long cylinders. After they reach a diameter of three to four inches, they are sliced into the thin wafers used to manufacture chips and other devices. In the microgravity environment, with the lack of buoyancy in the fluid and solidification processes, larger, more homogeneous crystals with far fewer dislocations are possible.

To cash in on materials processing, one needs more than the shuttle. As we progress to a larger and larger system in space, the demand for economy becomes paramount. If larger space structures are to be economically competitive with terrestrial systems their costs and costs of transporting them to low Earth or geosynchronous orbits must be minimized. Specified performance at minimum cost will dictate the way in which large space systems are designed and made. The large number of missions, and the growing proportion of these missions with a more or less "commercial" character, have made economic factors increasingly important. An infrastructure of support facilities and services is needed. The governments will provide the infrastructure, but on a long run the emphasis will shift to the private sector. The private sector has begun the manufacturing of small rockets that would launch satellites that the shuttle can't or shouldn't handle. Other parts of the private sector are developing small space laboratory platforms, which may be used on a lease basis for early production. These initial activities are precursor to the space station, and that remains a government project. A government-built space station may or may not be necessary, but clearly it would accelerate the development of space commerce—just as government subsidized railroads and highways have accelerated economic development in years past. A space station would make possible permanent crews of astronauts, whose activities will range from routine manufacturing operations to the repair of any malfunction, and retrieval for repair and upgrading of satellites. This will improve the economics of any investment dramatically. When discussions move from space stations and orbiting platforms to the next cycle of innovations, the true visionaries take over. Besides solar-powered satellites they dream of nuclear-waste disposal in space, mining operations on asteroids rich in precious minerals, and solar sails driven by photons from the sun. Such a trend would create a new wave of economic activity based on the production, launch, assembly of vast structures, and services necessary to support space manufacturing such as power generation, orbital refueling, food supplies, health care and so forth.

How real is the economic promise of space? According to the American Institute of Aeronautics and Astronautics (AIAA), space is a $22–23 billion industry today, coming from the satellite communication, remote sensing, military and other governmental agencies. Satellite-communication revenues are growing at a rate of 20 percent a year. Therefore, the projection of annual revenue of $40–100 billion serves reasonable by the turn of the century. The materials processing could be worth another \$10-20 billion, launch services \$5 billion, military \$25 billion and NASA \$8 billion, including remote sensing and other related activities, revenue of \$100-200 billion can be projected. In the end the important question may be not how abundant the fruits of space commerce will be but who will enjoy them, and how soon.

Space commercialization and technology now spans a vast range of complexity from fairly common and comparatively simple APT receiving stations to extremely complicated fully-automated space stations. It also encompasses equipment whose costs range from a few hundred dollars—for a television direct reception system, for example—to many millions of dollars. The investments required to develop and produce such equipment differ by many orders of magnitude, as do the infrastructure and technical skills involved. The variety of space activities now feasible for commercialization is quite substantial. It is within this almost bewildering array of possibilities that a country must make choices about what applications and which technologies it wishes to pursue. The choice must be determined by: the needs, priorities, feasibility to meet the needs and priorities with due regard to other needs, financial resources, technical and managerial infrastructure and human resources, and recognition of the rights of other countries. It is therefore obvious that there can be no fixed formula of universal validity, i.e., the benefits of application will vary from situation to situation and from country to country. Many of the developing countries may not themselves have all the expertise necessary for such interdisciplinary studies. In many cases, assistance by the developing country may range from problem analysis and system studies through fabrication of hardware. Most countries—developed as well as developing—face a number of difficulties with regard to space commercialization. These difficulties might spur innovative approaches. Some of these difficulties, and suggestions on how to overcome them, are given here, contributing to increased self-reliance and socio-economic advancement. These difficulties, in part, are: financial and industrial resources for development, fabrication and operation of space hardware; development of high-caliber manpower; equipment and its suitability thus assuming return on investment; the strategy of maximizing value-added profits; transfer of technology from laboratory to industry and from country to country; continuity, compatibility, and complimentary—a major impediment for developing countries participation; availability of data and information—access to needed information is an invaluable aid to the speed of appropriate space technology.

It has been stated that space technology is not the solution to all problems and that—depending on actual situations, needs and resources—conventional means might in many cases be a more suitable answer. Nevertheless, there is little doubt that space technology can lead to economic growth and overall development. If, however, it continues to be dominated by a few nations who invest in it, there is a danger that it may result in widening further the chasm between developing and developed nations. At the same time, this very stimulus of space technology has the potential to help the developing countries narrow the gap and to accelerate the process of development along the path of their choice. The countries must be encouraged to participate in various space applications so as to derive the fruit of space technology.

F. Shahrokhi
C. C. Chao
K. E. Harwell
September 1987

Chapter I. Space Platforms

Perspectives in Space Transportation

J. A. Vandenkerckhove
VDK System, 1150 Brussels, Belgium

INTRODUCTION

Many articles have recently been published comparing the existing or planned launch capabilities. This paper will take a more general approach. The existing or planned capabilities of the U.S. and of Europe have just been presented by the two previous speakers; the U.S.S.R. has a well diversified stable of expendable launchers and may soon flight test a reusable and manned vehicle in the SHUTTLE category, as well as a Heavy Lift Vehicle in the Saturn category. Both Japan and the Republic of China have successfully launched Medium Lift Vehicles which already take advantage of rather advanced technologies such as cryogenic propulsion and are obviously aiming at an operational capability which China has almost got already with LONG MARCH 3 and Japan is planning in 1993-1994 when the H-2 rocket will be available.

Other capabilities are likely to emerge: India has significant ambitions and other countries, in particular Pakistan, Corea, Brazil and Argentina are known to be interested. Commercial as well as military perspectives, exacerbated by the current shortage of launch capability, supplement national prestige, a situation which has been sarcastically characterized as follows by Mr. Beggs, a former NASA Administrator: "The launch vehicle business seems to be the 1980's equivalent of the steel mills back in the 1950's when every country of any size had a steel capacity. These days it appears that every country that wants to be in the forefront wants to have a launch vehicle capability".

The analogy between launchers and steel mills is most
pertinent. When the current backlog will have been resorbed,
the emergence of a definite overcapacity crisis: indeed
with ARIANE hopefully back to business soon, at least three
or four U.S. systems in operation by the turn of the decade,
China and later Japan most probably succeeding to capture a
share of the market and with the U.S.S.R. trying to do the
same it is clear that the capabilities will by far exceed
the commercial requirements which are estimated to be of
the order two dozens of satellites per year.

Mr. Beggs analogy stops here because steel is a much
more reliable product than rocket launchers. The formidable
industrial effort which in Europe is in support of the
ARIANE program and which unfortunately has not prevented
last year failure (similarly NASA's famous "marching army"
did not prevent the accident of Challenger), the magnitude
of the efforts required for maintaining a reliable operational capability is often underestimated and dispropor-
tionate with the modest launch rates which may result from
overcapacity.

Synthetically very high transportation expenditures
(not necessarily costs in the forthcoming competitive environment) could prevent the growth of space activities in
the medium term. This paper will discuss possible ways to
get out of this dilemma.

ROCKET PROPELLED LAUNCHERS

So far space activities can only be carried-out by
means of rocket propelled mutistage vehicles launched from
a very limited number of sites requiring huge facilities
and along strictly constrained trajectories. In addition
space transportation remains prohibitively expensive
(typically 4000 AU per kilogram to Low Exith Orbit: LEO)
and recent events have unfortunately shown that reliability
can still not be guaranteed. The objectives for future
developments must clearly be:

- RELIABILITY

- ECONOMY

- FLEXIBILITY.

Some prospects do definitely exist for improving substantially the next generation of rocket launchers through
simplification combined with utilization of more advanced
technology but simultaneously requirements on performance

become more and more demanding so that ultimately only a marginal improvement will likely be achieved. The following three approaches, which can be pursued for improving future transportation systems, have definite limitations because the above three objectives are conflicting with one another. They are

i) INCREASED REDUNDANCY, HIGHER MARGINS and MORE STRINGENT QUALITY CONTROL which as demonstrated by air transport can very much improve reliability but would also significantly penalize economy and flexability.

ii) HEAVIER LAUNCHERS which have the potential for significantly improving economy since as a rule of thumb the specific costs per kilogram of payload only increases as the cubic root of scale. Outside the U.S. and the U.S.S.R. a firm requirement for very heavy rockets has not yet been established since projected traffic remains limited.

iii) REUSABLE LAUNCHERS which in theory have the potential for significantly improving economy as well as flexibility since they are heading towards horizontal take-off and launching. In practice, however, this potential is very difficult to harness because reusability translates necessarily into additional masses for wings, landing gears, extra propellant and the like. Reusability of NASA's Shuttle has obviously not permitted to improve economy, at the contrary, and great caution must be exercised in assessing its benefits for the future, at least for only a modest launch rate.

In practice, however, the debate on reusability is somewhat academic since in the future a fair percentage of the missions, in particular all manned ones will require safe return to the ground.

Synthetically it appears that it is the MARGINALITY INHERENT TO ROCKET PROPELLED VEHICLES, in particular the rather low specific impulse which can be obtained event with the most energetic propellants, which severely limits the prospects for improving space transportation, and in particular would make rather risky the development of a Single-Stage-To-Orbit, otherwise a most attractive approach for its utmost simplicity.

In addition the issues discussed in the following paragraphs will also influence the characteristics of future transportation systems.

THE NUMBER OF DIFFERENT SYSTEMS: Whilst two or more systems, each optimized for covering different requirements, are probably well justified to support a very large

space program (i.e. that of the U.S.A. or the U.S.S.R.)
which in addition would reduce their vulnerability in case
of problem, more modest programs, one may hardly be able
to afford two systems.

GEOGRAPHY introduces significant restraints upon space
transportation systems. On the one hand only a few sites
are suitable for launching rocket vehicles and even so the
orbits which can be achieved from them are severely constrained by the disposal of the expended stages and other
safety limitations and on the other hand most countries
having a space transportation capability are located at
relatively high latitudes which forces either to accept a
substantial penalty in performance or to build and maintain
at rather high costs a launch site located near the equator. Futher, for reasons of economy, the reusable elements
should also, insofar as practicable, be capable of return-
ing directly to base at the end of the mission.

AIR-BREATHING LAUNCHERS

The development of a reliable, economical and flexible
reusable space transporation system sufficiently flexible
to carry-out most missions of importance and capable to
operate from a domestic base appears hardly within reach
with only rocket propulsion which requires to embark on-
board both the fuel and the oxydizer. Quite naturally,
therefore air-breathing propulsion currently arises much
interest for it offers the prospect of eliminating largely
the marginality inherent to rocket propelled vehicles
thanks to the oxygen freely available in air (for a rocket
the oxydizer represents typically 80 to 90% of the propell-
ent mass). Whilst the potential of breathing propulsion has
been recognized since the beginning of space flight, until
recently it was not considered as a practicable alternative
as long as the Mach number reached during the air-breathing
phase could hardly exceed three to four. Today technological advances, in particular in the fields of high temperature materials, ultra light structures and computational
fluid dynamics offer the prospect of extending considerably
this limit and to develop a rather efficient space transportation system, which explains the surge of interest of
the last three years (cf. studies on HOTOL and SANGER concepts in Europe and in the U.S. AEROSPACE PLANE project
through which U.S.A. is tending to dominate the scene
through a massive expenditive of resources; 450 M$ over a
initial feasibility assessment phase of 42 months).

Whilst the prospects of an air-breathing space transportation system, for the longer term beyond the turn of

this century, appears definitively promising, it is nevertheless necessary to recognize a few hard realities:
- at this point in time, the feasibility of developing
  an efficient air-breathing vehicle, cannot be taken
  for granted.
- much research, development and even perhaps feasibility demonstration will be necessary before committing
  a full scale operational development.
- fundamental options do exist, relying on widely different technologies and the selection of the best will
  be difficult and risky.

The rewards, however, could be enormous resulting from
easier and cheaper access to space as well as from important synergies with other sectors, in particular air transport (cf. ORIENT EXPRESS concept for hypersonic transportation but also improvements derived from applying advanced
technology to subsonic airplanes).

MAJOR AIR-BREATHING OPTIONS

GENERAL

To take advantage of air-breathing propulsion, it is,
by necessity, required to fly in the atmosphere at hypersonic speed which creates two severe problems:

- on the one hand the aerodynamic losses are much higher than the sum of gravity and aerodynamics on a
  rocket (for climb and acceleration only, they may represent typically 50% of the Delta V actually provided instead of 15%) and

- on the other hand aerodynamic heating becomes very
  severe, especially if sustained flight is contemplated.

In fact, the design of the first stage and of its
motors will be predominently influenced by thermal considerations: the temperature of the outer skin will be kept
acceptable by selecting a suitably high altitude allowing
radiation cooling to cope with kynetic heating but, of
course, this will require to increase the lifting surface
and the engine intakes, thereby making it more difficult
to achieve an ultralight structure. Further should sustained cruise be contemplated active cooling of critical elements will likely be required which will considerably
narrow down the choice of fuel (Hydrogen, Methane, ...).
The thermal problem is much more critical, still, for the
engines because radiation cooling is not available.

PROPULSION ASPECTS

The utilization of the following four classes of air-breathing motors, can be contmepleted for propelling a space transportation vehicle during the first phase(s) of its flight:

AIR-TURBO-ENGINES

This category which includes the TURBOFAN and the TURBOJET as well as the AIR-TURBO-ROCKET (Figure 1) is equipped with a rotary compressor and it is therefore capable of statically ingesting air, which allows for its use from take-off onwards.

In a TURBOFAN in which a ducted propeller is driven by a gas turbine and in a TURBOJET in which thrust is generated by the gas turbine exhaust itself the highest temperature is reached at the inlet of the turbine and increases with both flight Mach number and compression ratio. Figure 2 represents its variation with Mach number for rather low compression ratios. It shows that if, for instance, blade material does not allow to exceed $1850^{O}K$ at the turbine inlet (which likely represents today's state-of-the-art)

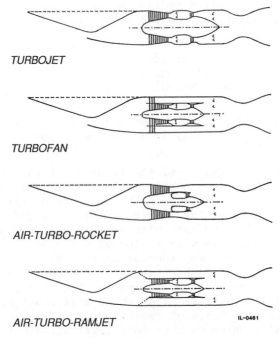

TURBOJET

TURBOFAN

AIR-TURBO-ROCKET

AIR-TURBO-RAMJET

IL–0461

Fig. 1 Propulsion Cycle for Launch.

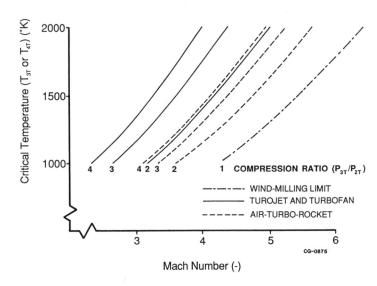

Fig. 2 Turbine Material Temperature Limit.

then it will not be feasible with a compression ratio of
only 2 to fly above Mach 4.75. Even this "rather modest"
speed would require to improve substantially the technology
of the compressor currently limited to 1100°K which would
correspond to Mach 3.88.

Two possibilities exist, however, for increasing the
flight Mach number:

i) "windmilling" the gas turbine above a certain speed,
   say Mach 3, in order to avoid transferring energy to
   the flow within the compressor and beyond it burning
   fuel to drive the turbine, or

ii) to use an AIR-TURBO-ROCKET, in which the compressor
    in the air flow is driven by a turbine fed by a
    rocket motor which allows to control turbine temper-
    ature independently of flight number

Figure 2 also represents the maximum cycle temperature, for
these cases which is the temperature at the compressor outlet, as a function of flight Mach number. It shows that,
for instance, with compressor technology allowing to go as
high as 1850°K (i.e. relying on turbine technology or using
very advanced materials) it would theoretically be possible
to reach Mach 5.40 with an AIR-TURBO-ROCKET with a compression ratio of 2 and Mach 6.15 through "windmilling" any
of the three types of AIR-TURBO-ENGINES.

Cooling of part of the whole air-flow, or of critical
elements of the gas turbine would of course permit to low-

er somewhat the temperatures and therefore to increase the
flight Mach number, but of the cost of a significant increase in complexity and mass.

RAMJET

In a RAMJET the air is slowed down to subsonic in the
inlet diffuser, with a consequent rise in pressure which is
sufficient, after combustion, to produce thrust through
expansion in the nozzle without need for a gas turbine.
However, the RAMJET, which is unable to ingest air statically, cannot operate at low subsonic speeds. The RAMJET,
without rotary elements (except the fuel pump), is well
suited for propulsion at supersonic and low hypersonic
velocities. However above Mach 6 and 7 the engine becomes
excessively bulky and heavy due to the dimensions of the
diffuser required to slow the air down to subsonic and to
the consequent rise in pressure.

SCRAMJET

In a SCRAMJET the slowing down of the air in the diffuser is limited and the combustion takes place at supersonic
velocity, which allows the burning of more fuel (before
reaching very high temperatures at which dissociation becomes significant) and ensure great compactness.

However a SCRAMJET cannot operate below the Mach 5 to 6
region.

Theoretically SCRAMJETs should be able to operate up to
very high Mach numbers, at least 10 to 15, and according to
some experts, perhaps even beyond 20 or more, but this can
only be demonstrated in flight since it is unlikely that
test facilities can be built to be capable of realistic
simulation much above Mach 8. The main objective of the
AEROSPACE PLANE in the U.S. is precisely to demonstrate in
flight the feasibility of using SCRAMJET's beyond the
conditions which can be simulated on the ground. SCRAMJETs
are still very much in research stage and require the satisfactory solution of some very difficult issues such as
the very short time available for combustion (of the order
of the millisecond), the extreme heat transfer to which the
combustor is subjected, the need to adapt internal geometry
to flight Mach number, the large air intakes required
(which may represent 50% of the vehicle cross-section) and
the integration of the engine in the frame.

COMBINED ENGINES

Each type of air-breathing engines being best suited or
simply capable to operate within a certain Mach region it

could be necessary to combine several cycles into a single engine for being able to cover efficiently the spectrum of flight conditions, from take-off to separation of the upper stages. The following three types can be identified.

The AIR-TURBO-RAMJET also represented on Figure 1, which operates as a TURBOJET or TURBOFAN up to a certain speed above which the gas turbine is bypassed and the engine operates in RAMJET mode, which should permit to increase up to 6 and perhaps 7 the maximum velocity.

The AIR-RAM-ROCKET or ducted rocket in which a core rocket motor provides most of the thrust for take-off and early acceleration but also works like an ejector for ingesting air, augmenting thrust and improving specific impulse taking advantage of the fact that the rocket exhaust is fuel rich.

The AIR-BREATHING-ROCKET (also call CRYOJET) in which air is cooled down to very low temperature in a heat exchanger, perhaps even liquified, and pumped directly into the combustor of the rocket for being burned with the fuel, almost certainly hydrogen which has first been used for cooling the air. Above the atmosphere liquid oxygen is injected in the chamber instead of air to operate in rocket mode.

Synthetically

A) a pure AIR-TURBO-ENGINE of very advanced design should be capable of propelling a vehicle up to Mach 5 or slightly beyond

B) an AIR-TURBO-RAMJET might be capable of propelling efficiently a vehicle up to Mach 6 or slightly beyond and

C) the SCRAMJET is the only air-breathing motor capable of propelling a vehicle beyond Mach 6 to 7.

In the later case propulsion during take-off and early acceleration would require either the use of a separate set of AIR-TURBO-ENGINES or that of a combined AIR-RAM-ROCKET capable to change its geometry for being adapted to each flight regime.

S.S.T.O. versus T.S.T.O.

An essential choice is between a Single-Stage-To-Orbit: S.S.T.O. which would be the least complex and a Two-Stage-To-Orbit which may in certain cases be much preferable to avoid marginality and reach great flexibility. A simplified parametric comparison already shows interesting results.

First a S.S.T.O. and a T.S.T.O. were considered both capable of bringing a payload into a 300km sunsynchroneous orbit and both having to transition from air-breathing to rocket mode at Mach 5 and 30km (which corresponds approximately to 800°K for the critical parts of the outer skin. The comparison was performed in terms of propellant ratio (i.e. ratio of lift-off mass to payload mass) the independent parameter being the extra ON DEMAND VELOCITY INCREMENT available beyond the Delta V required from the flexibility of the system. The design of a stage is characterized by its propellant ratio LAMDAi (i.e. the ratio of propellant mass to the sum of propellant and inert mass, payload mass excluded) and by its specific impulse ISi. The result is represented in Figure 3 which clearly shows that for a transition at Mach 5 a T.S.T.O. can be expected to be very flexible, capable in particular to provide an extra

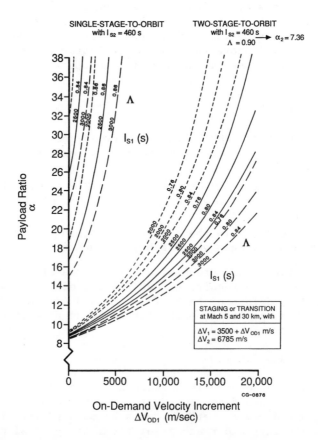

Fig. 3 Performance Summary for a Transition at Mach 5 and 30 Km.

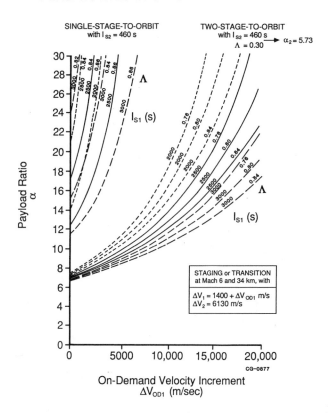

Fig. 4 Performance Summary for a Transition at Mach 6 and 34 Km.

Delta V of the order of 10 000 m/s sufficient for performing a very long cruise, whilst a S.S.T.O. would provide extremely little flexibility and could even be considered as
definitively marginal as indicated by the steepness of the
curves.

   The comparison can be repeated from a transition at
Mach 6 and 34km; the result is represented in Figure 4
which suggests the same conclusion but also shows that a
rather significant improvement in payload ratio can be expected from increasing the Mach number at transition.

   As flexibility of the second stage is as important,
perhaps even more, than for the first stage (since many
missions will require to go above 300km, preferably without
using an additional stage) the analysis was repeated for
several values of the on-demand velocity increment available
in the second stage; the result represented in Figure 5
shows that the flexibility of a T.S.T.O. is not limited to
its first stage.

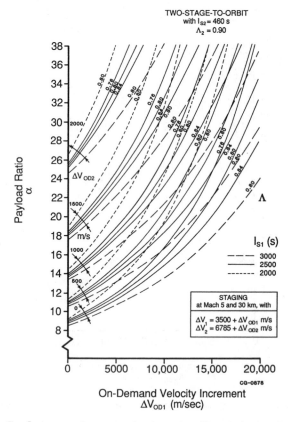

Fig. 5 Performance Summary showing the flexibility of a T.S.T.O.

These results suggest that an air-breathing S.S.T.O. would be rather marginal with its transition in the Mach 5 to 6 region require whilst a T.S.T.O. with similar characteristics carries the prospect of a significant improvement in both flexibility and performance.

Finally the case of a SCRAMJET propelled launcher can also be tentatively assessed. The performance was found to be rather sensitive to specific impulse and aerodynamic losses during the initial phase, which suggests to consider a high acceleration provided by air augmented rocket motors, followed by a pure air-breathing phase and finally a pure rocket phase beyond the atmosphere. Variable internal geometry might permit to build a suitable AIR-RAM-SCRAM-ROCKET.

Under the assumption that the pure air-breathing phase would extend from Mach 3.5 to 16.5 the performance of an

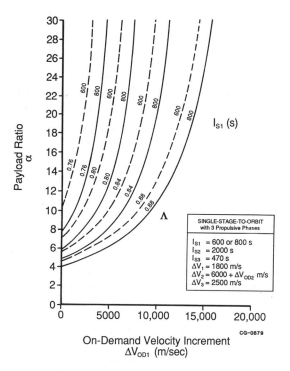

Fig. 6 Performance Summary of a hypothetical T.S.T.O.

hypothetical T.S.T.O. can than be assessed. The result is
represented on Figure 6 which shows that an S.T.T.O. powered by an AIR-RAM-SCRAM-ROCKET might provide superior performance and simplicity whilst remaining rather flexible.
It is no wonder that the ultimate objective of the AERO-
SPACE PLANE is the development of a SCRAMJET powered
S.S.T.O. but the challenge is formidable and the risks are
enormous.

DISCUSSION

The above analysis suggests to concentrate further
analysis on the following options, namely
   - a T.S.T.O. with transition between Mach 5 and 7,
     possibly similar to the Sanger concept, and
   - a S.S.T.O. capable to operate in pure air-breathing
     mode well beyond Mach 10.
The development of the former would be a considerably
less tisky undertaking than that of the later but the
successful development of the later, if at all feasible,
would be much more rewarding.

U.S. Laboratory Module: Its Capabilities and Accommodations to Support User Payloads

Luther E. Powell,* Walter V. Wood,† and Charles R. Baugher II‡
NASA Marshall Space Flight Center, Huntsville, Alabama

Abstract

This paper describes the U.S. Laboratory (US Lab) Module, as defined during the recently completed Phase B studies, with emphasis on user accomodation. The US Lab Module is one of five pressurized modules joined together by four resource nodes to form the manned core of the Space Station. The US Lab is to provide a permanently manned laboratory for interactive research and development activities in the low gravity enviro-ment of space. Accommodations and support for a wide variety of Materials and Life Sciences payloads and operations are envisaged. Materials experiments and process development activities to be conducted in the US Lab are expected to lead to a better understanding of material properties and to space-based commercial materials processing operations. Life sciences research will examine the effect of variable gravity on fundamental biological processes. A brief overview of the total Space Station configuration is presented to show the relationship of the US Lab to other elements. Configuration and subsystems capabilities including power, thermal, environmental control, data management, vacuum system, process materials management, and laboratory support equipment planned for the US Lab are described. An overview of experiment operations to be performed in the US Lab is described and contrasted with those

---

*Manager, Space Station Projects Office.
†Manager, Laboratory Module Project Office.
‡Study Manager, Microgravity and Materials Processing Facility.

16

in other manned space laboratories. Design and operational
approaches are presented.

Introduction

The United States has identified the construction and
operation of a permanently manned International Space Station (Fig. 1) as its next major goal in space research.
After more than two decades of consideration, study, and
conceptual design, the National Aeronautics and Space Administration (NASA) was formally committed, by President
Reagan in his January 25, 1984, State of the Union Address
to the U.S. Congress, to initiate the beginning phases of
the project. In the subsequent three years, NASA has worked
to focus the collective experience and talent of virtually
all elements of the U.S. Aerospace industry and space science community toward developing thepreliminary design of a
multiuser facility that could satisfy the requirements of
diverse disciplines (from Laboratory science to satellite
servicing) while maintaining the flexibility to accept new
and unforeseen missions over a thirty-year lifetime.

To accomplish this task the designers have been able to
draw on an evolutionary history of experience, which began
with Skylab, the world's first Space Station, progressed
through the use of the Shuttle for operations and experiments requiring manned involvement and quick implementation,
and finally achieved its highest level of accomplishment in
flights of the cooperative NASA and European Space Agency
(ESA) Spacelab. Each of these flight opportunities has contributed valuable and specific information, lessons, and
insight into the task of specifying and designing a functional and operational Space Station.

In particular, Spacelab demonstrated the absolutely
necessary utility of scientists and astronauts as interactive investigators in the weightless and isolated environment of an orbiting laboratory, both as scientists in re-
directing and adjusting experiment hardware. Also on Spacelab, it has been possible to further the evolution of the
research potential of the space environment through the
implementation of the "facility" approach to experiment
hardware development and investigator organization. In this
approach, a team of investigators defines and guides the
development of a major, general purpose apparatus that can
support a family of research objectives, and that subsequently can be utilized in succeeding flights by any qualified investigator who proposes a useful investigation. At
the same time, and in spite of its success as a scientific
research tool, the Spacelab experience has emphasized the

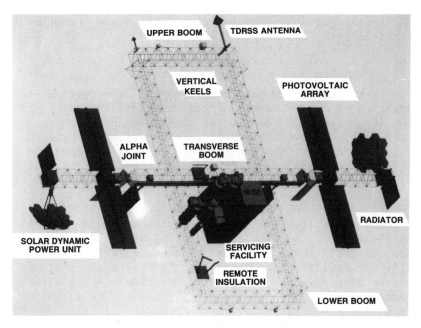

Fig. 1 Final assembly configuration.

necessity for a much longer operational interval in orbit
to achieve a reasonable level of return for the invested
effort.

Therefore, the Space Station has become a natural se-
quel to follow the research path laid down by Skylab, the
Space Shuttle and Spacelab. Two decades of experience have
demonstrated the utility of a facility that can support
continuous laboratory activities in the space environment,
serve as a platform for terrestrial and astronautical observations, and above all, allow for the presence of man to
operate, repair, replace, and utilize the scientific devices necessary to exploit the advantages of space.

Laboratory Design Approach

At the initiation of the Space Station design effort it
was envisioned that the United States portion of the Station
would contain two seperate laboratories, one dedicated to
material research and the other to life sciences. Because
of its lead in the development of Skylab, its central NASA
role in Spacelab operations and payload development, and a
predominant capability in the application of the microgravity enviroment into research in the science and

commercial applications of material processing in space, the
Marshall Space Flight Center (MSFC) at Huntsville, Alabama,
was assigned to lead the NASA Space Station design of the
manned materials research laboratory. The Goddard Space
Flight Center was assigned responsibility for the life-
science laboratory. Subsequently, and well into the initial
design of the elements, the decision was made to consolidate
the requirements of the life-science laboratory into a single United States' laboratory element under MSFC direction
and into the newly added European and Japanese elements
under their respective direction. The pattern for connecting these module elements is shown in Fig. 2.

To accommodate this addition discipline into the MSFC
design effort, the MSFC laboratory element was renamed from
the Material Technology Laboratory to the U.S. Laboratory
(US Lab) and the overall design was reviewed with the life-
science community to ascertain that it would satisfactorily
meet the requirements for life-science research. In general,
it was found that the design required only minor adjustments
to be fully compatible with this additional set of research
requirements and these were added. The major exception arose
from the life-sciences requirement for access to a variety
of centrifuge facilities for variable gravity studies. Two
separate facilities, a 4 m centrifuge and a 1.8 m centrifuge, were identified and specially studied during the design. It was concluded that the 4 m facility would require
separate accommodations in an attached node, while accommodations for the 1.8 m facility could possibly be accomplished in the US Lab Module if the system could be designed
with sufficiently small dynamic disturbances to adjacent

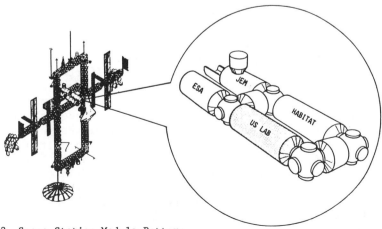

Fig. 2 Space Station Module Pattern.

experiments. Beyond this, and because the initial design of
the laboratory was derived directly from the requirements
of the material-sciences community, the discussion that
follows will be generally material-science oriented, even
though it is likely the US Lab will accommodate investigations outside of this discipline.

Within the design challenge of the Space Station, the
challenge of designing and implementing a permanently manned
research laboratory in space is possibly the greatest of
all. From the outset, the objective of the research community was to obtain an orbiting facility that would maintain a
microgravity environment while allowing a freedom of operation that closely approximated the freedom of a ground-based
laboratory facility. This laboratory must be flexible enough
to accommodate a diverse range of scientific instrumentation; it must provide a scientifically productive combination of support facilities; it must be able to evolve its
scientific capability during a thirty-year lifetime in a
fashion that presently cannot be predicted; its embedded
utilities must survive this thirty years without a major
overhaul and those utilities that are likely to require
modification due to technological evolution must be accessible for replacement; it must accommodate both large
multiuser science facilities and small experiments sponsored by low-budget university programs; it must maintain a
perfect safety record while not unduly compromising the
scientific return; it must provide a quick and certain implementation of investigations to support profit motivated
commercial research; and finally it must accomplish all of
these objectives within an established cost schedule, and
launch weight constraints, and formal program guidelines
established at the initiation of the effort (such as utilizing hardware components common with other Space Station
elements where at all possible).

Because of the complexity of the design task it was
obvious at the beginning that it could not be accomplished
without a concentrated and cooperative effort from a dedicated team of visionaries, planners, engineers, and research
scientists with all elements of the MSFC participating (Fig.
3). This team was organized and assembled under the leadership of the Space Station Laboratory Module Project Office.
This office assumed the responsibility for directing and
coordinating the several groups and integrating the efforts
of the design team into the overall Space Station Project.

The task of focusing and concentrating the scientific
requirements for the US Lab was assigned to the Commercial
Materials Processing in Space Office of the MSFC Program
Development Directorate. This Directorate has primary re-

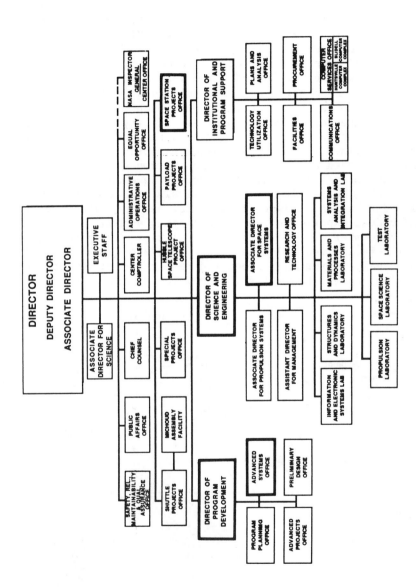

Fig. 3 Marshall Space Flight Center organization.

sponsibility at the Center for the early definition and
preliminary design of new projects, inlcuding scientific
payloads, and maintains active communications with virtually
all elements of the U.S. space science community. To implement this single point contact for investigator requirements, the organization established the Microgravity and
Material Processing Facility (MMPF) Study. The objective
of the study was to poll the entire materials-science
community within the United States, both scientific and
commercial, to obtain these potential user's "best estimate" of the resarch investigations which would benefit
from a manned microgravity orbital facility.

The information gathered at this initial phase of the
study provided a very large number of potential experiments
that were then analyzed for common hardware requirements in
the context of the "facility" concept developed during
Spacelab. The results of this analysis organized the envisioned research hardware requirement and indicated that the
individual experiments could be largely accomplished in a
finite number of core research facilities. These core research facilities (30 were identified, see Appendix A) were
then analyzed for their required resource requirements including; power, thermal rejection, logistics, crew operation

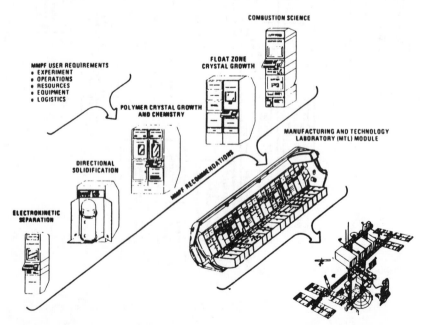

Fig. 4 Relationship of Microgravity and Materials Processing
Facility to Space Station.

time, volume, weight, fluid and gas utilization, microgravity level, waste disposal, and support equipment (Fig. 4).
The results of this study underwent a continuous review with
the user community and the appropriate commercial and scientific offices in NASA Headquarters and was communicated to
the MSFC engineering teams to serve as the foundation for
the design of the accommodations and outfitting for the US
Lab.

The primary engineering responsibility for the Laboratory was assigned to the MSFC Science and Engineering Directorate. This organizational element provides the Center's
primary engineering expertise and experience in system engineering, structures, electronics, communications systems,
materials requirements and applications, power systems, data
managements, hardware, software, and general design methodology through the Chief Engineer to determine the most
efficient means of accommodating the requirements derived
from the research community and translating those requirements into specific design concepts for the US Lab Module.
This included identifying subsystems required to service
the planned research facilities, sizing and specifying
those subsystems, and developing recommended design approaches consistent with cost and technology constraints,
accepted engineering practicies, and overall programmatic
considerations (see Fig. 5). The results of this design
effort were maintained by the Laboratory System Engineer
as a recommended US Lab design and an officially baselined

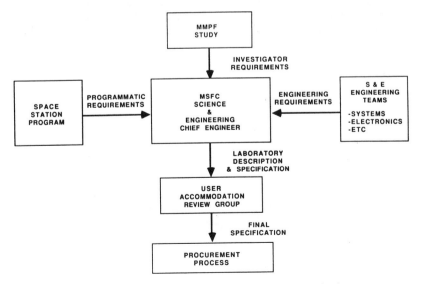

Fig. 5 Space Station engineering module design.

Contract End Item Specification; this latter document
serving as the basis for the contractual requirements for
the final design and implementation of the US Lab Module
by the prime industrial contractor to be selected to con-
struct the facility.

To provide for an overall check between the degree of
compliance of the recommended design with the requirements
of the user community, a review group comprised of noted
and potential investigators from both the scientific and
commercial communities was constituted under the MSFC Project Scientist, R. J. Naumann, to judge the progress of the
effort at critical intervals. The review activity was supported by the MMPF Study team, which maintained an active
and continuous representation of investigator interests
within the engineering group that was deriving the baselined documentation. The MMPF study team also directly
supported the design reviews by identifying areas that
would require special attention by the review group during
their periodic meetings. The composite process of definition, engineering assessment, and review is shown in Fig.5.

Laboratory Configuration

Because of certain Space Station program requirements
such as the utilization of common hardware and subsystem
elements whenever possible, certain of the design features
of the U.S. Laboratory Module were dictated from outside
of the dedicated Laboratory design team; although the
Laboratory requirements were one of the inputs that were
utilized in deriving the identification and design of these

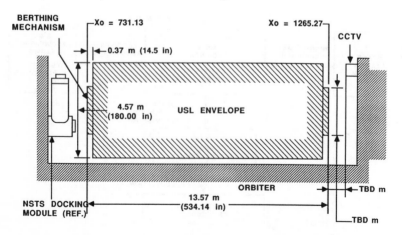

Fig. 6 Module standoff and rack geometry.

common elements. The prime example of the common component
approach is the basic structure and configuration of the
Laboratory pressure shell and its internal architecture.
The module (Fig. 6) is a cylinder 4.5 m in diameter and
13.6 m in length (the maximum size which will fit in the
cargo bay of the Shuttle) and contains a standard hatch in
both ends. Each hatch is large enough to pass a double rack
of laboratory experiment hardware. Internally, the module
is divided into quadrants by four "stand-off" structures
(Fig. 7) that are designed both to carry the necessary
utilities down the length of the module and to provide
structural support for standard racks of equipment and
subsystem facilities. This basic architecture and requirements for a large aisle to facilitate free passage through
the module by the crew while a rack is pulled away from the
wall, established the depth of the rack. The width of the
rack was chosen to accommodate the standard 19 in. panel
width common in Spacelab and ground-based laboratories.
While this approach dictates a certain allowabl size for
experiment racks and subsystem containers, it also establishes a situation of maximum flexibility because all
units will be interchangeable to various locations within
the module(s). In each U.S. module there are 44 double rack
locations - 11 in. each wall - for experiment or subsystem
units. Generally the subsystems and elements requiring only
occasional manned interaction will be located in the floor
and ceiling. The experiment facilities will tend to be

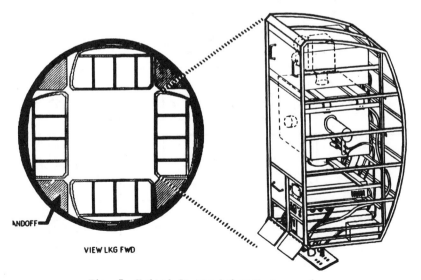

VIEW LKG FWD

STANDOFF

Fig. 7 United States Laboratory envelope.

located in the walls to provide the crew with a visual reference which simulates an earth based laboratory (Fig. 8).

Other subsystems located in the Laboratory and derived from common design considerations are the Laboratory environmental control and life support system, the fire detection and suppression system, emergency provisions, housekeeping systems, the element control workstation, and the internal audio and video communication systems. The design of the electrical power and thermal control system is also common with those in the other module; however, in the Laboratory version the size of these two systems was based on an accommodation analysis of the potential research facilities projected by the MMPF study and the systems have substantially larger capability.

Further analysis of the MMPF data by the engineering team indicated that the research facilities to be accommodated in the Laboratory tended to require a number of services which would be most efficiently supplied by special subsystems and support eqiupment incorporated directly in

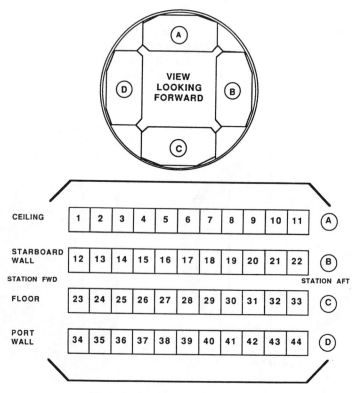

Fig. 8 Standard rack locations.

the Laboratory, rather than supplied by the individual facilities themselves. In particular, it was determined that the majority of the research facilities required substantial supplies of various process fluids and gases if they were to be operated in a continuous and interactive mode. Usages range from purge gases to flush furnace chambers of potentially harmful vapors, to ultrapure water for cleaning biological equipment. It was also determined that continuous and interactive operations of the large number of experiments planned for the Laboratory will, at one time or another, involve various amounts of virtually every chemical or reagent used in earth-based laboratories.

To meet the challenge of accommodating the maximum variety of materials and chemicals, and therefore maximizing the potential scientific return from the Laboratory, the Laboratory will contain as a major subsystem a centralized fluid and gas supply system with specific provisions for managing, controlling, containing, and disposing of any noxious material or by-product of potential experiments. This utility, the Process Materials Management Subsystem (PMMS), will be accessible to every experiment rack location and will potentially supply argon, oxygen, carbon dioxide, hydrogen, helium, nitrogen (liquid and gas), and ultrapure water and subsequently collect and dispose of the return from the experiment facilities. In addition, the subsystem will provide facilities for controlled storage of any chemical or material required by the experiments, but whose accidental release into the Laboratory module would create a hazard or discomfort for the crew (Fig. 9).

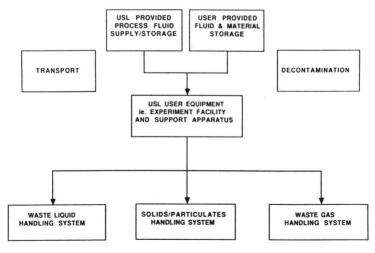

Fig. 9 Process Materials Management Subsystem.

A second critical area in which it became apparent that centralized Laboratory services are required is the data and communication interface between the experiment and investigator. A major objective of potential science investigators on Space Station will be to significantly improve and evolve their ability to monitor and control experiments from ground-based locations. Accomplishing this goal requires substantial knowledge by the investigator of the status of processes within the experiment facility, the status of the material sample itself, and information on the microgravity acceleration environment within the Laboratory during the process procedure. Because the nature of amny material-science studies is such that substantial information can be gained by visual inspection of the material sample, the requirement for in-depth investigator knowledge of the experiment process translates into a requirement for thorough video monitoring of the various phases of the experiment.

Although in most individual cases the application of video monitoring is reasonably straightforward and its implementation within a particular facility would be the responsibility of the builder of that facility, the multiple and simultaneous operations of a number of experiment facilities (a likely situation) will quickly exceed the data downlink capability of the Station. This will be particularly true if many of the facilities require high-resolution pictures. The indicated solution is to provide a centralized-video-management facility as a part of the Laboratory outfitting. This unit will be capable of acquiring a number of channels from the complement of experiments and processing or compressing the data into a manageable bandwidth for downlink and subsequent unpacking and distribution to the various investigators on the ground. It is anticipated that this capability, together with a reasonable uplink command capability to on-board experiment facilities, will provide the basic tools needed by ground-based scientists both to control many routine experiment processes from home facilities and to interact with the on-board crew in a significant and cooperative fashion which will maximize the research return. To enhance the interaction the Laboratory will contain an embedded accelerometer system for monitoring and communicating the Laboratory microgravity environment as required by the investigators during experiment operations and on a routine basis.

The third responsibility of the Laboratory and one whose accomplishment will rely heavily on the materials management and data control subsystems previously described, is the capability to reconfigure and service experiment hardware for multiple-sample runs and to implement small

experiments that require manned operation. To facilitate
this end, the Laboratory will be supplied with both a sci-
entific-experiment workbench and a glove box. Both of
these units will have access to the PMMS and contain embedded video monitoring. It is envisioned that it will be
possible for a crew member to remove internal portions of
experiment facilities for repair, refurbishment and/or sample exchange at the workbench (or within the glove box if
the potential for a hazardous leak is present), with the
ground-based investigator monitoring the procedure via
video and assisting with instructions where appropriate.
Similarly, a small apparatus, perhaps one designed as a
precusor to a larger facility, can be operated by the crew
with power and utility connections at the workbench or
glove box and with the close involvement of a ground-based
scientist.

Finally, the Laboratory will provide for a limited cap-
abaility to examine and characterize samples from experiment runs. Although an extensive characterization capability
within the orbital facility would require a prohibitive
amount of crew time, a measured amount of sample inspection
and manipulation will be necessary to ascertain that the
experiment processes are indeed proceeding as appropriate.
This capability will be supported by tools within the glove

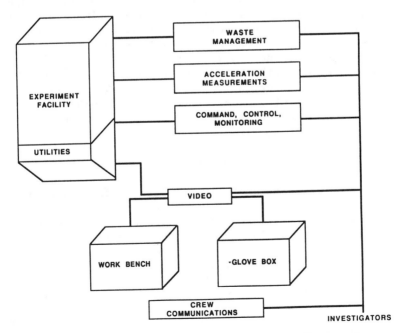

Fig. 10 United States Laboratory operational concept.

box for operations such as the cutting and polishing of crystals and other sample preparation procedures, and implemented using a system of microscopes and X-ray units. These units will also be coupled to the video system and will possibly allow for sample manipulation via remote control from ground investigators.

In summary, the US Lab has been designed to support an operational concept (Fig. 10), which will provide the highest level of crew and scientist interaction with the scientific process. The US Lab has been designed to provide a set of utilities to installed experiments, which studies indicate will be sufficient to provide for the operation of any envisioned research facility. To directly support the facilities, the US Lab will provide for waste management, acceleration measurements, and command control and monitoring by investigators from operations centers or their home laboratories. To provide the interactive environment necessary for productive scientific research and the implementation of small experiments, the Lab will furnish glove boxes and a multipurpose workbench, complete with utility hookups. Stanard and comprehensive utilities at all experiment locations will remove from the experiments the burden of supplying their own resources and eliminate the potential for the duplication of such support systems among experiment facilities. Extensive video and data acquisition capability will allow investigators to retain intimate and interactive relations with their experiment over the long intervals of orbital operations. Comprehensive working facilities will provide the means to manipulate experiment facilities in orbit with a freedom similar to that found in ground-based laboratories. Tools and facilities will be available to inspect and test the results of experiments while carfully protecting the crew and the Laboratory from from potential accidents.

Conclusion

As a result of a carefully considered design approach, the MSFC was able to efficiently focus its wide range of engineering experience toward the US Lab design, while taking advantage of the insights of the investigator community into its future research directions and objectives. In addition, close communication links were established between the investigators and the engineering representatives, which provided both sides with confidence that requirements were recognized, considered, and included whereever possible. Although the preliminary design of the Laboratory has been accomplished, the details of the implementation will occupy the full attention of the engineering teams for the next several years, and it is obvious that this implementation will continue to benefit from close communications with the investigator community. In several cases (one example being the video interface to the experiments and support equipment) it will be necessary for both the investigators and the engineers to explore and learn through joint testbed activities the interaction of the hardware and operational procedures to arrive at optimum configurations. The end result of these efforts, to be achieved sometime in the next decade, will be an unprecedented laboratory facility maintained in earth orbit and dedicated to an entire generation of researchers for the commercial and scientific exploitation and understanding of the effects of the microgravity environment.

Acknowledgments

The information content of this paper was derived not by the authors, who merely had the task of reporting the accomplishments of the Laboratory design effort, but by the numerous members of the design team. Within the U.S. Laboratory Module Project Office these individuals include: C. C. Priest, the Deputy Project Manager, W. R. Bowen, J. R. Graves, H. B. Hester, and D. Xenofos. A. Boese provided engineering leadership from the Chief Engineers Office, and N. C. Parker served as the Lab Module System Engineer. Among the companies participating in the effort were Boeing Aerospace Company, Martin-Marietta Corporation, McDonnell Douglas Astronautics Company, Teledyne Brown Engineering, and Wyle Laboratories.

Appendix A

ACOUSTIC LEVITATOR FACILITY: Acoustic levitators and associated furnaces for processing materials without contact with container walls.

ALLOY SOLIDIFICATION FACILITY: An array of various experimental furnaces (such as an isothermal unit with rapid quench) for research into the metalurgical properties of alloys processed in microgravity under various circumstances.

ATMOSPHERIC MICROPHYSICS FACILITY: A combination of chambers for research and parametric studies into the formation of clouds and ice particles under controlled conditions.

AUTOIGNITION FURNACE FACILITY: Integrated electric furnaces and instrumentated pressure vessels for controlled studies of ignition and burning of solids and liquids in microgravity.

BIOREACTOR/INCUBATOR FACILITY: Growth cells designed to propagate and grow biological cells in microgravity for studies of the growth process and as a potential source of material for separation by electrokintic techniques.

BRIDGMAN FURNACE FACILITY, LARGE: High-temperature tube furnace for directional solidification processing of metal and semiconductor samples several centimeters in diameter.

BRIDGMAN FURNACE FACILITY, SMALL: Several high-temperature furnaces for directional solidification processing of a large variety of material samples and for controlled studies of the directional so-lidification process itself in microgravity.

BULK CRYSTAL FACILITY: A large isothermal furnace for experimental and preproduction processing of crystals of 10-30 cm in diameter.

CONTINUOUS FLOW ELECTROPHORESIS FACILITY: Research system for stu-dies of high resolution separation of material using electrophore-tic techniques in microgravity.

CRITICAL POINT PHENOMENA FACILITY: A comprehensive set of tools and equipment for conducting controlled studies of liquified gases near their critical point.

DROPLET/SPRAY BURNING FACILITY: Pressure vessels and associated equipment for studying the combustion of droplets and small parti-cles in microgravity under controlled and varied conditions.

ELECTROEPITAXY FACILITY: Research system for growth of high quality commercial and research electronic crystals of various diameter to develop electroepitaxy growth techniques and theory in the micro-gravity environment.

ELECTROSTATIC LEVITATOR FACILITY: System for positioning samples by electric fields in the hot zone of a furnace for processing without contacting the walls of the container.

ELECTROMAGNATIC LEVITATOR FACILITY: System for the containerless processing of materials positioned within a furnace cell by the Lorentz forces induced in the sample from RF coils.

FLOAT ZONE FACILITY: A hot-wall furnace facility capable of produc-ing a hot zone with a carefully taylored gradient that can be tran-slated along the length of the sample.

FLUID PHYSICS FACILITY: A multipurpose fluid system for research in basic fluid phenomena, such as convection, phase transition, and buble behavior.

FREE FLOAT FACILITY: Containerless processing facility that takes advantage of the low-gravity environment to process a sample in free fall within an enclosed chamber with position corrections be-ing provided by gas jets.

HIGH-TEMPERATURE FURNACE FACILITY: Extremely high temperature iso-thermal furnace for experimental processing of materials with high melting points such as ceramics.

ISOELECTRIC FOCUSING FACILITY: Research facility for studying al-ternative electrokinetic techniques for the separation of biologi-cal materials.

LATEX REACTOR FACILITY: Research facility for studying the poly-merization process and the related morphology of the products.

MEMBRANE PRODUCTION FACILITY: System for producing inorganix poly-meric Langmuir-Blodgett type membrane in microgravity.

OPTICAL FIBER PULLING FACILITY: Research system to develop tech-niques for pulling ultrapure fibers from levitated melts.

ORGANIC AND POLYMER CRYSTAL GROWTH FACILITY: Support unit to accom-modate a number of small, specilized devices designed to research all aspects of polymer crystal growth in microgravity.

PREMIXED GAS COMBUSTION FACILITY: Pressure vessl capable of igniting premixed gas and liquid system and recording the evolution of the combustion process in microgravity.

PROTEIN CRYSTAL GROWTH FACILITY: System of a large number of growth cells designed to provide the accommodations necessary for the growth of protein crystals, along with interfaces to allow for in-teraction of the ground-based investigator with the analysis pro-cess.

ROTATING SPHERICAL CONVECTION FACILITY: An apparatus designed to study convection flows in planetary and terrestrial system in a working spherical model.

SOLID SURFACE BURNING FACILITY: Pressurized apparatus that can gen-erate a laminar flow profile across burning samples and record the results of varing the parameters of the reaction.

SOLUTION CRYSTAL FACILITY: System designed to evaluate several approachs for producing crystals from liquid solution in the micro-gravity environment.

VAPOR CRYSTAL FACILITY: System consisting of several special gra-dient furnaces for studying vapor phase and thin film crystal growth of both organic and inorganic compounds.

VARIABLE FLOW SHELL GENERATOR FACILITY: Apparatus for producing near-perfect spheres and/or shells by special techniques that eli-minate the effects of levitator positioners.

ARIES: The Ariane 5 Extended Stage for Orbital Transfer and Rendezvous

Pierre Molette
MATRA Espace, 31077 Toulouse, France

Abstract

The operations related to the In-Orbit Infrastructure
will require the launch of large payloads to a Space Station
or platform, typically pressurized modules for either logistics resupply or station buildup, or large payload instruments (or instrument complement). These large payloads are
not usually provided with resources such as power generation, communication equipment, attitude control, which are
required for the autonomous life of the payloads before they
are docked to the platform or station. There is therefore
the need for vehicle which will provide those resources during that short phase of the mission and provide the transfer
and rendezvous capabilities, which are currently beyond the
definition of Ariane 5. However it appears to extend the
capabilities of the Ariane 5 Vehicle Equipment Bay (VEB) and
L5 upper stage, which are already used for the launch and
can be upgraded for a typical 48 h mission. The resulting
expendable vehicle, called ARIES (ARIane Extended Stage) can
be configured in a basic vehicle and additional extension
packages for rendezvous and docking capabilities for different mission types. A review of the potential missions requiring ARIES and their corresponding mission timelines allow to derive the ARIES configuration. The modifications
brought to the standard L5 and VEB to allow them to fulfill
the various mission requirements are as follows:
1) The upgraded L5 stage provides the main propulsion
functions for injection into orbit (Ariane 5 mission), orbit
transfer and homing boosts. The main modifications are due
to lifetime and reliability extension, and particular attention has been paid to the safety aspects;

2) The upgraded VEB provides guidance, navigation and control for launch, coast and transfer phases; it ensures also up and down link interfaces with ground and Data Relay Satellites for both ARIES and its payload.

The adaptations to the ARIES missions have been defined so as to minimize the modifications, and thus maintain the impacts on development and recurring costs within an acceptable level. Major modifications, such as the rendezvous equipment, are configured in kits to be mounted on the payload adaptor. Same is valid for the extra batteries required for the 48 h mission to power the ARIES and payload equipment as well as for the thermal control heaters: a provision of 500 W continous has thus been allocated to the payload.

1. Introduction

The operations related to the In-Orbit Infrastructure will require the launch of large payloads to Space Station or platform, typically large payload instruments (or instrument complement) or pressurized module for either logistics resupply or station buildup. These large payloads are not usually provided with resources such as power generation, communication equipment, attitude control, which are required for the autonomous life of the payloads before they are docked to the platform or station. There is therefore the need for a vehicle which will provide those resources during that short phase of the mission and provide the transfer and rendezvous capabilities, which are currently beyond the definition of Ariane 5.

Instead of developing a completely new vehicle, it appears much more interesting (on the technical as well as on the development and recurring costs points of view) to extend the capabilities of the Ariane 5 Vehicle Equipment Bay (VEB) and L5 upper stage, which are already used for the launch and can be upgraded for a typical 48 h mission. This vehicle, called ARIES (ARIane Extended Stage) can be configured in a basic vehicle and additional extension packages for rendezvous and docking capabilities for different mission types. A review of the potential missions requiring ARIES leads to identify three types of ARIES missions; they differ by the level of involvement of ARIES in the rendezvous operations:

1) The first type is the less demanding one for ARIES, which has to provide resources to the payload and a minimum orbit transfer capability; the vehicle is passive in the rendezvous and docking operations which are performed by another vehicle (e.g. OMV);

2) The second mission type requires in addition a rendezvous homing capability, to bring the payload in the vicinity of the station or platform;

3) The third mission type requires the proximity operations and docking capabilities to perform the whole sequence in the active mode.

2. Mission Scenarios

2.1 ARIES missions

In the frame of the European In-Orbit Infrastructure operations, Ariane 5 will have to launch large modules or elements towards spacecraft already in space. That is mainly the case for:

1) The buildup of a space station: launch of a pressurized module (PM) towards a resource module (initial construction) or an operational space station (growth),

2) The launch of logistics modules for the resupply of a space station.

3) The servicing of automatic platforms: launch of a large payload module (construction or growth) or of fuel tanks (refuelling) towards an automatic platform.

Table 1 summarizes the three classes of ARIES missions which are detailed below and illustrated in Fig. 1.

2.2 Mission type (a)

The typical case is the launch of a pressurized module (PM) towards an operational Space Station and its recovery by a space based OMV. ARIES will support the PM once in orbit while waiting for the OMV. The main features of this scenario are as follows:

1) ARIES and its payload are injected in phasing orbit; this orbit depends on the spacecraft orbit and drives the mission duration and   V requirements.

2) ARIES performs the transfer to the rendezvous orbit (if different from the phasing orbit).

3) Then, ARIES will not perform any orbit maneuver: all the rendezvous (homing, proximity operations) and docking maneuvers are performed by the space based vehicle or by the spacecraft itself (e.g., Resource Module).

4) The arriving vehicle will dock on the payload side; once the interfaces between this vehicle and the payload have been activated, the separation between the payload and ARIES is done. ARIES is de-orbited, while the payload is transferred to the Space Station or the assembled spacecraft moves to its operational orbit.

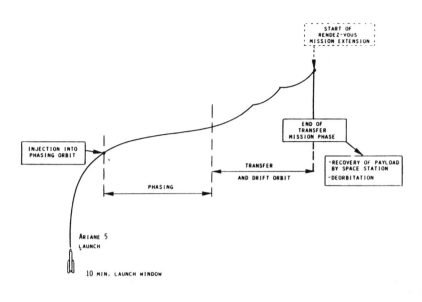

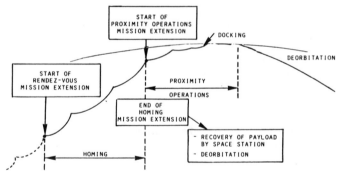

Fig. 1   Illustration of possible mission scenario.

2.3 Mission type (b)

The typical case is the launch and transfer of a pressurized module to the proximity of a space station (a few
km), where it will be picked up by a specific station vehicle, such as an OMV or a Proximity Operation Module (POM).
The main features of this mission type are as follows:

1) After the phasing period, ARIES transfers its payload to a drift orbit (some km lower than the Space Station
or spacecraft orbit).

2) Once it has acquired the Space Station by its sensors, long or medium range, ARIES performs a homing phase so
as to reach a point located at proximity of the Space Sta-

Table 1   Typical mission for ARIES

Type of mission	Spacecraft in orbit	Aries payload	Class of Aries mission
Build-up of a Space Station	Non operational Space Station (or Resource Module)	Pressurized module (P.M.)	a, b, c
Build-up or Logistic Resupply of a Space Station	Operational Space Station without space based vehicle	P.M.	c
	Operational Space Station with specific vehicle (OMV, SV,-)	P.M.	a, b
Assembly, Growth or Servicing of automatic PF	Large platforms	Payload Module Fuel Tanks	c

a) ARIES provides resources and attitude. Recovery by space based
   vehicle or spacecraft.
b) ARIES provides resources and rendez-vous to a safety area.
   Recovery by space based vehicle or spacecraft.
c) ARIES provides resources and rendez-vous and docking to the
   spacecraft.

tion. Navigation from GPS or ground control center data,
channelled through the Delay Relay Satellite (DRS), may reduce or avoid the use of homing sensors.
    3) Then, as for the mission type (a), ARIES remains
passive until a Space Station dedicated vehicle performs the
docking with the payload; after separation, ARIES is de-orbited.

2.4 Mission type (c)

    This type of mission is the most demanding one for
ARIES. The typical case is the launch of a pressurized module towards a Space Station (not equipped with a specific
vehicle) or of a payload module towards a space platform.
The main features of this mission type are as follows:
    1) Same sequence as for mission type (b) until ARIES
has reached the proximity of the Space Station.
    2) The proximity operations are carried out by ARIES.
    3) Last final meters and docking operations are driv-
en by ARIES, the docking port being on the payload side.

4) If a berthing procedure is used (case of the Space
Station equipped with a manipulator system), ARIES will perform a stationkeeping at a few meters from the space station
and the capture of the ARIES/payload composite will be done
via the payload; the separation may be done by de-docking
procedures, or alternatively by the manipulator. Proximity
operations, necessary for moving away before de-orbitation,
will be ensured by ARIES.

2.5 ARIES mission timeline

An overall ARIES mission timeline is presented in Fig.
2, in the case of the launch of a module towards a Space
Station located at 500 km, 28 5. This timeline is roughly
applicable to the various mission scenarios, whichever is
the chaser during the rendezvous (and thus the mission type).
The driving parameter of the overall mission duration is the
phasing period for the rendezvous. This phasing period depends on the true anomaly being corrected (this parameter is
driven by the potential launch delay, the evolution of the
Space Station orbit altitude and the launch window) and the
altitude of the phasing orbit, which drives the V required
for the next orbit transfers. The duration of the overall
rendezvous sequence may be reduced by using the DRS.

Taking into account the above parameters and a margin
for rendezvous re-entry, the total mission duration will be
up to 48 h. Analyses have shown that this maximum mission
duration is also applicable to ARIES missions to other orbits (sunsynchronous for instance).

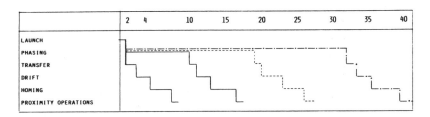

—— CASE 1 : USE OF DRS, NO PHASING   8 H

—— CASE 2 : USE OF DRS, H_0 = 300 KM, $\theta \in$ [0.90]   ≤ 17 H

---- CASE 3 : IDEM CASE 2, H_0 = 400 KM   ≤ 26 H

—·— CASE 4 : NO DRS, H_0 = 300 KM, $\theta \in$ [0.360]   ≤ 42 H

THESE OVERALL MISSION DURATIONS ARE APPLICABLE TO ALL SCENARIOS (WHATEVER IS THE CHASER DURING
THE RENDEZ-VOUS).

THESE DURATIONS DO NOT INCLUDE MARGIN FOR RENDEZ-VOUS RE-ΓRY

Fig. 2 Overall mission timeline.

3. The ARIES Concept

3.1 General architecture

ARIES is an expendable vehicle directly derived from
the Ariane 5 Vehicle Equipment Bay (VEB) and L5 upper stage.
Therefore, as shown in Fig. 3, ARIES is composed of 3 main
elements: an upgraded Ariane 5 L5 stage, which will perform
the required boosts for orbit transfers and homing maneuvers; an upgraded Ariane 5 VEB, which ensures the control-
command of the vehicle, and the payload adaptor, ensuring
interfaces with the payload, and supporting most of the
ARIES dedicated addtional equipments.

Being based on the extension of the AR5 upper s
VEB capabilities, ARIES will provide the same functioɳ
the standard L5 and VEB during launch; upgrading of the
two elements has been carried out so as to allow ARIES tᴜ
fulfill the various mission requirements; however , in order to minimize the impacts of the ARIES mission on the
standard AR5 elements, their overall architecture and equipments have been kept unchanged as far as possible, most of
additional equipments or subsystems being mounted on the
payload adaptor or configured as a kit.

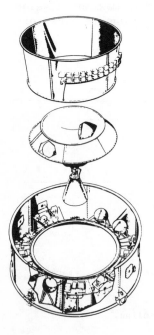

Fig. 3 Overall configuration of ARIES.

3.2 The L5 stage

The L5 stage upgraded for the ARIES mission provides
the main propulsion functions for injection into orbit
(Ariane 5 mission), orbit transfer and homing boosts. The
actual L5-stage concept fulfills ARIES requirements with
regards to its general propulsion capabilities but not concerning lifetime, reliability, and safety for man-involved
missions. The main modifications concern the propulsion
system, the electrical system and the thermal control system and are due to lifetime and reliability extension.

A particular attention has been paid to the safety requirements which become a critical design rule when ARIES
has to perform co-orbiting or proximity operations towards
a manned or man-tended space element(Space Station, COLUM-
BUS elements). The extended mission of Ariane 5 (48 h including reignition of main engine) has a basic impact on
the L5 stage system performance in opposite to the short
lifetime of a normal transportation rocket. The ARIES propulsion system architecture has to be considered under the
failsafe/safelife philosophy to avoid any impact on payload
(i.e. COLUMBUS PM) through failure propagation. Another
important aspect is that Pressurized Module transportation
(Ariane 5 PM) cannot be considered strictly as unmanned
space flight, because the stage will perform proximity operations or will be subsequently docked to manned modules.
During these phases the main engine must be safe (non-oper-
ational). These safety aspects are considered for the definition of the hazard categories and safety critical functions of the L5 stage.

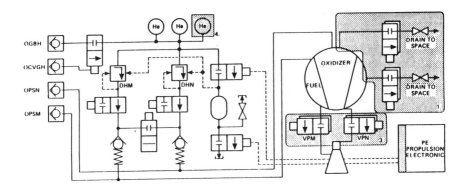

Fig. 4 L5 propulsion system (simplified).

3.2.1 Propulsion system (Fig. 4)

The main modifications are as follows:
1) Additional main tank emptying device to meet safety
requirements before beginning of the proximity operations.
After fulfilling its operational propulsion tasks, the L5
stage must be set to safe status regarding the main tanks
content of storable hypergolic propellants via defuelling
and depressurization of the outer compartment via equatorial
connected bleed lines to space and depressurization of the
inner compartment.
2) The 20kN main engine can be taken in its baseline L5
status, since it will be qualified for long mission duration
(100 days) and multistarting capability (20 cycles) for HER-
MES application.
3) Redundancies in engine shutoff devices (VPN, VPM) to
guarantee reliable function.
4) Additional 3rd He High-Pressure tank to cover increased consumption for additional engine head purging cycles, long time He-leakages and for improvment of the blow-
down operating performance.

3.2.2 Electrical system (Fig. 5)

The main modifications are as follows:
1) Functional system upgrading by implementation of additional drivers for the increased number of commands to the
propulsion system and by implementation of redundancies for
general reliability improvement. Extension of Electronique
Sequentielle to provide means for the additional commands
and for signal conditioning.
2) Additional disarming possibility for the destruction
system (i.e., make safe the BAM) during safety-critical mission phases.
3) Intermittent operating mode for TM system (active/
standby) meaning extended (additional) switching capability

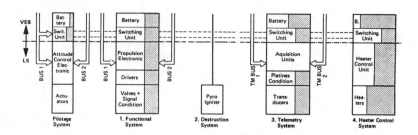

Fig. 5 L5 electrical system.

for BMO. Additional redundant PC's (Platine Conditionneur)
for monitoring of vital functions/statii.
    4) Additional circuits for Heater Control System with
redundancies for vital parts and switching via Data-Bus-in-
terface from VEB computer.

3.2.3 Thermal control system

    The 48 h mission duration leads to additional insula-
tion/redundant heating for propellant pipes between tank and
engine to guarantee restarting capability within 48 h; additional redundant insulation/heating for the electrical
equipment to keep it within storage/operational temperature
limits; and additional insulation for different critical
parts.

3.3 VEB adaptation

    The adaptation of the VEB to the ARIES mission has been
done so as to minimize the modifications of the standard
VEB, and thus to maintain the impacts on development and recurring costs within an acceptable level.
    The upgraded VEB provides guidance, navigation and con-
trol for launch, coast and transfer phases; it ensures also
up and sown link interface with ground and European DRS for
both ARIES and its payload. The upgraded VEB configuration
is illustrated in Fig. 6. The main modifications concern
the electrical system, the attitude control system, and a
particular attention has been paid to thermal aspects.

3.3.1 Electrical system

    It reuses the basic electrical architecture: the separ-
ate chains for GNC (see Fig. 7), for TTC (see Fig. 8), and
for safety, as well as the standard fault tolerant architecture. The functional electronics (on board computer, inertial reference system and command electronics) are kept
identical; redundant units are already part of the standard
Ariane 5 VEB configuration and provide the necessary level
of redundance for ARIES without the need for additional
equipment. Use of GPS for navigation is considered, to ensure compatibility with HERMES orbital operations.
    If necessary, these equipments are connected to the
ARIES electrical system via a harness in a gutter along the
payload. In addition, ARIES may play an active role in the
rendezvous operations and two rendezvous extension packages
may be used (see generic diagram in Fig. 9).

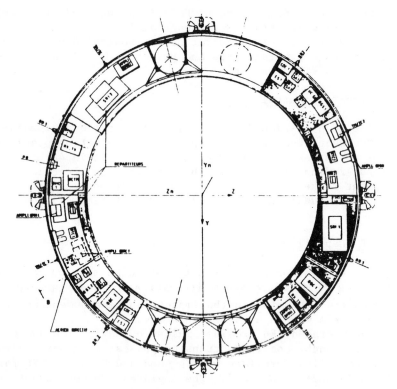

Fig. 6 Veb configuration.

1) Homing extension package. The ARIES navigation may
be ensured by long range sensors (RF sensors) with one (or
two) S-band antenna(s) installed on the payload or on a deployable mast on ARIES, and an S-band transponder mounted on
ARIES (typically on payload adaptor). However, in order to
ensure commonality of equipments with HERMES, the selected
solution would consist in bringing ARIES to a point as close
as possible (below 10 km) to the receiving spacecraft by
using navigation data from GPS (GPS antenna on ARIESO or by
the ground control center (through DRS system). Thus, either this point is close enough to the spacecraft (end of
homing phase), or medium range sensor (i.e., laser) is necessary.

2) Proximity operations package. Short range sensors
(CCD camera) are necessary to allow ARIES to acquire rela-
tive parameters and attitude during the proximity operations.
They are mounted on the front side of the payload, close to

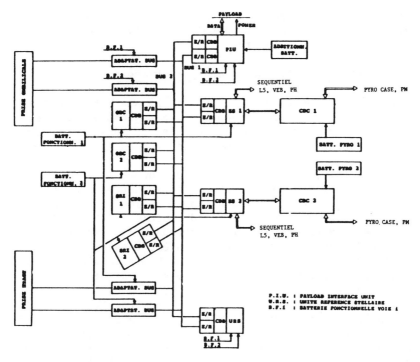

Fig. 7 GNC chain.

the docking port. The associated electronic boxes (for processing of camera data) are implemented close to the cameras.
An electrical link via an external gutter allows to power
and monitor these equipments from the ARIES avionic system.
Medium range sensors (laser) may also be required, according
to the overall spacecraft/ARIES configuration (size of safety area), and especially if no long range sensors are used.
     Additional propulsion equipment will support the prox-
imity maneuvers and the braking. They are biased in order
to minimize the plume effects on the payload and station.
The need for a cold gas system during the last final meters
to avoid the contamination of the Space Station (or other receiving spacecraft) is to be evaluated in relation with the
selected strategy (i.e., docking vs berthing). A dedicated
set of processors are used to extend the capabilities of the
central processor for the monitoring of the proximity operations; the architecture of these processors allows fully
fault tolerant operations within the Space Station safety
area; however the level of safety requirements (and of relevant fault tolerance) depends whether the receiving spacecraft is manned, man-tended, or automatic. These processors

are mounted on the payload adaptor and connected to the VEB
basic avionic system. An abort capability is implemented in
these dedicated processors.

3.5 General characteristics and performances

The mass budget of the upgraded L5 and VEB is indicated
in Table 2. The payload adaptor is not included, as being
part of the payload in the Ariane payload mass definition,
but the masses of the equipments mounted on this adaptor are
included in the VEB dry mass.

Therefore, with respect to the standard L5 and VEB, the
utilization of the basic ARIES vehicle induces a mass penalty of about 550 kg on the Ariane 5 payload; that is much
lower than the use of a dedicated vehicle which would provide the same functions. The major part of this mass penalty is due to the power supply system, composed of batteries (lithium cells) and sized for 48 h mission: about 300 kg
of batteries are mounted around the payload adaptor. ARIES
will provide the following services to the payload:

1) Power supply. 500 W may be permanently (during 48
h) supplied to the payload; this power may be necessary for
thermal control (heaters), but also for the monitoring and
checkout of the payload. On a case by case basis, it is
possible to increase the level of power available for the
payload (for instance if a shorter mission is planned).

2) Monitoring and control. All the telemetry and tele-
commands necessary to allow the ground control center to
monitor and control the payload will be exchanged via the

Table 2   ARIES mass for the basic configuration

L5	Stage baseline	675	
	Additional mass (propulsion, electrical system, thermal control)		80
	Veb baseline	920	
VEB	Additional mass (functional, TM-TC, batteries, harness, propulsion, thermal control, secondary structure)		471
	Total (dry)	1 595	+ 551

ARIES electrical system and the DRS system. The payload is
connected to the data handling system of ARIES via an interface unit. The resources allocated to the payload are: 10
kbps for downlink, 2 kbps for uplink.

3) The barbecue mode, in slow spin rate, allows minimal
use of heaters for the thermal control of the payload and
avoidance of any critical hot case. In its basic configuration, as well as in case of mission (b), ARIES acting as a
target, is capable of fulfilling the attitude control requirements (in position and stability) necessary to allow
the arriving space based vehicle to perform the proximity
operations.

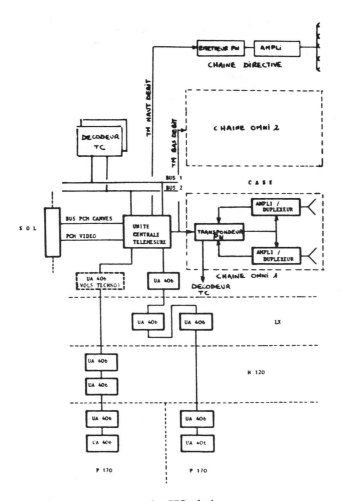

Fig. 8 TTC chain.

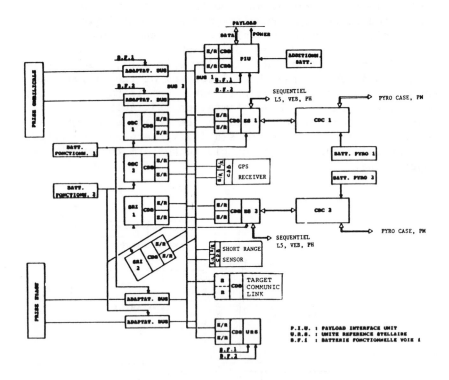

Fig. 9   Rendevous extension.

4. Conclusion

The combination of the Vehicle Equipment Bay and the L5 stage of Ariane 5 constitutes a very sound basis for the definition of an expendable transfer vehicle. The maximum reuse of the Ariane 5 capabilities and the minimal modifications and implementation of additional equipment lead to a most effective vehicle in terms of additional development and recurring costs.

ARIES will perform such missions as payload or module transfers to Space Stations or platforms while meeting all safety requirements, including final rendezvous and docking with a manned space station.

Additional missions such as the placement of multiple payloads in different orbits may also be performed with the basic version of ARIES.

ARIES sill therefore dramatically increase the Ariane 5 capabilities for operations in Low Earth Orbit and as such will become an important element of the In-Orbit Infrastructure.

Columbus Operations: Planning and Execution

Joachim Kehr*

Deutsche Forschungs und Versuchsanstalt für Luft und Raumfahrt (DFVLR), Oberpfaffenhofen, Federal Republic of Germany

Abstract

The paper presents the latest status on the COLUMBUS Operations Planning concept as proposed by the COLUMBUS Flight Operations Organization (FOO) study group. Planning and execution interfaces with the International Space Station and appropriate partner organizations will be identified. The paper will be centered towards the operations planning concept for the European Attached Pressurized Module (APM) and the associated interfaces. Following the description of operations planning principles, the major planning products and their logical interconnection will be identified. The rest of the paper will describe in some detail the activities, outputs and interfaces on the various planning levels (strategic, tactical and execution) with special emphasis on the USER involvement.
As a result of the (still ongoing) study centralized, international strategic planning is proposed with respect to payload manifesting, resource allocation and on all safety critical issues.
DECENTRALIZED tactical planning is promoted by Europe as the operations concept suiting best the goal to achieve operational autonomy within the international frame of Space Station utilization.

Nomenclature

APM   = Attached Pressurized Module
AR-5   = Ariane 5
CMC   = Combined Mission Center

*Study Manager, Columbus Flight Operations Organization.

```
COOPF        = Coorbiting Platform (=enhanced EURECA)
DRS          = European Data Relay Satellite
D-2          = National Spacelab Mission No. 2
ECS          = Environment Control System
EMCC         = European Mission Control Center
ESOC         = European Spaceflight Operations Center
EURECA       = European Retrieveable Carrier (Platform)
EVA          = Extra Vehicular Activity
FCC          = Flight Control Center
FOO          = Flight Operations Organization (COLUMBUS)
GSOC         = German Space Flight Operations Center
IOI          = In-Orbit Infrastructure
IVA          = Intra Vehicular Activity
JSC          = Johnson Space Center (Houston)
KSC          = Kennedy Space Center
MCC          = Mission Control Center
MTFF         = Man Tended Free Flyer
POCC         = Payload Operations Coordination Center
PPF          = Polar Platform
STS          = Space Transportation System (Shuttle)
TROP         = Tactical Resource and Operations Plan
TDRSS        = U.S. Tracking and Data Relay Satellite
UOC          = User Operations Center
USC          = User Support Center
US SSSC      = U.S. Space Station Support Center
```

1. Introduction

The following paper gives a summary of the COLUMBUS Operations Planning and Execution Concept from the European point of view, as developed by the Flight Operations Organization (FOO) Phase B Study Team, formed by European Space Flight Operation Center (ESOC) and German Space Flight Operations Center (GSOC) engineers as well as engineers from the Italian Space Organization Piano Spaziale Nazionale (PSN). Credit has to be given to all team members from ESOC, GSOC and PSN in particular to Mr. Antonio Sesma, FOO study manager at ESOC and Mr. Mel Brooks, consultant to the study team.
The presented European IOI Flight Operations System Implementation proposal in chapter 4 is that one being promoted by the German ESA delegation.
The final decision on the actual implementation is pending and will be resolved by the ESA member states ministerial conference planned for July 1987.

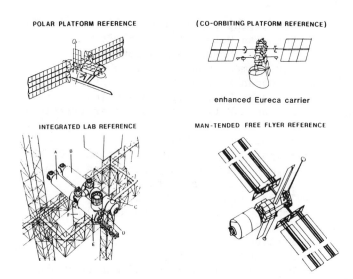

POLAR PLATFORM REFERENCE   (CO-ORBITING PLATFORM REFERENCE)

enhanced Eureca carrier

INTEGRATED LAB REFERENCE   MAN-TENDED FREE FLYER REFERENCE

Fig. 1 Elements under Study in Phase B2.

2. The Columbus Programme

Columbus is the European contribution to the International
Space Station Programme to develop a manufacturing and
research infrastructure in low earth orbit. This will comprise a manned space station in an equatorial orbit and a
number of unmanned or man-tended platforms in equatorial or
polar orbits. Construction is due to start in space 1994.
Europe under the umbrella of the European Space Organisation (ESA) is expected to provide the following flight
elements to the programme (see also Figure 1)

Pressurised Module   (PM) a 4 segment pressurised module
                          similar to Spacelab which will be
                          permanently attached   to the manned
                          space station; it will house materi-
                          als and life science payloads which
                          will be tended by the station crew;
                          resupplies will be transported to the
                          station by the shuttle.

Man Tended Free Flyer (MTFF) a 2 segment pressurised module
                          flying unmanned in the same orbit as
                          the space station; it will house
                          payloads requiring   micro-gravity
                          environment lower than that of the
                          space station; it will be serviced

through the European space transportation system Hermes or at the
space station. The MTFF will comprise
a short pressurised module similar to
Spacelab and a module providing
resources and propulsion (RM, Resources Module) to be launched by Aria-
ne-5.

Polar Platform   (PPF) an unmanned platform in polar
orbit; it will carry earth observa-
tion payloads, and it will be serviced every four years by the Shuttle
or Hermes. The launch will be carried
out by Ariane 5.

The main European participants in Columbus are West Germany
(prime Contractorship and RM), Italy (PM), France and the
UK (PPF), with significant contributions from other European countries. ESA is providing overall programme management.
The flight elements are intended to remain in orbit inde-
finitely and operations will therefore be a significant and
long term activity. An operations infrastructure will be
developed on the ground and in orbit the main aim of which
is to provide users with facilities to operate 'their' payloads. The Flight Operations Organisation (FOO) has already
been setup with the task of studying the operations infrastructure. Three organisations contribute to the FOO: ESOC
(ESA), GSOC (DFVLR/BRD) and PSN (Italy).
The Columbus Programme is just about to complete Phase B2
during which the FOO developed a concept for the operation
of the flight elements and derived from this a preliminary
design for a flight operations infrastructure on the ground
The distribution of these centres amongst the various
European countries has yet to be determinated. They will
comprise extensive data processing facilities for mission
planning, control, evaluation etc. as well as communication
links. In order to minimise costs, as a groundrule as many
existing facilities as possible will be used.
Through their use a number of words and phrases have
acquired a meaning specific to Columbus flight operations.
This usage is carried over in this paper:

Operations   This encompasses three major activit-
ties: "Ground Processing" "Logistics"
and "Flight Operations"

Ground Processing   Covers   all   activities performed pre-
                      launch and post-landing for subsystems
                      and payloads, such as transport to
                      launch site, launch readiness verifi-
                      cation and   required Ground Support
                      Equipment

Logistics   Covers all sizing, placing and creating
                      of stocks of consumables, spare parts,
                      ORU's, etc. before, during and after a
                      mission.

Flight Operations   This involves System Operations inclu-
                      ding launch support, in orbit assembly,
                      deployment, orbit and attitude maneu-
                      vering, servicing   and   maintenance
                      operations and Payload Operations Coor-
                      dination for Columbus space elements in
                      the defined Flight Operations Phases.

3. Operations Planning

Operations Planning for the Columbus mission can be divided
into three major categories: strategic planning, tactical
planning and that planning which occurs during the actual
execution of the operational mission. This section will
address the strategic and tactical planning; the execution
level management will be covered in chapter 4. The concepts
discussed in this section primarily address the Attached
Pressurised Module (APM) as it is by far the most complex.
This complexity derives from the fact that most of the
resources for the APM come from the International Space
Station (ISS) which is managed by an Organization formed by
the ISS participants. The Space Station resources must be
shared by all modules which make up the Space Station,
which in turn dictates a need for some integrated planning.
The MTFF and PPF, on the other hand, only require an interface with NASA when NASA elements are involved in servicing
operations. With the exception of the NASA interface, the
concept is generally applicable to the MTFF and PPF as
well.

Operations planning is guided by the following principles:

o Operations planning will be conducted from Europe to the
   maximum extent possible
o Payload operations will be planned and executed directly
   by the user to the extent possible, and when applicable

by the on board crew. It is the goal of the planning process to provide as much freedom as possible for the user to plan and operate his payload.

It is understood that the intended high degree of decentralisation and autonomy in the planning for and conduct of the operations, may lead to an initial non-optimal utilisation of the Columbus resources. This reduced utilisation of Columbus resources should be partly or totally compensated by the improvements in other areas (such as reduced mangement interfaces, replanning flexibility, user in user environment, European personnel in European environment, faster decision taking, equitable cost sharing and industrial return).

3.1. Strategic Planning

Strategic planning covers a period of time from 1-5 years ahead, and the strategic decisions are taken at very high levels of management. The planning concerns primarily programmatic and funding matters and establishes the mechanism, framework and resources for the following levels of planning to commerce, and for the planned mission to be implemented. Some initial decisions and agreements may already be established by such top level documents as a memorandum of understanding between agencies, or internal agreements within Europe. These agreements will address such things as:

o Commitments from all participants for flight and ground elements to be developed
o Commitments from NASA to provide Space Station resources to the APM and formulas for calculating the amount of resources
o Commitments regarding use of other NASA or European resources such as Shuttle/Hermes flight for resupply and servicing flights, KSC or Ariane ground facilities for ground processing and launch support, TDRSS/DRS usage and sharing etc.
o Costing policies for provided services
o Agreements on combined logistics planning for Shuttle resupply flights, lead times for manifesting of Shuttle cargo, etc.
o Program directives regarding product assurance and safety requiements, change control policy, system test and verification requirements, rules for management decisions, major program milestones and review schedule etc.

Within the framework of these international or inter-project agreements, the Columbus strategic planning takes place and covers such things as:

o Utilization planning which results in a manifest for each element, including payload definition and system configuration.
o Definition of the ground segment and communications systems configurations.
o Plans for upgrading flight or ground systems and new technology development.
o Allocations of funding for development of the above capabilities, and for improving existing facilities or building new ones.
o The negotiation of support agreements from the System and Element Contractors for sustaining engineering operations support, and the use of contractor facilities such as simulators, engineering models etc.
o The negotiation of agreements with the Logistic Organisation for Logistics Operational Support and Servicing Mission Planning support.

The output of the strategic planning is a set of documentation and mangement instructions which defines the mission and which contains all of the necessary information for the various implementing organisations to prepare and conduct the mission. The mission defined by this strategic Plan would probably cover a mission period of 1-5 years, and would include such things as:

o Configuration and manifest of the flight segment throughout the period of the mission.
o Configuration of the ground segment and communications system throughout the period of the mission.
o Top level mission objectives, including payload objectives, payload priorities and success criteria, and system performance objectives.
o Top level schedule, showing mission start and end dates, servicing/resupply flights, major required systems operations such as re-boost manoeuvres, etc.
o Flight crew complement and top level functions.
o Gross allocation of Space Station resources to the APM throughout the mission period (Element Resource Plan).
o Key milestones and decision points such as design reviews, acceptance reviews, readiness reviews, etc.
o Management Plan and Policy, showing rules for decision making, program requirements and directives

3.2 Tactical Planning

Tactical Planning beginns with the implementation phase of the mission. Tactical planning as it applies to Flight Operations is related to the distribution of resources to users/consumers within the APM (or any other flight element) and to the production of a schedule of activities for operations in flight and on the ground.

During Phase C/D, the mission definition documentation from the strategic planning will be analysed and a mission plan generated which schedules the operations activities throughout the period of the mission. This plan will be iterated and tested until the tactical planners are satisfied that it is a feasible plan that can be executed within the envelope of resources allocated to the APM by the strategic plan, and which will achieve the objectives of the mission. It will then form the basis for more detailed planning for the first execution period and will serve as a model for subsequent tactical planning activity.

As the mission progresses, tactical planning will always address the next execution period. Currently, execution periods are tied to Shuttle resupply flights and are 90 day periods. The tactical planning phase will always consider the entire mission period defined in the strategic plan, but will concentrate on the next 90 day period. It is necessary to consider the entire mission in order to evaluate achievement of objectives to date and to look ahead for opportunities, or requirements still to be achieved.

Since this schedule resulting from the planning also includes ground segment and communication systems, it may include certain pre-mission activities such as verification and testing of new flight or ground systems, personnel training, development of procedures, etc.

The objectives of the planning is to allocate available resources to systems and users and the output, therefore is called a Tactical Resources and Operations Plan, detailed enough to show a schedule of systems operations and blocks of time where specific resources are allocated to UOC's.

Examples of the resources scheduled by the Tactical Resource and Operations Plan are:

o Electrical energy
o Electrical power
o Thermal control/heat rejection
o IVA crew time
o EVA crew time
o Special crew skills

o Data management systems
o Communication network
o Environment
   - Micro gravity
   - Vibration
   - Electromagnetic
   - Contamination
     * Between experiments
     * Gas or liquid dumps
o Consumables
   - Oxygene, nitrogen (for ECS, life support, airlock)
   - Food and drinking water
   Personal hygiene supplies
o Stowage space
o Resupply/logistics
   - Volume
   - Mass
   - Stowage space
o Ground equipment
   - Ground processing facilities
   - Simulators
   - Mock ups
   - Trainers

The starting point for the tactical planning is the strategic level allocation of resources to the APM. It is this envelope of resources which has been allocated to each module which enables decentralised tactical planning. As the mission progresses and the performance of the Space Station systems becomes better known, the resources could also be necessitated by failures or anomalies in the Space Station systems. These resource envelopes will be routinely taken into consideration by the distributed tactical planners. Furthermore, Lab Modules may trade resources for specific execution periods within the constraints of system limitations, of course.

The first step in the tactical planning process is to assign the required resources to the APM systems, based on routine operations. The remaining resources are then allocated to the payload. The users decide on the distribution of resources to the various UOC's based on the strategic mission plan and coordinated by the EPOCC. UOC's may barter among themselves for resources and the decentralised plans are integrated by the EPOCC. The EPOCC may also discuss and evaluate trades of resources with other POCC's. The EMCC evaluates the systems capabilities to accommodate the negotiated trades, and then integrates the payload and system plans. The plan at this time consists of a 90 day

schedule showing envelopes of resources allocated to UOC's
for blocks of time, as well as systems operations and any
known Space Station systems operations such as re-boost
manoeuvres.
This APM plan is then sent to the SSSC. The SSSC evaluates
any trades of resources negotiated between modules against
the Space Station system capabilities and then merges the
TROPs from the various Lab Module MCC's into an integrated
Tactical Resource and Operations Plan, which now shows the
same level of detail as the APM plan for all modules and
the Space Station itself. Each of these planning activities
may require reiteration and coordination. During the sub-

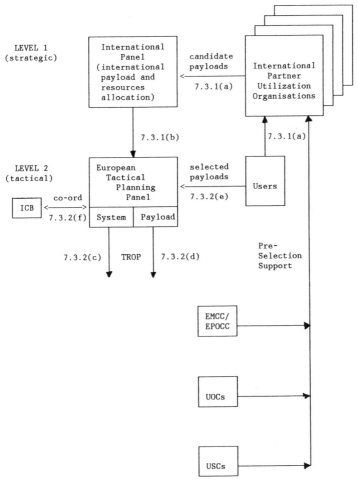

Fig. 2 Strategic and Tactical Planning Structure.

sequent detailed planning and execution, the EMCC and EPOCC
ensure that the module and payload remain within their resource envelope. During the execution of the 90 day plan,
tactical planning for the next 90 day period is going on in
parallel. The TROP must be available with sufficient lead
time to enable the nect level of planning. The lead time is
presently TBD, but is currently considered to be 2-4 weeks.
     The main point of the concept for tactical planning is
that it is decentralised; each module does its own planning. This is made possible by the allocation of resources
to the individual modules at the strategic planning level,
and is based on a willingness to accept less than 100%
efficiency in utilisation of Space Station resources.
Allowing modules to trade resources will provide sufficient
effciency. It is, therefore, not neccessary that resources
be allocated to each element by the Space Station at the
beginning of each planning cycle; they are already known
from the Strategic Plan, plus updates during the mission –
certainly, communication and coordination will be required
between the SSSC and module MCC's between POCC's and
between UOC's. But, there is not a need to colocate the
people, as electronic means can be used.
The other main point of the tactical planning is that it is
the objective to give as much freedom as possible to the
users to assign resources to the payload and to plan their
own operations, just as it is for each module to manage its
own resources and plan its own operations. The Strategic
Plan not only provides resources to individual modules, but
also provides rules for decision making and guidelines for
conflict resolution.
Figure 2 shows the strategic and tactical planning structure.

3.3 Execution Planning

The TROP is intended to allow decentralised detailed
planning and operations execution. From the TROP executable
timelines and crew activity plans (Composite Timeline) will
be generated for weekly increments of time. This time lines
will be more detailed, but the level of detail for the
payload operations timeline will be at the discretion of
the UOCs.
Operations will normally have to be planned in detail owing
to internal EMCC, EPOCC or UOCs requirements to use properly the allocated resources. The access to the last
planning information available by all parties concerned
will allow an improvement in the utilisation of resources.
The operations concept therefore includes a common data

base and tools for operations planning. For some scarce
resources, it can be expected that detailed planning and
iterations between UOCs and EPOCC or between EPOCC and EMCC
will be unavoidable. The common data base and planning
tools will improve the response time of these iterations.
In any case, the flexibility for last minute replanning
within allocated envelopes requires that the control
concept is based on resource monitoring (and not only on
command screening). Some implications on resource monitoring are described in the following section. Figure 3 shows
the Execution planning structure, Figure 4 summarizes all
Operations planning products.

4. Execution Management

The requirements on the management structure for operations
execution derive from the decisions to be taken during
operations. These decisions rely on earlier decisions and
management procedures approved at higher management levels
(strategic and tactical).
To be effective, the strategic and tactical decisions require that the various affected parties are properly represented in the decision process. The various parties in the
Columbus operations derive from the relationship customer-
supplier, as well as from the funding entities. In this re-

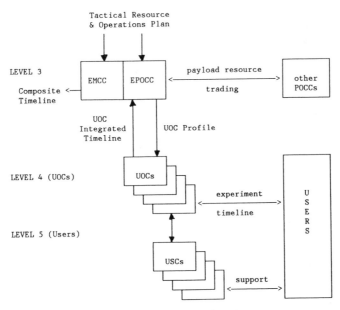

Fig. 3 Execution Planning Structure.

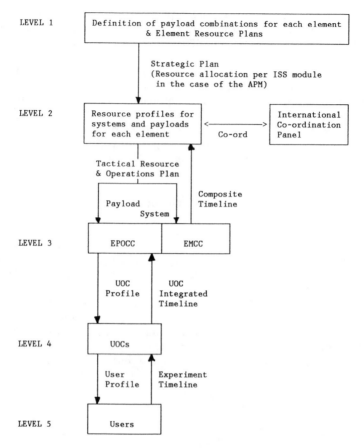

Fig. 4   Operations Planning Products for the Columbus Elements.

spect, we have to consider on one side:

o   Columbus Utilization Organisation
o   Flight Operations Organisation
o   Support functions (Logistics, Ground Processing, Engi-
     neering Support, NSTS, HERMES, ARIANE-5, TDRSS, EDRS,
     etc).

and on the other side

o   the European interests,
o   the international partners' interests

The principle for the mangement of Columbus Operations is
that most situations shall be resolved by predefined

decisions (i.e. Mission Rules) without any need for management intervention of the next higher level. Nevertheless,
the execution level management structure must be such that
response to exceptional situations can be taken in a timely
fashion.
At execution level, the various management interfaces are
represented in Figure 5.
Only Japan and Europe are shown in the figure as international partners, but it should be possible to exchange
any of them by the United States, for example.
The Flight Operations provides a service to the Columbus
users and is supported by the various operations support
functions. In these three groups of functions, there are
interests of the various international partners and situations may arise where the individual partners' interests
are in conflict. When those interests cannot be easily
separated, an international management body is needed.
Therefore, the need for all international partners to be
represented at the Space Station Support Centre is foreseen, although the size or staffing of this international
management body is still to be defined. For the Polar Platform and MTFF, there is no need for an international body
at execution level, the resources are allocated by European
strategic and tactical planning levels.

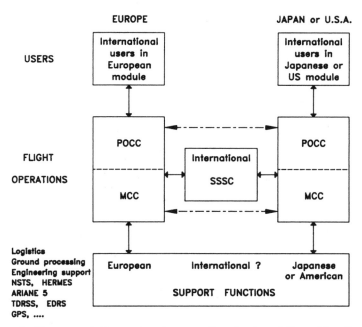

Fig. 5 Management Communications at Tactical and Execution Level.

The Flight Operations management structure at execution
level will be as follows: All interfaces of the flight
operations function with the Columbus flight element mani-
fested Users will be through the EPOCC. Conflicting issues
between UOC's should be decided by the Payload Operations
Coordination Manager who might be supported by some user
representation on site. The Payload Operations Coordination
Manager will therefore represent the Users' interests via
the Flight Director, located at the EMCC. The Flight
Director is responsible for systems operations coordination, as well as achieving the overall mission objectives
at execution level and will decide on issues between the
EPOCC and the EMCC.

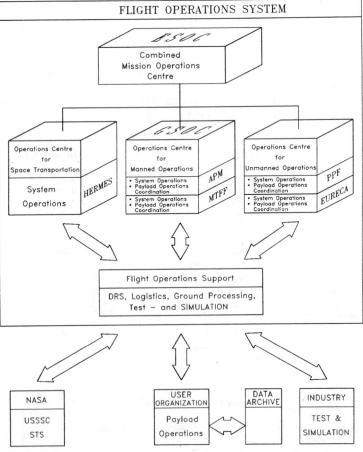

Fig. 6 European IOI Flight Operations System Implementation
Proposal.

Direct communications   between   EPOCC's can be desirable in
order   to discuss possible trades of resources between   the
APM and other attached modules.
Issues between the APM EMCC and the SSSC are expected to be
decided   at the SSSC, with the participation of the   inter-
national representatives.
Execution level issues between the MTFF or PPF-EMCC and the
Support Fuctions are expected to be decided by the EMCC.
For combined European mission activities involving a manned
vehicle (MTFF or PPF), the   final management decisions will
be taken by a combined Mission Operations Center.
Figure 6 shows an implementation proposal for the described
European   In-Orbit-Infrastructure   (IOI) driven by   opera-
tions-oriented seperation of tasks.

Computer-Aided Conceptual Design and Cost Modeling of Space Transportation Systems

Malcolm G. Wolfe*

The Aerospace Corporation, Los Angeles, California

Abstract

A primary key to the increased utilization of space by all nations is low-cost space transportation. The problem is that a vast number of competing transportation system concepts are being proposed, and a system that might appear cost-effective under one scenario could be completely unacceptable economically under another. Each country needs to be able to examine its own space transportation requirements and to answer such questions as, for example, whether to depend on other nations or whether to develop its own domestic capability, or if it intends to use another nation's transportation system, which system best meets its needs. Modern computer technology can aid in this activity by permitting the synthesis of micro- or minicomputer-based design/cost/economics models. A computer-aided procedure is described which aids in creating and analyzing conceptual space transportation systems. The program assists the designer by developing an optimal transportation vehicle design, analyzing the vehicle's performance characteristics, performing a mass properties analysis, and estimating the total life-cycle cost of the system. The program architecture is outlined. The individual program modules and their input and output parameters are described. The model is designed to deal with most kinds of systems (up to four stage vehicles), including expendable, partially reusable and fully reusable vehicle concepts.

This paper represents the opinions of the author only and does not necessarily reflect any official endorsement by either The Aerospace Corporation or the United States Air Force.

*Senior Engineer, Space Launch Operations.

66

Keywords

Launch vehicle, transfer vehicle, space transportation, computer-aided design, microcomputer, design, mass properties, costing, modeling.

Introduction

The space community is searching for new plateaus in the exploration and exploitation of space with ambitious plans in telecommunications, space travel, space manufacturing and processing, permanent manned space stations, and the use of space for military purposes (Paine et al)[1]. Virtually all nations wish to participate. Several countries own and operate their own satellites, and many more countries are expected to do so in the future. More than 100 nation-states have joined the International Telecommunications Satellite Organization, for instance, and new alliances and agreements are being made continually.

An important key to the increased utilization of space by all nations is low-cost space transportation. However, future space transportation architectures to support the kind of activities described by Paine et al[1] are likely to be quite complex (see Fig. 1).

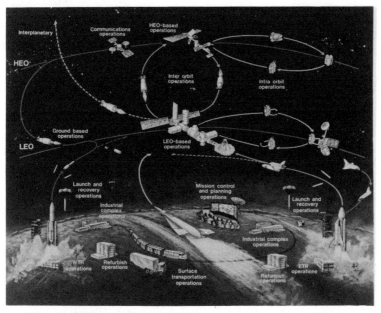

Fig. 1 Typical Advanced Space Transportation Scenario

There are currently at least nine countries with
funded space launch programs (Brazil, France, India,
Japan, the Peoples Republic of China, the Soviet Union,
the United Kingdom, the United States, and West Germany,
see Fig. 2), and all countries need to be able to examine
their own space transportation requirements in the light
of their own national goals. A large number of
transportation system concepts are being developed or
studied, from current technology, multi-stage expendables
to high-technology, single-stage fully reusables. Each of
these competing systems have ill-defined advantages,
disadvantages, costs, and risks. In addition, it is very
unclear what mission requirements will be beyond about a
five-year future, while advanced transportation systems
are likely to take twelve years or more to bring to an
operational capability. A transportation system that
might appear very cost effective under one operational
scenario could be economically unacceptable under another.
Crucial questions that each nation must face are whether
to depend on other nations for space transportation,
whether to develop its own domestic capability, or, if it
intends to use another nation's transportation system,
which system best meets its needs.
    Wilhite, Johnson, and Crisp[2] and Wolfe and Knight[3]
have demonstrated that modern computer technology can be

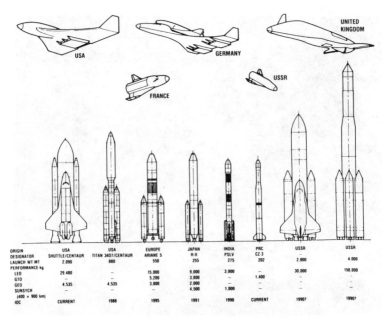

ORIGIN	USA	USA	EUROPE	JAPAN	INDIA	PRC	USSR	USSR
DESIGNATOR	SHUTTLE/CENTAUR	TITAN 34D7/CENTAUR	ARIANE 5	H-II	PSLV	CZ-3		
LAUNCH WT MT	2.090	880	550	255	275	202	2.000	4.000
PERFORMANCE-kg								
LEO	29.480	–	15.000	9.000	3.000	–	30.000	150.000
GTO	–	–	5.200	3.800	–	1.400	–	–
GEO	4.535	4.535	3.800	2.000	–	–	–	–
SUNSYCH	–	–	–	4.500	1.000	–	–	–
(400 × 900 km)								
IOC	CURRENT	1988	1995	1991	1990	CURRENT	1990?	1990?

Fig. 2 Global Space Transportation Concepts

used to synthesize computer-based design/cost/economics models which permit the experienced designer to investigate a wide variety of transportation system options. The computer can quickly process large numbers of transportation system scenarios, normalize information originating at a multitude of different sources, and store the information in an accessible data base. This permits tradeoff and sensitivity analyses to be performed and intelligent economic decisions to be made.

The model described herein encompasses the transportation system segment of a space transportation architecture (see Fig. 3). Complementary models are expected to be developed for the other two space transportation architecture segments: that is, the logistics support system and the mission control system. This model is designed to generate a computerized transportation system data base consisting of the basic transportation system design, performance, weight, cost, and economic characteristics to the element, stage, and subsystem level. Together with logistics system, mission control system, mission requirements, technology, and operations data bases, it can be used to aid in conducting broad transportation architecture, technology definition, economic, and budgetary studies.

The program is menu-driven, and inputs such as payload weight and dimensions; orbital destination parameters; launch site parameters; number of stages; desired stage and engine design characteristics; operational characteristics; and desired development, vehicle purchase, and operational schedules are requested of the user as the system design proceeds. Windowing is provided to permit a review of the assumed and calculated data at

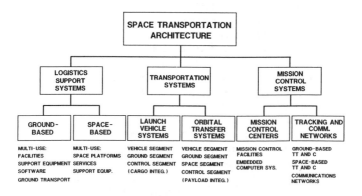

Fig. 3 Space Transportation Architecture

strategic points in the design. The windows can be used
to perform tradeoff and sensitivity analyses by permitting
original inputs or calculated data to be modified or to
permit initial entry into the program at selected points
downstream in the program. Typical transportation issues
that the program is intended to aid in attacking are
listed in Table 1; transportation-related technologies are
assumed to be lumped into the categories listed in Table 2
to permit gross technology tradeoffs to be made. A
computerized technology cost/benefit model could, if
developed, feed information into the program through this
entry point.

Table 1   Typical Transportation Issues

- STRUCTURES AND MATERIALS
- PROPULSION
- AVIONICS, COMPUTERS AND SOFTWARE
- AERO/THERMODYNAMICS
- FACILITIES/OPERATIONS
    - MANUFACTURING
    - SERVICING AND MAINTENANCE
    - RECOVERY
    - AUTOMATION
    - OPERATIONS
    - SURFACE TRANSPORTATION
    - POWER SYSTEMS

Table 2   Space Transportation
Technology Categories

- MANNED vs UNMANNED
- REUSABLE vs EXPENDABLE
- AIRBREATHING vs ROCKET
- VERTICAL TAKEOFF vs HORIZONTAL TAKEOFF
- BALLISTIC vs LIFTING REENTRY
- PROPULSIVE vs AEROASSIST RECOVERY
- GLIDE BACK vs FLY BACK RETURN
- LIQUID vs SOLID BOOSTER
- LOX/H2 vs LOX/HC PROPELLANTS
- MODULAR vs MONOLITHIC CONSTRUCTION
- LOW vs HIGH TECHNOLOGY
- SINGLE FLEET vs MIXED FLEET
- MANY SMALL VEHICLES vs FEW LARGE VEHICLES
- GROUND OPERATIONS vs FLIGHT OPERATIONS

Design Approach

Previous approaches to the problem of examining transportation vehicle design with computer aids have generally taken one of two possible approaches, each achieving some limited measure of success. The first is to utilize a number of stand-alone computer programs for performing very specific portions of the design process. The programs often are written by the engineering expert himself in whatever language he may be familiar with and utilizing whatever programming skills he may possess. The expert maintains the computer code that performs his specialty task and is responsible for making design improvements.

From the wide range of disciplines needed to design and cost transportation systems, a variety of programs are written. For example, sizing programs are used to estimate performance, to obtain the general size of the vehicle, and to develop a geometry. Then this information is passed on to a mass properties analyst who uses analytically and empirically derived data, in the form of Weight Estimating Relationships (WERs), to estimate the vehicle subsystem weights. Finally, these data are passed on, together with other programmatic inputs, to a cost analyst who uses historical data from previous systems in the form of Cost Estimating Relationships (CERs) to estimate the total cost of the system.

The problem with the above approach is that the stand-alone nature of the computer tools creates islands of automation in which there are no links between the sources of computerized expertise. The vehicle designer cannot get rapid feedback on the cost ramifications of design changes and thus is not able to quickly evaluate design options from a cost or economic standpoint. From a practical point of view, this limits the number of design alternatives that can be examined in a reasonable period of time and so can potentially result in designs that are considerably less than optimal. Without real-time guidance from other experts in each of the areas which need to be considered, an expert in one discipline cannot best perform the task for which his expertise is valued.

Another approach is to create an integrated computer tool by having the experts describe the essential aspects of their design or analysis task to a group of dedicated computer programmers. These programmers then code the design algorithms into one large program which bridges the islands of automation. A common programming language is used, allowing for a smooth transition from one task to another. Code readability and maintainability are improved, owing to the advanced programming techniques used by the programmers (as compared to the techniques used by the engineering experts).

There are a number of inherent problems in this approach, however, which have limited the usefulness of such projects. First, it requires a large program development team with a diverse mix of engineering and programming skills. The application engineers must accurately relate the design algorithms involved to the programmers. Then the computer model must be "exercised" and debugged to ensure that the design algorithms accurately reflect the design process. If there is a conflict, there is no single person who completely understands both the engineering and programming details involved. The heuristic algorithms that are a significant part of the design process are difficult to program and therefore are oversimplified or neglected. Finally, design is a complex and dynamic skill, and adding to or improving the program's design algorithms requires that the information exchange process be repeated each time a modification is made.

To avoid some of these difficulties, a pragmatic design philosophy was adopted which allows a computer model to be produced with a minimum of manpower and a maximum of flexibility. The technical judgments involved in the development of each of the technical modules which constitute the total space transportation model are left to the specialist in that particular area. The cognizant technical specialist creates the design or analysis model within the framework of the model architecture and, where possible, creates, maintains, and improves the computational algorithms. The resulting modules are integrated into a single user-friendly integrated package by the executive module. Within certain constraints, the technical specialist is then free to refine a particular portion of the model without necessitating extensive changes to the rest of the program. The model will improve with use, since exercising it as an operational tool will identify within which modules increased fidelity is desirable.

The user interacts with the model via the computer terminal. The model is intended, of course, to emulate the procedure used by a designer when he/she designs. Its objective is to put the analytical load and data management responsibility on the computer, leaving the designer to concentrate on the heuristic component of design. In spite of valiant (if sporadic) efforts, no one

yet understands how education, training, experience,
information, knowledge, imagination, curiosity,
creativity, and judgment are used by the designer to
produce a "good" design, and the complete design process
cannot at this time be fully automated. The Artificial
Intelligence (AI) community has recently initiated a
renewed and concentrated attack on this generic problem,
and progress is anticipated.

Model Anatomy

General Program Architecture

The general program architecture is shown in Fig. 4.
The program consists of an executive module and a number
of application modules. Each of these modules is
described below. The executive allows the user to
initiate the design and costing process from any of a
number of windows. The windows provide a method of
entering the model with little more than a general
specification of the vehicle performance requirements and
general characteristics, or alternately, a completely
defined vehicle can be costed. In this way, a
transportation system can be developed from the beginning,
or designs which already exist can be analyzed.
The program presently consists of three main logic
components: Design/Sizing (using a vehicle optimization
or a vehicle synthesis approach), Mass Properties, and

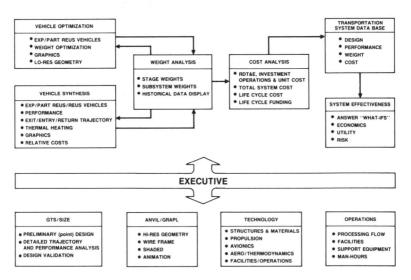

Fig. 4 Model Schematic

Costs. In addition, other useful functions, such as design
graphics (for verifying the reasonableness of program
inputs), and interfaces to external programs are provided.
The executive module links these programs together as
described below. The program is intended to deal with
both launch vehicle systems and transfer vehicle systems.
The launch vehicle design morphology is illustrated in
Fig. 5; the transfer vehicle design morphology is
illustrated in Fig. 6. The general input data format and
the windowed data format for a launch vehicle are outlined
in Figs. 7 and 8, respectively.The general input data
format and the windowed data format for a transfer vehicle
are outlined in Figs. 9 and 10, respectively. The output
data format outlined in Figs. 11, 12, and 13 is common to
both a launch vehicle and a transfer vehicle.

Executive Module

The executive module controls program information flow
between each of the application modules. The various
modules are accessed through the executive's selection
menus. The executive serves other functions, such as data
collection, error checking, help information, and model
storage and retrieval.
Before the sizing modules are run, the transportation
system must be defined. This is done through data
collection menus by inputting the general transportation
system characteristics. The user can easily move around
from field to field in the menu and edit any input. A
context-sensitive help screen, showing more detail on the
menu items (and/or instructions on using the menus) is

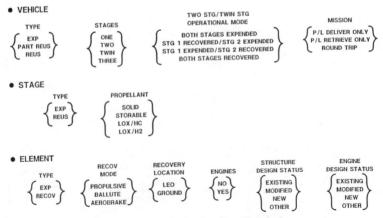

Fig. 5 Launch Vehicle Design Morphology

provided. When the user has specified the inputs, a key stroke is used to accept the menu and enter the data into the system model. Menus can always be redisplayed and changes inputted.

Menus such as this not only provide a convenient mechanism for system specification but are also used to check the data for validity. A mechanism is provided to check for nonsensical or conflicting inputs, and some default capability is provided to aid the less expert user. The program thus has some degree of inherent expertise built into its logic. However, it requires that an expert also be on the <u>outside</u> of the computer (i.e., the user must have at least a minimum level of expertise).

The data collection menus also serve as the previously mentioned program windows. This is because output data from the various application modules are inserted into what are actually data collection menus. The user thus may simply review the results or may change them. This might be desired if the user would like to perform parametric trade studies. Another possible application is when the user knows from previous experience that a particular quantity should be changed. Finally, this allows the model to be entered at any point in the procedure. For example, a set of weight data could be used to estimate vehicle costs without running the design and sizing portion of the model.

Design/Sizing Modules

Basic sizing of the vehicle is performed, based on the design characteristics specified, the payload weight, the launch site, and the mission orbital parameters.

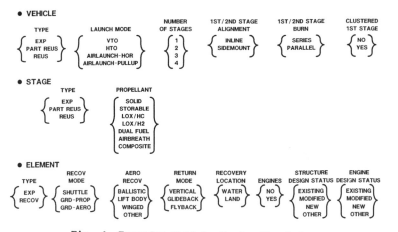

Fig. 6   Transfer Vehicle Design Morphology

LV ELEMENT DATA INPUT   VEHICLE: NAME: _____
CODE: _____

LV STAGE DATA INPUT   VEHICLE: NAME: _____
CODE: _____

LV VEHICLE DATA INPUT   VEHICLE: NAME: _____
CODE: _____

NAME OF ORIGINATOR: _____ DATE: _____
ORIGIN OF VEHICLE: _____
QUANTITY FLOWN: _____ ;YEAR OF CONSTANT DOLLARS:_____
YEAR OF ATP_____ ;YEAR OF 1ST FLT: _____ ;YEAR OF LAST FLT: ____
LAUNCH SITE: _____ (1-ETR; 2-WTR; 3-OTHER); LAUNCH AZIMUTH (deg) :____
LAUNCH SITE SELECTION 3: LATITUDE (deg) : _____ ;LONGITUDE (deg) : _____
TOTAL DELTA VELOCITY : (ft/sec) : _____ ; ORBITAL VELOCITY (ft/sec) : _____
DRAG LOSS (ft/sec) : ____; GRAV LOSS (ft/sec) :____; THRUST LOSS (ft/sec) :____
PAYLOAD: WEIGHT (lb) : _____ ;LENGTH (ft) : ____ ;DIAMETER (ft) :____
    FAIRING: _____ (SELECT CONFIGURATION: 1 THRU 8)
    ORBIT: INCLIN (deg) : _____ - ; APOGEE (nm) : ____ PERIGEE (nm) : ____
LAUNCH VEHICLE: TYPE: ____ (1-EXP; 2-PART REUS; 3-REUS)
    LIFTOFF MODE: ____ (1-VTO; 2-HTO; 3-AIRLAUNCH)
    GROSS WEIGHT (lb) : ____
    NUMBER STAGES:_____; INITIAL T/W:_____
    FIRST/SECOND STAGE ALIGNMENT: ____ (1 = INLINE; 2 = SIDEMOUNT)
    FIRST/SECOND STAGE BURN: _____ (2 = SERIES; 2 = PARALLEL)
    FIRST/SECOND STAGE THRUST RATIO: _____
    CLUSTERED FIRST STAGE: _____ (Y/N)
    NUMBER COMMON ELEMENTS: CLUSTER NO.1: _____
                     CLUSTER NO.2: _____
NOTE: _____

Fig. 7   Launch Vehicle Input Format

LV WEIGHT DATA BY STAGE (lb)   VEHICLE: NAME: ____ CODE: ____

LV DESIGN DATA BY STAGE   VEHICLE: NAME: ____ CODE: ____

LV COMMONALITY DATA BY ELEMENT   VEHICLE: NAME: ____ CODE: ____

LV WEIGHT DATA BY ELEMENT (lb)   VEHICLE: NAME: ____ CODE: ____

LV DESIGN DATA BY ELEMENT   VEHICLE: NAME: ____ CODE: ____

	ELEMENT	1	2	3	4	5	6
STAGE							
ELEMENT	– TYPE						
	– QUANTITY						
	– DESIGN STATUS						
	– LENGTH (ft)						
	– WIDTH (ft)						
	– DIA (ft)						
	– FUEL VOL (ft³)						
	– OXID VOL (ft³)						
	– X SECTION AREA (ft²)						
	– PLANFORM AREA (ft²)						
	– WING AREA (ft²)						
	– ASPECT RATIO						
	– RECOV MODE						
	– RECOV TYPE						
	– RETURN MODE						
ENGINE	– TYPE						
	– DESIGNATOR						
	– DESIGN STATUS						
	– QUANTITY/ELEMENT						
	– THRUST/ENGINE (lbf)						
	– SPEC IMPULSE (sec)						
P/L FAIR	– TYPE (1 thru 6)						
	– LENGTH (ft)						
	– DIA (ft)						
	– WEIGHT (lb)						

Fig. 8   Launch Vehicle Window Format

TV ELEMENT DATA INPUT   VEHICLE: NAME: _____
                                                CODE: _____

TV STAGE DATA INPUT   VEHICLE: NAME: _____
                                          CODE: _____

TV VEHICLE DATA INPUT   VEHICLE: NAME: _____
                                            CODE: _____

NAME OF ORIGINATOR: _____ DATE: _____
ORIGIN OF VEHICLE: _____
QUANTITY FLOWN: _____ ; YEAR OF CONSTANT DOLLARS: _____
YEAR OF ATP _____ ; YEAR OF 1ST FLT: _____ ; YEAR OF LAST FLT: _____
PARK ORBIT: INCLIN (deg): _____ ; APOGEE (nm): _____ ; PERIGEE (nm): _____
MISSION ORBIT: INCLIN (deg): _____ ; APOGEE (nm): _____ ; PERIGEE (nm): _____
DELTA VEL (ft/sec): DESIGN: _____ ;1ST BURN: _____ ; 2ND BURN: _____
PAYLOAD: DELIVERED: WEIGHT (lb): _____ ; LENGTH (ft): _____ ; DIA (ft): _____
            RETRIEVED: WEIGHT (lb): _____ ; LENGTH (ft): _____ ; DIA (ft): _____
TRANSFER VEHICLE: TYPE: _____ ; OPTIMIZE DV: _____ (Y/N)
            NUMBER STAGES: _____ ; INITIAL T/W: _____
            FLIGHT MODE: _____ ; RECOV TYPE: _____
            GROSS WEIGHT (lb): _____
            FIRST/SECOND STAGE ALIGNMENT: _____ (1 = INLINE; 2 = SIDEMOUNT)
            FIRST/SECOND STAGE BURN: _____ (1 = SERIES; 2 = PARALLEL)
            FIRST/SECOND STAGE THRUST RATIO: _____
            CLUSTERED FIRST STAGE: _____ (Y/N)
            NUMBER COMMON ELEMENTS: CLUSTER NO. 1: _____
                                               CLUSTER NO. 2: _____
NOTE: _____
        _____

Fig. 9   Transfer Vehicle Input Format

TV WEIGHT DATA BY STAGE (lb)   VEHICLE: NAME: _____
                                               CODE: _____

TV DESIGN DATA BY STAGE   VEHICLE: NAME: _____
                                        CODE: _____

TV COMMONALITY DATA BY ELEMENT   VEHICLE: NAME: _____
                                             CODE: _____

TV WEIGHT DATA BY ELEMENT (lb)   VEHICLE: NAME: _____
                                               CODE: _____

TV WEIGHT DATA BY STAGE (lb)   VEHICLE: NAME: _____
                                            CODE: _____

	ELEMENT	1	2	3	4
STAGE					
ELEMENT	– TYPE	___	___	___	___
	– QUANTITY	___	___	___	___
	– DESIGN STATUS	___	___	___	___
	– LENGTH (ft)	___	___	___	___
	– WIDTH (ft)	___	___	___	___
	– DIA (ft)	___	___	___	___
	– X SECTION AREA (ft^2)	___	___	___	___
	– PLANFORM AREA (ft^2)	___	___	___	___
	– RECOV TYPE	___	___	___	___
ENGINE	– TYPE	___	___	___	___
	– DESIGNATOR	___	___	___	___
	– DESIGN STATUS	___	___	___	___
	– QUANTITY/ELEMENT	___	___	___	___
	– THRUST/ENGINE (lbf)	___	___	___	___
	– SPEC IMPULSE (sec)	___	___	___	___

Fig. 10   Transfer Vehicle Window Format

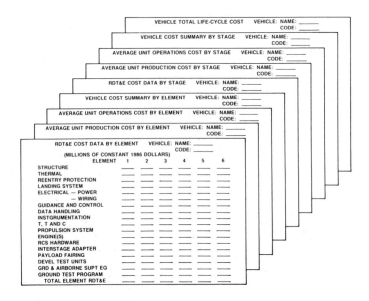

Fig. 11   Output Data Format

Algorithms are used to calculate the optimal propulsive
size of each element of the system from classical
equations which define the contribution of each stage to
the vehicle's total propulsive capability (incremental
velocity, or "delta V"). The weight of the specified
propellant for each stage and the structure factor are
iteratively calculated to determine the optimal overall
vehicle size. Performance equations are used to determine
thrust, drag, and gravity losses that need to be accounted
for in the sizing process. These losses are added to the
orbital velocity called for by the mission requirements.
The procedure can also determine the size of a system
where the various stages have fixed delta-V properties.
This is useful for examining transportation systems in
which one or more stages already exist.

      The modules are integrated together by the executive
in such a fashion as to provide a flexible method of
sizing vehicles. The user can input the orbital
requirements and determine the basic size of the vehicle.
Then the vehicle components can be modified in various
ways to determine the effect of a change in one stage on
the rest of the transportation system. A wide variety of
stages designs, propellant types, and technology
assumptions can be examined.

VEHICLE COST SUMMARY BY STAGE — VEHICLE: NAME: _____
                                     CODE: _____

(MILLIONS OF CONSTANT 1986 DOLLARS)

	RDT&E	(AVG UNIT)	INVESTMENT	OPERATIONS	TOTAL
STAGE 1	____	(____)	____	____	____
STAGE 2	____	(____)	____	____	____
STAGE 3	____	(____)	____	____	____
STAGE 4	____	(____)	____	____	____
PAYLOAD FAIRING	____	(____)	____	____	____
LAUNCH OPERATIONS	____	(____)	____	____	____
FLIGHT OPERATIONS	____	(____)	____	____	____
RECOVERY OPERATIONS	____	(____)	____	____	____
SPARES	____	(____)	____	____	____
SOFTWARE	____	(____)	____	____	____
FACILITIES	____	(____)	____	____	____
SE AND TECH SUPPORT	____	(____)	____	____	____
GOVERNMENT SUPPORT	____	(____)	____	____	____
CONTINGENCY (0%)	____	(____)	____	____	____
TOTAL COST	____	(____)	____	____	____

Fig. 12   Stage Cost Summary Format

VEHICLE TOTAL LIFE-CYCLE COST — VEHICLE: NAME: _____
                                 CODE: _____

(MILLIONS OF CONSTANT 1986 DOLLARS)

FUND TYPE	FY 1987	1988	1989	1990	1991	1992	1993	BALANCE TO GO	TOTAL
3600 (RDT&E)	____	____	____	____	____	____	____	____	____
3020 (Procurement)	____	____	____	____	____	____	____	____	____
3080 (Grnd Equip)	____	____	____	____	____	____	____	____	____
3300 (Facilities)	____	____	____	____	____	____	____	____	____
3400 (Operations)	____	____	____	____	____	____	____	____	____
3500 (MIL Pay)	____	____	____	____	____	____	____	____	____
TOTAL	____	____	____	____	____	____	____	____	____

(MILLIONS OF THEN-YEAR DOLLARS)

FUND TYPE	FY 1987	1988	1989	1990	1991	1992	1993	BALANCE TO GO	TOTAL
3600 (RDT&E)	____	____	____	____	____	____	____	____	____
3020 (Procurement)	____	____	____	____	____	____	____	____	____
3080 (Grnd Equip)	____	____	____	____	____	____	____	____	____
3300 (Facilities)	____	____	____	____	____	____	____	____	____
3400 (Operations)	____	____	____	____	____	____	____	____	____
3500 (MIL Pay)	____	____	____	____	____	____	____	____	____
TOTAL	____	____	____	____	____	____	____	____	____

Fig. 13   Vehicle Life-Cycle Cost Format

Vehicle Geometry

A simple graphics module is provided to show how the system looks geometrically. This provides visual feedback to the designer as to the design's validity and ensures that the vehicle design parameters have been accurately entered and the vehicle is indeed what is expected. An example of this graphical feedback is shown in Fig. 14. The screen displays four versions of a specific

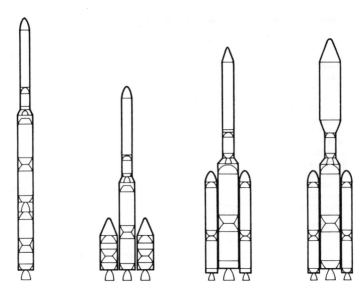

Fig. 14   Configuration Development

configuration at one time. However, scrolling is provided
to permit more than four versions to be compared.

Subsystem Weights

Once the vehicle has been sized, a breakdown of the
vehicle element subsystem weights is needed to aid in
estimating costs. The mass properties module provides an
estimate of these weights based on the overall size and
the design characteristics of each element or stage. WERs
are used that have been analytically and empirically
developed and refined over the years.

Costing Module

The cost module treats each element of a given vehicle
separately and aggregates the costs to obtain appropriate
element, stage, and vehicle subtotal and total data. Each
vehicle element, in turn, is estimated by each of its
major subsystems. CERs that are subsystem-oriented are
used to estimate the Research, Development, Technology,
and Engineering (RDT&E) and unit production costs for each
subsystem. Weight and quantity produced are the primary
determinants of cost and, thus, constitute the principal
inputs that are needed. In addition, the type and design
status of each stage and of its engine(s) is required to

determine the extent to which RDT&E must be included.
Programmatic inputs such as fiscal year when the program
will commence, first and last launches, and year of
constant dollars (for price index effects) are also
inputted.

The major output of the cost model is a life-cycle
cost estimate for the complete vehicle, broken down into
the principal categories of RDT&E, investment, and
operations. These costs are further broken down by stage
(or element), fairing, launch operations, flight
operations, recovery and refurbishment operations, spares,
software, facilities, system engineering and technical
support, and government support. Other output formats
provide details on subsystem and element RDT&E and average
unit production cost. Operations cost estimates are shown
for each important category of cost. For funding
purposes, the total life-cycle cost is automatically
spread by fiscal year by funding type; that is, RDT&E,
Vehicle Investment, and Operations. Finally,
cost-quantity curves for the total launch vehicle and for
each stage are generated. The costs can be displayed
either in tabular form or in the form of pie or bar charts.

The cost module can estimate the cost of launch
vehicle concepts utilizing stages that are: (1)
propelled by liquid-oxygen and hydrocarbon or
liquid-hydrogen fuels, by storable fuels, or by solid
rocket motors; (2) expendable, partially reusable, or
entirely reusable; or (3) ground recoverable in a winged,
lifting body or ballistic mode or on-orbit recoverable
(via the Shuttle Orbiter, for example). It can estimate
the cost of orbit transfer systems that are: (1)
recoverable or expendable or (2) propulsively recovered
or aeroassist recovered. The cost of airbreathing stages
can also be estimated (but with low fidelity at this point
in time).

The output from the cost module can be displayed in a
number of ways. A variety of tabular outputs of various
cost parameters of the system can be requested, or the
user may choose to show the cost data graphically. For
example, a pie chart can be generated for any column of
the tabular data, or bar charts may also be requested.
These graphical forms of the cost data allow the user to
easily convey the relative costs for a particular system
to others.

External Interfaces

The program described herein will permit the user to
test the validity of a design concept, optimize the

concept, and rapidly examine a large number of variants of
the concept in a short period of time. It will not
originate the concept itself; that is the responsibility
of the designer. Some level of default capability is
incorporated, but a limited level of expertise is assumed
of the user. Although the program is sufficiently
user-friendly so that a reasonably skilled user can enter
input and generate output, the model is intended to be
most useful in the hands of a skilled expert engineering
user. At times it may be necessary for several experts (a
designer, a mass properties analyst, and a cost analyst,
for instance) to cooperate with the aid of off-line
calculations in processing a vehicle concept through the
model.

The model is intended as a conceptual design tool and
not as a substitute for more detailed vehicle sizing
models such as the NASA/Langley Research Center's
Aerospace Vehicle Interactive Design (AVID) program
(Wilhite, Johnson and Crisp)[2] or a similar program
(designated GTS/SIZE) being developed by The Aerospace
Corporation. These are intended to be operated by
engineering specialists and are essentially point design
rather than parametric design tools. The program described
herein however will have an immediately useful interface
with the models mentioned above. It will perform the
parametric design analysis necessary to identify favored
vehicle system candidates for more detailed study using,
for instance, GTS/SIZE. An automated interface between the
program and other such programs is currently being
considered.

The geometry output of the type illustrated in Fig. 14
is produced and used in an iterative way during the design
procedure to configure the vehicle. Once a vehicle
configuration is decided on, a computer-aided design (CAD)
input data package can be assembled and a minicomputer
used to produce the type of wire frame or shaded image
drawing output illustrated in Fig. 15. The model can be
used to validate and normalize data originating from a
wide variety of sources. It can be used to populate data
bases for other models which need transportation system
information.

Future Work

The program described herein is now at a prototype
development stage, and individual modules have been
validated. Improvements are currently being made to
refine the algorithms necessary for a wider range of
vehicles to be examined, perform more accurate analyses,

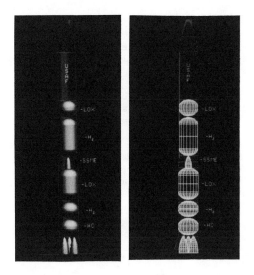

Fig. 15   ANVIL/GRAPL Output

and provide more sophisticated graphical output. The
second generation of the model will be hosted on a
micro-VAX minicomputer and will incorporate more efficient
programming procedures, an audit trail capability, and a
more sophisticated data base management system.

Summary/Conclusions

An integrated system for conceptualizing, sizing, and
costing future space transportation vehicles has been
described. This system has been found to quickly give
results which are almost as accurate as much larger models
which require detailed analysis and lengthy computer
processing. The model can therefore be used to provide
sufficiently accurate design and costing data to enable
parametric analyses to be performed while a system is
still in the conceptual design stage. The resulting
savings in computer time and manpower combined with the
capability to process a large number of options is a
valuable tool for selecting cost-effective future space
transportation systems.

Acknowledgments

The procedure described in this paper is the result of
the cooperative efforts of many individuals at The
Aerospace Corporation, including A. Arnold, J. Bennett,
R. Casten, H. Campbell, D. Ettlin, T. Hoang, J. Kimble,
F. Knight, and S. Wray.

References

[1]Paine, T.O., et al., "Pioneering the Space Frontier," The Report of the National Commission on Space, Bantam Books, Toronto-New York-London-Sydney-Auckland, 1986.

[2]Wilhite, A., Johnson, S. C., and Crisp, V., "Integrated Computer Programs for Engineering Analysis and Design," AIAA Paper 83-0597, Proceedings of the 21st AIAA Aerospace Sciences Meeting, Nevada, 1983.

[3]Wolfe, M. G. and Knight, F. L., "Space Transportation System Interactive Conceptual Design/Cost/Economics Model," IAA Paper 86-435, 37th Congress of the International Astronautical Federation, Innsbruck, Austria, 1986.

The Techniques of Manned On-Orbit Assembly

Leon B. Weaver*

Weaver Enterprises, Aptos, California

ABSTRACT

Manned on-orbit assembly may be defined as work performed by astronauts outside the habitable volume of an orbiting spacecraft to assemble and place in the operational environment spacecraft elements, which for one reason or another it is not practical to launch in the assembled configuration. This paper discusses the spectrum of activities required to design, test, emplace, and activate representative large space systems. Development of the assembly concept for any large space system begins with an analysis that characterizes the physical elements and defines options, and progresses through trade studies that ultimately provide a basis for selecting a specific design solution. Representative major trade issues inherent in many phases of this process are discussed. In general, manned simulations, particularly of the extravehicular activity (EVA) tasks, are required to support the decision-making process. The five basic reasons why manned simulation may be required are examined, as are the three requirements for "valid" manned simulations. The scenario for a representative manned simulation is discussed, and the four major mediums available for the conduct of manned EVA simulations are examined and compared. Manned on-orbit assembly is a practical technique for the emplacement and activation of large space systems. Although the actual in-flight experience base is still quite limited, a large body of experience exists with potential assembly tasks in the four earth-

    *Consulting Pilot-Engineer; President, Weaver Enterprises.

based simulation mediums to substantiate this conclusion. Finally, techniques for the analysis and simulation of EVA tasks are available and well understood.

INTRODUCTION

A wide variety of techniques have been analyzed, and many explored through manned simulation exercises, to accomplish the various assembly activities inherent in any large space system. Currently, one of the most visible examples of a planned large space system requiring a substantial degree of on-orbit assembly is the International Space Station, initiated by the United States. The techniques used to analyze, simulate, and ultimately assemble this large space system are generic in nature, i.e., are equally applicable to any other large space system.

The assembly of any large space system may emphasize either an erectable or a deployable design approach, or a hybrid combination of these. As used here, the word erectable refers to the piece-by-piece assembly in-flight of simple components not launched in a joined-together state. Deployable refers to spacecraft elements that are preassembled before flight, folded for launch, and unfolded and rigidized after achieving orbit.

In general, the design and development effort associated with the erectable approach is substantially lower than the analogous effort required for a deployable interpretation of the same hardware system, and the reliability of the emplacement operation is increased. The erectable approach, however, requires more on-orbit crew activity. The erectable vs. deployable issue represents a significant trade that must be considered in the development of a large space system. The present objective is not to explore this subject, but to show the potential on the erectable side of the issue.

ASSEMBLY FUNCTIONS

Assembly and activation functions span a wide spectrum of activities, each of which must be considered and treated with a specific design solution. Large space systems requiring multiple launches require, after the first flight, that each arriving load rendezvous and dock with a partially assembled spacecraft. Specific docking sites must be preplanned for each successive arrival, considering the specific assembly functions to be accomplished on that flight, as well as other significant issues such as constraints on the approach corridor, the stabilization and control of a unique and constantly changing configuration with each visit, the desire not to replicate or relocate the docking interface any more often than necessary, etc. Subsequent to docking, the newly arrived cargo must be unpacked from the launch vehicle. Attention is required to handling hardpoints on each cargo element, manipulator access to each element in the desired sequence, and the operation and backup of launch locks. Following release from the launch vehicle, each arriving cargo element must be transported to the assembly site and positioned for assembly. A trade at this point is whether an intermediate staging area is used to permit rapid unloading and release of the transport vehicle, or whether the cargo elements are unloaded only as required to support the assembly process.

The actual assembly process can be expected to require assembly support equipment (ASE); this may involve items ranging from simple portable foot restraints to complex robotically augmented extravehicular activity (EVA) workstations. Following installation, positioning, and any necessary checkout of the ASE, the actual assembly process may begin. For truss-based systems such as the International Space Station, the truss assembly itself is a necessary first step. However, many options must be considered for the positioning, installation, and activation of subsystem equipment on the truss; these options range on a continuum from installation essentially in parallel with truss assembly, to installation subsequent to completion of truss assembly. Each option presents a different set of requirements on the assembly support equipment, entails different timeline requirements, and offers different potential for the sequence of completion/activation of elements of the spacecraft capability. This is one of the most complex, significant, and difficult to scope assembly-related trade studies required in the development of a large space system.

Even within the elements of a given subsystem, many installation details and options must be considered. Subsystems typically include distributed modular equipment items interconnected by a variety of lines, including power, signal, fluid, gas, etc. Utility lines between protected areas are likely to require insulation or other environmental protection; the lines themselves may include both rigid and "flexible" types. The flexible lines may in some cases be large diameter items with substantial stiffness and/or memory, which may present handling difficulties to EVA crewmembers and their support equipment. Again, many aspects must be considered: Flexible vs rigid lines? Lines to be pre-installed in protected conduit trays, or installed on-orbit to achieve better launch packaging efficiency? Single or ganged deployment of flexible line sections? Types of deployment aids? Not surprisingly, the installation, checkout, and activation of utilities on a large distributed space system is a much more complex and time-consuming task than the assembly of the structure itself.

If the space system under study involves large modules or other equipment items (for example, the laboratory and/or habitability modules on the International Space Station), the transfer and positioning of these modules presents another challenge. In this case, the problem may well be more one of mobility and reach envelope, than mass handling. The solution may require a mobile manipulator system that can traverse the space structure. The problem may also involve significant stabilization and control issues resulting from the shifting mass distribution as the large object is transported across the structure.

The large space systems currently under study normally are intended for long operational lifetimes, and as such pose a new requirement on the designer—essentially all aspects of the system, including the primary structure itself, must be designed for on-orbit repair or replacement. This is a very significant requirement with far-reaching implications--some obvious, such as the need for physical and visual access to all system components that may one day require replacement, and some that are likely to be overlooked—for example, designing to accommodate the system impacts of the operations necessary to remove and replace each specific component. In this regard, a significant advantage of an erectable system is that, in general, maintenance and repair requirements are more easily accommodated on an erectable system—which was assembled on-orbit in the first place—than on a deployable system, which, in general, experiences strong design pressure to be compact and pre-integrated before launch to the maximum degree possible.

A final significant consideration in the assembly of a large space system is the effective utilization of both the EVA and IVA (intravehicular activity) crewmembers, and the recognition of the significant overhead which is inherent in EVA operations—as a minimum, prebreathing, extravehicular mobility unit (EMU) donning and doffing activities, EMU and life-support system, tool staging and stowing, etc. Extravehicular activity overhead may also include EMU resizing, EMU cleaning, Manned Maneuvering Unit (MMU) donning, doffing, checkout, and servicing, and a host of other activities.

ANALYSIS, TECHNIQUES, AND TOOLS

The development of an assembly concept for any large space system begins with analysis, which characterizes elements and defines options, and progresses through trade studies, which compare options and ultimately provide a basis for selecting the most desirable design solution. In general, manned simulations, particularly of the EVA tasks, are required to support the decision-making process.

The first step in assembly analysis is to identify and physically characterize each individual component in the system or subsystem under study. For example, if the subject is a distributed thermal control system on a large space structure, the study begins by identifying each separate item that must be handled in the course of the assembly process, and determining the physical characteristics and handling constraints associated with each item. An installation sequence must be developed and interacted with the launch manifest analysis to ensure that each element can be made accessible in the sequence and at the time needed. Specific installation requirements must be identified for each separate component—i.e., details of transportation, positioning, conditioning, and mechanical, electrical, and fluid interfaces must be defined, and those requirements must be accommodated by the combined resources and capabilities of the crewmembers and the ASE.

With the system components characterized, a step-by-step assembly scenario is developed for each of the subsystems or major hardware elements that comprise the system. For a distributed thermal control system, this list may include radiators, heat-transport lines, utility trays, cold plates, heat exchangers, signal and power lines, remotely located performance sensors, accumulators, and the operating fluid itself.

Development of the scenario will result in the identification of a multitude of assembly support requirements, in the areas of EVA tasks, EVA support equipment, IVA support, control functions and locations, reach envelopes, visibility and/or feedback, lighting, potential robotic or teleoperated support, etc.

With individual components characterized and major component or subsystem installation scenarios analyzed, the stage is set to develop an integrated assembly scenario. At this level, the objective is to sequence the individual component installation activities in the most resource-efficient manner, consistent with real-world requirements such as assembly support equipment utilization limits or constraints, the probable need to activate selected portions of the total system in a phased manner, natural break points in the required activities, etc. At this point, postassembly checkout functions may be considered. Selected checkout functions may best be performed at various points during the overall system assembly flow; others of necessity must follow completion of the assembly activities. Those checkout functions requiring EVA must be added to the integrated assembly activities to provide the basis for development of the EVA crew activity plan that will accomplish the assembly activity.

The analysis of the assembly requirements for a large space system that relies on EVA is certain to identify tasks that require further evaluation through full-scale manned tests under simulated EVA conditions. Specific simulation techniques and considerations will be discussed later. At this point, the key concern is that these simulations be identified, properly scoped, and conducted in a timely manner to support the selection and development of hardware designs that are amenable to assembly. In the author's experience, the proper scope and timing of manned simulations is often seriously miscalculated; the result is either needless design effort expended in the development of design approaches that ultimately will prove unworkable; or premature simulations that cannot answer the objectives of the test—again, a needless expense.

A wide variety of techniques have been examined for accomplishing the physical activities of assembly; for example, the efficient emplacement of truss structure, the assembly and installation of utility lines (considering both rigid and flexible elements), the grouping and installation of modular system components, etc. Techniques can be developed for essentially any desired assembly activity. Space and time limitations prohibit significant discussion of specific examples of each concept. The practical question, which ultimately must be answered for each technique considered, is "How long does it take?" The answer is almost always highly dependent on the design details of the flight-crew interfaces.

Space and time limitations also prohibit a significant treatment of the tools available for assembly. Suffice it to say that: (1) tools in this case may be broadly defined to include a wide spectrum of equipment, including space suits, EVA hand tools, assembly support equipment (both major and minor), etc. (2) in most cases significant equipment already exists, and (3) also in most cases significant additional development and refinement is desirable and may be anticipated. In particular, with respect to EVA handtools, many tools in the current United States inventory were developed for specific applications in specific situations, and subsequently retained in the inventory for general purpose use. What is needed now is the development of a well-rounded generic EVA tool set, with emphasis on the flight-crew interfaces.

UNITED STATES FLIGHT EXPERIENCE

The majority of the United States actual on-orbit experience with erectable assembly was gathered by NASA astronauts Sherwood Spring and Jerry Ross during STS 61B in late 1985. Two planned EVA assembly experiments were flown on this mission—EASE (Experimental Assembly of Structures in EVA) and ACCESS (Assembly Concept for Construction of Space Structures). These experiments successfully explored the on-orbit assembly of two types of erectable structures, using various restrained and free floating work positions. A significant objective of the EASE experiment was to develop a correlation between the results of the preflight neutral buoyancy assembly tests and the actual on-orbit assembly performance.[1]

The previous United States experience with on-orbit assembly was gained in the Skylab program in August of 1973 with the successful emplacement of an improvised fabric thermal protection shield deployed on a V-shaped erectable structure composed of two 55 foot poles.[2] This operation marked the world's first experience with erectable structures in space.

MANNED SIMULATION OF EXTRAVEHICULAR ACTIVITY

There are five basic reasons why a manned simulation may be required in the development of hardware for an extravehicular task. The first is simply to verify the feasibility of the projected crew task, i.e., sufficient physical access exists, the visibility is adequate, the force application and repetition requirements fall within acceptable limits for the EVA crewmembers and their support equipment, and the support equipment functions and requirements are understood. A second significant reason to do manned simulations is to ensure early elimination of design approaches that are inherently unsatisfactory from the crew interface point of view. A third reason is to determine realistic task timelines—excessive crew time requirements can be the death of a design solution that may look good on paper. A fourth reason is that it may be necessary or prudent to demonstrate the feasibility of the projected crew tasks to the customer. Finally, as designs move into the later stages of development, add-on features that require excessive crew time or energy expenditure may appear as designers attempt to satisfy belatedly-recognized requirements. Additional full-scale or part-task manned simulations may be required to develop design refinements.

There are three requirements for "valid" manned simulations, i.e., simulations worthy of supporting design decisions that may cost hundreds of thousands or millions of dollars, or even more significantly, the competitive edge. The first is that mockups be of the appropriate fidelity to answer the question at hand. "Fidelity" is not synonymous with "complexity"—rather, it means that the relevant features of the situation are included in the mockup, and that they are represented in the simulation environment in a way that accurately represents the crew task as it will be on-orbit. The second requirement is that the simulation medium—one-G, neutral buoyancy, Keplerian trajectory zero-G atmospheric flight, or computer-driven motion base—be appropriate to the task at hand. Each mode has specific attributes, and each is appropriate for some tasks while totally inappropriate for others. The third requirement is that the test crewmembers for the simulation be properly qualified—they must be individuals who can accurately interpret their actions and decisions in terms of the corresponding action in the orbital situation. They should be experienced across all simulation mediums, since each represents the orbital flight situation differently, and ideally they should have seen hardware that they have personally been involved in shaping actually perform in the space environment. Using available engineers as test crewmembers, because they happen to fit the suit and are interested, may be informative to the individuals involved but it does not satisfy the third criterion for a valid simulation.

Manned simulation of extravehicular activity may be defined as manned evaluations of tasks that are later to be conducted outside the pressurized environment of a spacecraft. The scenario for defining and conducting a manned simulation will include the following steps: (1) Select the appropriate medium for the test, (2) Design and build appropriate mockups, (3) Prepare a test plan and test procedures specific to the simulation being planned, (4) Conduct a safety review and obtain approval on a per test basis, both for the hardware elements and for the proposed operations, (5) Install and checkout the mockups in the simulation facility, and conduct a walkthrough, (6) Conduct a pretest briefing, (7) Conduct the test, (8) Conduct a posttest debriefing, and (9) Analyze the results, formulate design inputs, and document the test as appropriate through written reports.

There are four major simulation mediums in use for manned extravehicular simulations—one-G, neutral buoyancy, zero-G or reduced-G atmospheric flight, and computer-driven motion base. One-G simulations are conducted wearing a pressurized spacesuit in a normal one-G environment. They are appropriate for tasks characterized by a fixed body location and upright position, equipment item exchange or transfer limited to small items, "quick-look" evaluations with foamcore mockups, and, since the test crewmember carries much of the 160-pound weight of the suit, test durations limited to 20-30 minutes. The neutral buoyancy technique is widely used for EVA simulation; it is appropriate for tasks that require crewmember translation, multiple body positions or extended work envelopes, equipment exchange or transfer involving large items, tests involving mobility or stability aid development, and suited durations greater than 20-30 minutes. Neutral buoyancy tests frequently run 2-3 hours in length, and can run much longer. Zero-G flight in the atmosphere, achieved by flying Keplerian trajectories in subsonic transport-class jet aircraft, represents a higher-fidelity simulation of the orbital EVA situation, but in short (20-30 second) intervals, achieved at far greater expense than equivalent periods of neutral buoyancy. Zero-G flight is appropriate for tasks such as space-suit donning/doffing, operations requiring high-fidelity cable or flexible conduit behavior or other dynamic motions, operations involving fluid flow or behavior operations, etc. Modified trajectories can be flown to produce reduced gravity for simulation of lunar, Martian, or other extraterrestrial surface operations. Multiple trajectories (typically 40 or more) can be obtained in a given flight.

The fourth major earth-based simulation medium, the computer-driven motion base technique, is appropriate for selected tasks not amenable to simulation in the first three mediums. Typical tasks that have been or are being successfully studied in computer-driven motion base simulators include Lunar Roving Vehicle (LRV) handling characteristics evaluations, which required simulating vehicle responses under lunar gravity conditions, and Manned Maneuvering Unit (MMU) handling characteristics and operations, which require high-fidelity dynamic response for much longer periods of time than a zero-G aircraft can provide.

The ultimate in simulation fidelity, of course, is the orbital flight environment. As shuttle operations mature, it can be expected that selected hardware evaluations, in cases where the results have major significance and none of the earth-based simulations mediums adequately simulate the tasks required, will be conducted in space itself. The EASE/ACCESS experiments, in this regard, were simulations conducted in space to evaluate potential space station assembly concepts under high-fidelity conditions.

SUMMARY

Manned on-orbit assembly is a practical technique for the emplacement and activation of large space systems. Although the actual in-flight experience base is still quite limited, a large body of experience exists with the simulation of potential assembly tasks in the four earth-based simulation mediums to substantiate this conclusion.

Although EVA is considered an operational capability, it is not trivial. The available EVA time on any mission is a limited commodity whose use must be planned and budgeted carefully. Flight-crew interface design is critical to achieving maximum productivity in an EVA, and flight crew interface considerations must be treated as an integral part of the design process.

Finally, techniques for the analysis and simulation of EVA tasks are available and well understood. Missions that rely on EVA can be launched with a high degree of confidence that the EVA tasks will be accomplished as planned.

REFERENCES

[1]Levin, George M., "EASE/ACCESS Postmission Management Report," NASA Headquarters, Washington, D.C., undated.

[2]"MSFC Skylab Crew Systems Mission Evaluation," Marshall Space Flight Center, Alabama, NASA TM X-64825, August 1974, pp. 329-337.

On the Orbiter Based Construction of the Space Station and Associated Dynamics

V. J. Modi* and A. M. Ibrahim†
*The University of British Columbia, Vancouver,
British Columbia, Canada*

Abstract

Using a relatively general formulation procedure, the paper reviews complex interactions between deployment, attitude dynamics and flexural rigidity for two configurations representing deployment of beam and tether type appendages. The governing nonlinear, nonautonomous and coupled hybrid set of equations are extremely difficult to solve even with the help of a computer, not to mention the cost involved. Results suggest substantial influence of the flexibility, deployment velocity, initial conditions, and appendage orientation on the response, and under critical combinations of parameters the system can become unstable. The information has relevance to the design of control systems for: (i) the next generation of communications satellites; (ii) the orbiter based experiments such as SAFE (Solar Array Flight Experiments), SCOLE (Structural COntrol Laboratory Experiment), STEP (Structural Technology Experiment Program), and the NASA/CNR tethered subsatellite system; as well as (iii) the evolutionary transient and postconstruction operational phases of the proposed space station.

Introduction

In the early stages of space exploration, satellites tended to be relatively small, mechanically simple and essentially rigid. However, for a modern space vehicle carrying light deployable members, which are inherently flexible, this is no longer true. The space shuttle based

*Professor, Department of Mechanical Engineering.
†Graduate Research Assistant, Department of Mechanical Engineering.

experiments as well as construction of the proposed space
station with gigantic trusses, power booms and solar
panels, which will have to be errected in space, dramatically emphasize the role of deployment, slewing maneuvers
and flexibility on dynamics, stability and control[1-3].

This being the case, flexibility effects on satellite
attitude motion and its control have become topics of
considerable importance. Over the years, a large body of
literature pertaining to the various aspects of satellite
system response, stability and control has appeared. A
recent issue of the Journal of Guidance, Control, and
Dynamics published by the AIAA (American Institute of Aeronautics and Astronautics) contains a series of articles
reviewing the state of the art in the general area of large
space structures[4].

Attention is also directed towards planning of on-
orbit experiments such as SCOLE, the Orbiter Mounted Large
Platform Assembler Experiment, NASA/Lockheed Solar Array
Flight Experiment and a host of others to check, calibrate
and improve algorithms. It is generally concluded that
on-orbit information acquired during the construction phase
of a space station is the only dependable procedure for its
overall design. Obviously, this promises to open up an
exciting area of in-flight measurements of structural
dynamics, stability and control parameters necessary for
design. With the U.S. commitment to a space station in
mid 1990's, the need for understanding structural response
and control characteristics of such time varying, highly
flexible sytems is further emphasized. This paper briefly
reviews representative results concerning dynamics of
spacecraft having flexible deployable members with the
attention focused on two classes of problems of contempor-
ary interest:

(i)   the Orbiter based deployment of a beam-type appendage
       in and out of the orbital plane;
(ii)   deployment and retrieval of tethered subsatellite
       systems.

Details of the mathematical formulations and analyses being
extremely lengthy are omitted here, however, appropriate
references cited. Emphasis throughout is on the analysis
of results and corresponding conclusions.

Deployment of Flexible Structural Members - A General
Formulation

A relatively general formulation of the nonlinear,
nonautonomous and coupled equations of motion applicable to

a large class of spacecraft with flexible plate/beam-type of members has been discussed by the authors earlier[5]. It has been extended to account for membrane, shell and tether type of appendages with viscous/structural damping and momentum/reaction wheels[6]. Essential features of the general formulation may be summarized as follows:

- arbitrary number, type (tether, membrane, beam, plate, shell) and orientation of flexible members in orbit, connected so as to form a topological tree configuration, deploying independently at specified velocities and accelerations (Fig. 1);
- the appendage is permitted to have variable mass density, flexural rigidity and cross-sectional area along its length;
- governing equations account for gravitational effects, shifting center of mass, changing inertia and appendage offset together with transverse oscillations;
- modified Eulerian rotations (roll, yaw, pitch, respectively) are so chosen as to make the governing equations applicable to both spin stabilized and gravity gradient orientations;
- the equations are programmed in nonlinear as well as linearized forms to permit the study of:
  (i) large angle maneuvers;
  (ii) nonlinear effects.

The program is written in a modular fashion to help isolate the effects of flexibility, deployment, character and orientation of the appendages, inertia and orbital param-

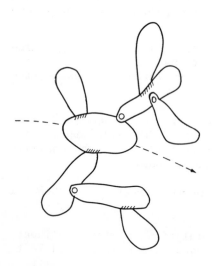

Fig. 1 A system of arbitrarily connected flexible bodies in orbit.

eters, number and type of admissible functions, etc.
Environmental effects due to solar radiation pressure aerodynamic forces, Earth's magnetic field interaction, etc.,
can be incorporated easily through generalized forces. The
same is true with internal and external energy dissipation
mechanisms.

The Orbiter Based Deployment of a Beam in an Arbitrary
Orientation

    A particular configuration selected for study corres-
ponds to the Orbiter Mounted Large Platform Assembler
Experiment once proposed by Grumman Aerospace Corporation
(Fig. 2). Its objective is to establish capability of
manufacturing beams in space which would serve as one of
the fundamental structural elements in construction of the
future space station. The assembler is fully collapsible
and automatically deployed.
    For analysis, the flexibility and deployment rate
parameters were taken to be of the same order of magnitude
as used or likely to be employed in practice. In the

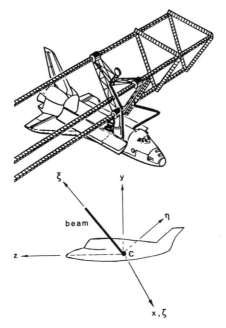

Fig. 2 A schematic diagram showing the Oribiter based construction
of beam-type structural members. The principal coordinates x, y, z
with their origin at the instantaneous center of mass and beam
coordinates ζ, η, ξ with the origin at the attachment point are
also indicated. In general the two origins are not coincident.

diagrams e represents orbital eccentricity; α, β, γ (roll, yaw, pitch, respectively) are the librational angles; EI is the beam flexural rigidity, assumed constant over the length in this example; and $\dot{L}$ corresponds to the deployment rate. λ_{in} and λ_{out} denote beam inclinations to the local vertical in and normal to the orbital plane, respectively. The perigee was taken to be 331 km. The truss or beam vibrations were represented by a maximum of the first four modes, ψ_i, of a cantilever. P_ℓ, Q_ℓ represent generalized coordinates associated with the admissible functions used to represent beam-type appendage oscillations in the ℓth mode in ζ and η directions, respectively. $\overline{P}_\ell$ and $\overline{Q}_\ell$ represent transverse generalized coordinates normalized with respect to the total length.

Numerical values for some of the more important parameters used in the computation are given below:

Orbiter:
    Mass =   79,710 kg
    Ixx = 8,286,760 kg m^2 Ixy =   27,116 kg m^2
    Iyy = 8,646,050 kg m^2 Iyz = 328,108 kg m^2
    Izz = 1,091,430 kg m^2 Izx =   -8,135 kg m^2

Here x,y,z are the principal body coordinates of the Orbiter with the origin coinciding with the center of mass. In the nominal configuration x is along the orbit normal, y coincides with the local vertical and z is aligned with the local horizontal in the direction of motion (Fig. 2). γ (roll), β (yaw), and α (pitch) refer to rotations about the local horizontal, local vertical and orbit normal, respectively.

Beam:
    Mass (M_b) = 129 kg
    Length (L)   =   33 m
    Flexural Rigidity (EI) = 436 kg m^2

Figure 3 shows tip response of the beam for two different orientations in the plane defined by the local vertical and the orbit normal, λ_{out} = 20° and 90°. The initial tip disturbance was taken to be as 4% of the beam length. Note, the two transverse motions, ζ and η, are completely coupled with the plane of vibration precessing, due to the Coriolis force, at a uniform speed which is governed by the beam inclination angle λ_{out}. For the case of λ_{out} = 0, the uncoupled motion showed no precession. On the other hand, the precessional velocity increased with λ_{out} and reached a maximum value at λ_{out} = 90°. The plane of vibration of the

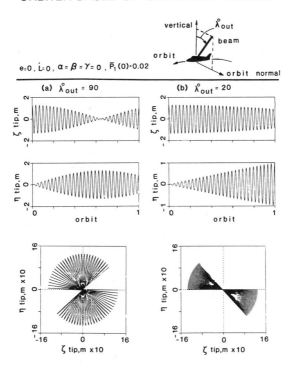

Fig. 3   Response of the beam to a tip disturbance when located out of the orbital plane.   Note the transverse motions ζ and η are coupled and the vibration plane of the beam tip precesses at a uniform rate.

beam precessed in one direction only (in this case clock-wise) for a given λ_{out}.

Figure 4 attempts to assess the effect of eccentricity on the tip deflection of the beam and first two modes' contribution to it.   Two orientations of the beam are considered, $\lambda_{in} = 30°$ and $\lambda_{out} = 90°$, to facilitate compar-ison with the earlier responses.   The orbital eccentricity tends to impart a periodic increase in the stiffness, and hence a rise in the natural frequency of the beam, which is a minimum at perigee and reaches a maximum at apogee.   This is apparent in the tip time history as well as the modes. Such modulations of the frequency, although shown here for two orientations, were found to be to be present at all λ. Furthermore, it is interesting to observe, for the planar configuration of the beam ($\lambda_{in} = 30°$), modulation of the inplane vibrations (ζ, Q_1, Q_2) around the changing equili-brium position of the beam in orbit due to eccentricity.

Effect of beam deployment on the tip dynamics is studied in Fig. 5. Initial tip deflection is the same as before (4% of the length, i.e., 1.32 m). Two time histories with the same duration of deployment are considered. As can be expected, the frequency of oscillations in and out of the orbital plane gradually decreases with deployment finally attaining a steady state value upon its termination. It is of interest to recognize that they reach the same steady state amplitude, although it is much larger during deployment compared to the deployed case.

In practice the Orbiter's librations will be controlled to a specified tolerance limit. A typical time history[7] of the controlled Space Shuttle librations during an orbit is shown in Fig. 6. In the following results attention is focused on response of the deployed beam during such forced excitation of the Orbiter in the Lagrange configuration.

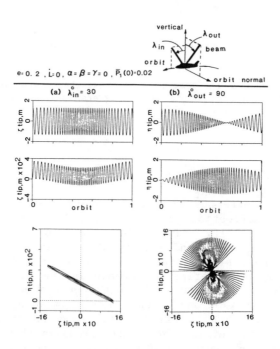

Fig. 4 Effect of an orbital eccentricity on vibratory response of the tip for two orientations of the beam. Note the frequency condensation effect particularly pronounced near the apogee. It was found to be present in both the modes used in the analysis.

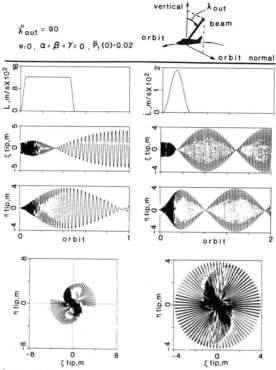

Fig. 5 Effect of deployment strategies on tip response of a beam deploying normal to the orbital plane. Note a reduction in beam frequency during deployment. The steady state amplitude is essentially independent of the strategy for a given time of deployment.

Figure 7 shows the forced tip response as well as the first two modes contributing to it for a beam deployed along the orbit normal with the Orbiter in the Lagrange configuration. At the outset it should be recognized that, for this out-of-plane configuration of the beam, the out-of-plane motion ζ and inplane response η are coupled as seen before (Fig. 3). Hence one would expect the Orbiter's yaw and roll to be reflected in both η (inplane) and ζ (out-of-plane) motions. The response shown in Fig. 7 precisely reveals these trends. However, the roll disturbance being at a higher frequency, and hence with a higher acceleration, appears to be dominant as apparent from the amplitude modulation of the response at the roll frequency (around 13 cycles per orbit).

A word concerning computational accuracy would be appropriate. To this end permissible error during numerical integration of the governing nondimensional differen-

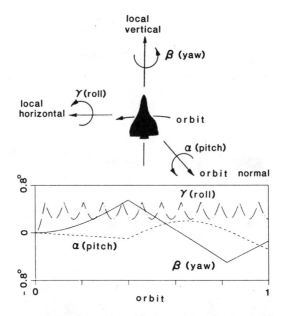

Fig. 6 A representative controlled motion of the Orbiter during a typical orbit. Note the roll, yaw and pitch motions are with reference to the local horizontal, local vertical and orbit normal, respectively.

tial equations was varied systematically to assure reliable data. The case corresponding to the Lagrange configuration with λ_{out} = 90° was studied using error tolerance of 10^{-4}, 10^{-6}, and 10^{-8}. Both components of tip deflections and generalized coordinates were compared. The tolerance level of 10^{-4} was found to be inadequate and gave misleading response. However, the results obtained using the tolerance levels of 10^{-6} and 10^{-8} were essentially the same. Hence during the numerical integration the error tolerance of 10^{-8} or lower was used.

The Oribiter Based Tethered Subsatellite System

A vast potential of the Shuttle based tethered systems (Fig. 8) has led to many investigations concerning their dynamics during operational (station-keeping), deployment and retrieval phases. In its utmost generality, the problem is indeed quite challenging as the system dynamics is governed by a set of ordinary and partial, nonlinear, nonautonomous and coupled equations which account for:

• three dimensional rigid body dynamics (librational motion) of the Shuttle and the subsatellite;

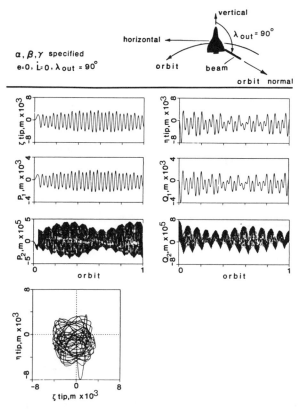

Fig. 7 Forced response of a beam aligned with the orbit normal. Note, the effect of the dominant roll excitation at 13 cycles per orbit which is particularly clear in the second mode.

- swinging inplane and out-of-plane motions of the tether, of finite mass and elasticity, with longitudinal and transverse vibrations superimposed on them;
- offset of the tether attachment point from the Shuttle's center of mass;
- aerodynamic drag in a rotating atmosphere.

Over the years, investigators have attempted to obtain some insight into the complex dynamics of the system using a variety of models which have been summarized by Misra and Modi[8]. In general, the studies show that the dynamics of the system during deployment is stable, however, the retrieval dynamics is basically unstable. The system involves a negative damping approximately proportional to $\dot{L}/L$, where L represents the unstretched tether length. This suggests a need for an active control strategy partic-

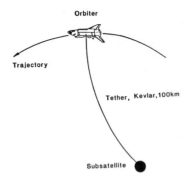

Fig. 8   The NASA/CNR
proposed Orbiter based
tethered subsatellite
system.

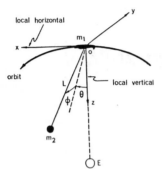

Fig. 9   Typical unstable
response of the tethered
system during uncontrolled
retrieval.

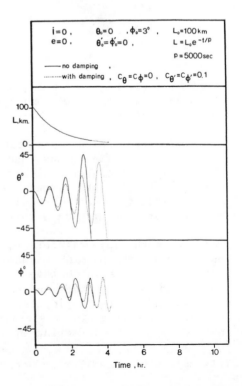

ularly to limit inplane (θ) and out-of-plane (ϕ) swing motions of the tether.

The pioneering contribution that may help realize this objective is due to Rupp[9] who introduced a tether tension control law for the system. Librational motion in the orbital plane was analyzed and growth of the pitch oscillations (θ) during the retrieval phase noted. The system was further studied in detail by Baker et al. taking into consideration the three dimensional character (i.e., θ and ϕ degrees of freedom) of the dynamics and the aerodynamic drag in a rotating atmosphere[10]. Several more sophisticated models have followed since[11,12], however, one of the major conclusions of all the analyses remains essentially the same: even when the various tension control schemes are used, large amplitude motion can result under certain conditions, particularly during retrieval, which may not be acceptable.

Figure 9 shows inplane and out-of-plane response of a 100 km long tether during retrieval of a 170 kg subsatel-

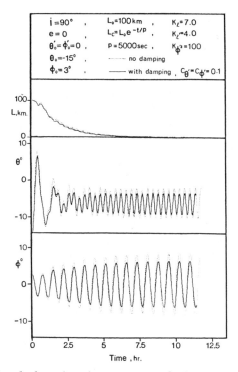

Fig. 10 Retrieval dynamics in presence of the proposed nonlinear control showing dramatic reductions in the limit cycle amplitudes.

lite to the Shuttle in a circular equatorial orbit at a height of 220 km. Physical properties of the tether, subsatellite, atmosphere as well as the exponential retrieval procedure are taken to be the same as those used by Baker and others[10]. Both the inplane and out-of-plane tether librations become unstable even in the presence of dissipation (C_θ, = C_ϕ, = 0.1). Depending upon the retrieval rate used, one can easily calculate the damping level required to guarantee stability, however, it may not always be possible in practice to provide the required level of damping. Clearly this suggests a need to evolve an effective control procedure.

Attention was now turned to assess effectiveness of a number of nonlinear control strategies. A procedure where the tether tension is controlled as

$$T = K_\ell\, \ell + K_\ell,\, \ell' + K_\phi,\, \phi'^2 \qquad (1)$$

appeared to be the most promising. Here: T = tether tension level, ℓ = non-dimensional difference between the actual and the commanded tether length. Primes denote differentiation with respect to the non-dimensional time. Figure 10 presents representative results for the tether response during retrieval from 100 km to 500 m using the control strategy mentioned in equation (1). Note, the amplitudes in both pitch and roll are substantially reduced. The controller continued to remain effective under a wide range of diverse situations involving different initial conditions, orbital elements, physical properties of the system, retrieval rate, etc.

In the actual practice the tether material is indeed elastic causing a longitudinal stretch (ζ) of the tether. Any realistic analysis must account for it. Hence it was essential to check the controller's effectiveness when the tether is elastic. To this end appropriate stretch equation was added to the system of equations representing θ and ϕ degrees of tether rotations. Figure 11 summarizes controlled retrieval dynamics for such a tethered subsatellite system for the Space Shuttle in a circular polar orbit at a height of 220 km. The high frequency longitudinal oscillations of the tethered system made the numerical integration of the governing nonlinear coupled equations quite expensive. Clearly the nonlinear control strategy remains effective even when the longitudinal stretch is accounted for.

Transverse vibrations of the tether would further complicate the problem. They are excited by the Coriolis

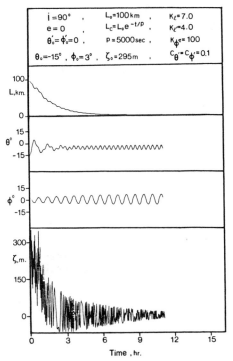

Fig. 11 Plots showing effectiveness of the proposed nonlinear control strategy accounting for longitudinal flexibility of the tether.

forces during deployment and retrieval as well as the aerodynamic forces when a part of the tether is at a relatively low altitude. The transverse and longitudinal vibrations are strongly coupled, particularly at the terminal phase of retrieval. The transverse vibrations are also unstable during uncontrolled retrieval. Although growing transverse vibrations have negligible effect on the tension at the beginning of retrieval, it can make the tether slack towards the end[11]. This can be avoided in two ways: one is to speed up the retrieval when the length is small, to avoid very slow retrieval rate due to exponential character. Alternatively, thrusters may be used in conjunction with a tension control or a length rate law. Using three mutually perpendicular thrusters, one along the nominal tetherline and the other two opposing in-plane and out-of-plane rotations, Xu et al. have shown a remarkable success in controlling all the degrees of freedom[12].

Results of a typical example with three orthogonal thrusters T_{α}, T_{γ}, T_c along pitch, roll and tether direc-

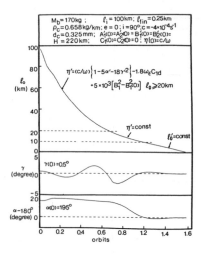

a) variation of length
and rotations.

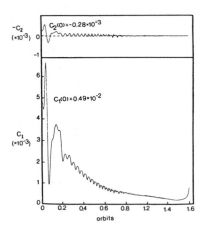

b) response of longitudinal
generalized coordinates.

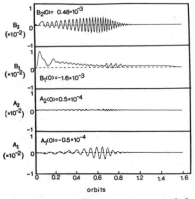

c) behavior of transverse modal
coordinates.

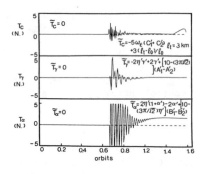

d) time history of thrusters.

**Fig. 12   Retrieval dynamics using
mixed control with feedback of
transverse vibrations.**

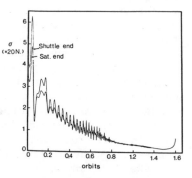

e) variation in tension.

tions, respectively using the following physical parameters, are shown in Figure 12:

- circular polar orbit;
- spherical satellite with projected area of 1 m^2, mass = 170 kg;
- 1 mm diameter Kevlar tether;
- Initial Conditions
  All the generalized coordinates excited, i.e., transverse as well as longitudinal disturbances with pitch
  $\alpha - 180° = 15°$ and $\gamma = 0.5°$;
- thrust vector $\bar{T} = i\bar{T}_\gamma + j\bar{T}_c + k\bar{T}_\alpha$, maximum thrust limited to ± 5N.

Note, here $\eta' = L'/L$ and A_i, B_i, C_i represent generalized
coordinates associated with transverse and longitudinal
oscillations, respectively. The subsatellite is retrieved
from 100 km to 250 m with an exponential retrieval phase
over the first 90 km followed by an uniform retrieval rate.
The thrusters are activated only over the terminal 20 km of
the retrieval maneuver.

It is apparent from Figures 12a-c that pitch and roll
rotations as well as the transverse and longitudinal vibrations are damped, rather effectively, in less than two
hours during the proposed thruster augmented retrieval of
the 100 km tether. Note, a finite value of tension exists
even at a small tether length (Fig. 12d). Variation in the
thruster time history over the last 20 km is shown in
Figure 12e. Even a relatively low magnitude thrusters of
±5N are able to damp the tether motions rather rapidly.
The total thrust impulse required is around 10^4 Ns with a
substantial reduction in the retrieval time.

Concluding Comments

With a relatively general formulation in hand and the
program operational, efforts are in progress to develop a
comprehensive data bank for spacecraft with flexible
appendages. Not only will it prove useful to design engineers involved in planning of future communications satellites but also help in assessing dynamical, stability and
control considerations associated with the Orbiter based
construction of space-platforms.

The entire field is wide open to innovative contributions. Dynamics and control of such nonlinear, time-
dependent systems accounting for damping and environmental
forces remains virtually untouched. Their applications to
the construction of space station has received attention

only recently. Development of an algorithm to predict the effect of mass, inertia and stiffness of the station as it evolves on the dynamics and control parameters represents an exciting challenge. Application of the tether and associated dynamics presents an area of enormous potential.

Acknowledgment

The investigation reported here was supported by the Natural Sciences and Engineering Research Council of Canada, Grant No. 67-1547.

References

[1] Modi, V.J., "Attitude Dynamics of Satellites with Flexible Appendages - A Brief Review", Journal of Spacecraft and Rockets, Vol. 11, November 1974, pp. 743-751.

[2] Williams, C.J.H., "Dynamics Modelling and Formulation Techniques for Non-Rigid Spacecraft, Proceedings of the ESA Symposium on Dyanamics and Control of Non-Rigid Spacecraft, ESA SP 117, Frascati, Italy, May 1976, pp. 53-70.

[3] Lips, K.W., "Dynamics of a Large Class of Satellites with Deploying Flexible Appendages", Ph.D. Dissertation, University of British Columbia, September 1980.

[4] Special Section, "Large Space Structure Control: Early Experiments", Journal of Guidance, Control, and Dynamics, Vol. 7, No. 5, September-October 1984, pp. 513-562.

[5] Modi, V.J., and Ibrahim, A.M., "A General Formulation for Libration Dynamics of Spacecraft with Deploying Appendages", Journal of Guidance, Control, and Dynamics, Vol. 7, No. 5, September-October 1984, pp. 563-569.

[6] Ibrahim, A.M., and Modi, V.J., "A Formulation for Studying Dynamics of N Connected Flexible Deployable Members", 37th International Astronautical Congress, Innsbruck, Austria, October 1986, Paper No. IAF-86-240; also Acta Astronautica, in press.

[7] Budica, R.J., and Tong, K.L., "Shuttle On-Orbit Attitude Dynamics Simulation", AIAA/AAS Astrodynamics Conference, San Diego, California, August 1982, Paper No. AIAA-82- 1452.

[8] Misra, A.K., and Modi, V.J., "Dynamics and control of Tether connected Two-Body System", Invited Address, 33rd Congress of the International Astronautical Federation, Paris, France, September 1982, Paper No. IAF-82-316; also Space 2000, Selected Papers from the 33rd IAF Congress, Editor: L.G. Napolitano, AIAA Publisher, pp. 473-514.

[9] Rupp, C.C., "A Tether Tension control Law for Tethered Subsatellites Deployed Along Local Vertical", <u>NASA</u> TM X-64963, September 1975.

[10] Baker, P.L., et al., "Tethered Subsatellite Study", <u>NASA</u> TM X-73314, March 1976.

[11] Modi, V.J., Zu, D.M., and Misra, A.K., "Influence of Dynamical Modeling of a Tethered Satellite System on its control System", <u>AIAA 23rd Aerospace Sciences Meeting</u>, Reno, Nevada, January 1985, Paper No. AIAA-85-0025.

[12] Xu, D.M., Misra, A.K., and Modi, V.J., "On Thruster Augmented Active Control of a Tethered Subsatellite System During its Retrieval", <u>AIAA/AAS Astrodynamics Conference</u>, Seattle Washington, August 1984, Paper No. 84-1993.

Space Transportation: Options and Opportunities

J. P. Loftus,* Jr., R. C. Ried,† and R. B. Bristow‡
NASA Lyndon B. Johnson Space Center, Houston, Texas

Abstract

The transportation options to Earth orbit and beyond that are expected to be available in the period 1990 and beyond are summarized. The basic performance of available systems is characterized, and some significant issues in payload interfaces are summarized. The objective is to provide a synoptic summary of available capability and some assessment of the most significant design and operations issues for new systems.

Current launch capability is a combination of the continued production of systems designed 20 yr ago and their derivatives and a family of more recent designs. New designs range from the partially reusable Space Shuttle systems, new design expendables such as the Ariane and the H-II liquid-oxygen/hydrogen systems to new simple systems such as the Space Services Incorporated solid-propellant Conestoga and the American Rocket Company hybrid industrial launch vehicle. The systems available today are adequate for near-term performance demands; availability and reliability are the critical issues.

---

* Assistant Director.
† Manager, Advanced Programs Office.
‡ Aerospace Technologist, Advanced Programs Office.

114

Any substantial increase in space activity will require a new generation of vehicles that must have significant improvement over current systems. Among the more significant improvements required are significantly increased reliability, reduced unit cost, reduced lead time on production, and improved payload accommodations. Systems definition and technology efforts are in progress to define the manner in which these goals can be achieved.

Introduction

Space launch vehicles serve no useful purpose in and of themselves; they are a means to achieve other objectives. The only useful function of a space launch vehicle is to place a useful payload in a productive environment; i.e., to convert kinetic into potential energy.

This is a very significant point because the flame and thunder of launch are noticeable to the press and the public and thus launch capability becomes in the public mind synonymous with space capability. This perception is misleading and potentially counterproductive. The true significance of space is what occurs out of sight but not necessarily out of mind. The real value of space operations is not in launch but in the assets on orbit for their contribution to information acquisition and distribution and other process operations such as materials formulation.

Space launch vehicles are unique among all transportation systems. Space launch requires a system to accelerate continuously throughout its effective life, approximately 600 sec. No other transportation system makes such demands on its components. Most transportation systems with which we are familiar require a few seconds of acceleration and many hours of sustained but significantly derated operation to overcome resistance - air drag for aircraft, rolling friction and drag for other systems.

Because space launch systems are in essentially fundamental opposition to a basic law of nature - gravity - they are energy intensive beyond the bounds of more customary systems and easily misunderstood. The fact that the fuel pump for a single main engine on the Space Shuttle has a greater shaft horsepower than any ship afloat is a direct consequence of the mass ejection - 11 tons/sec - that is required of the Space Shuttle vehicle to combat the

force of gravity, which is an acceleration term. All
launch vehicles are similarly energy intensive.

The value of an asset in space is that it has a unique
environment.

- Minimal gravitational influence: $10^{-5}g$ or $10^{-4}g$

- Potentially constant lighting

- Constancy or periodic constancy of position relative
  to a point on Earth

- Predictable position

- A one-time investment in translating a spacecraft
  from the Earth surface through a gravity well to a
  point at which it possesses essentially indefinite
  use of the energy investment in converting kinetic
  to potential energy

As a practical matter, only a few meters per second
per annum velocity increment are required to maintain a
given state as contrasted to the 8 km/sec required to
attain a low Earth orbit (LEO) or to the 10.66 km/sec
required to achieve a geosynchronous orbit (GEO).

Because space launch is highly energy intensive, it is
inherently hazardous. Effective space launch requires high
energy density, low mass fraction, and consequently minimum
structural margins of safety; but, to be economically
useful, there must be some reasonable reliability based on
the assets at risk. When the payload cost is less than the
cost of launch (one-third to one-half), a reliability of
0.90 may be acceptable; when the payload cost equals the
launch cost, a reliability of at least 0.95 is needed.
When payload costs, or assets at risk, are 10 times the
cost of launch, a reliability of better than 0.99 is
needed.

All original space launch vehicles were derived from
military missile systems. When they were converted to
space launch systems, the inherent 80% to 90% reliability
proved unacceptable because the payload to launch cost
ratio rapidly exceeded unity and continued to grow. Space
transportation advocates focused on the cost of a pound of
mass to orbit; users and insurers focus on the cost of
transportation in relation to the cost of the payload and
other assets at risk. Increasing capability and costs of
payloads require significant increases in the reliability
of transportation. By systematic control of components and
processes, reliabilities of 93% to 96% have been achieved

for the most frequently used space launch systems, but they
are sustained only by maintaining tight process and
component control and a certain minimal rate of use. A
significant aspect of recent U.S. expendable launch vehicle
failures can be potentially ascribed to the use of long-
shelf-life items; i.e., these failures likely would not
have occurred had the vehicle been flown 2 years earlier
when the hardware was in a greater state of readiness and
when manufacturing and operations personnel were at higher
training levels.

     Through experience, we have arrived at a condition in
which a 95% to 96% reliability of the launch vehicle is
consistent with the cost and operational life of a
spacecraft having value that approximates the cost of
launch. Less costly spacecraft can be flown economically
on less reliable launch vehicles, and much more costly
spacecraft or partially reusable launch systems require
significantly higher reliability to be economically
effective. In all cases, higher reliability in launch
activities benefits all users.

     During the last 25 yr, much has been learned about
both space launch and space-based systems and their
economic potential. There are potential opportunities in
Earth surveillance systems for weather, crops, and
minerals; for microgravity processing of glasses, ceramics,
solid-state devices, and biologic materials; and of course
in exploiting many more aspects of communication than the
point-to-point systems currently in use (e.g., point-to-
points and broadcast systems). Many forms of scientific
study in astronomy, astrophysics, solar dynamics, and
planetary exploration can only be done with space-based
instruments.

     In such a context, the natural question is, "What are
the launch system alternatives and what issues should users
consider as they plan for such systems?"

Earth to Orbit Vehicles

Existing

     Figure 1 illustrates the currently available launch
systems. All are mature systems for the state of the art
at the time of their commitment to production. A brief
characterization of each of the vehicles follows. The
information is based on internationally recorded data.
Success and failure are as reported by the sponsoring
agency; failures can be characterized generically as

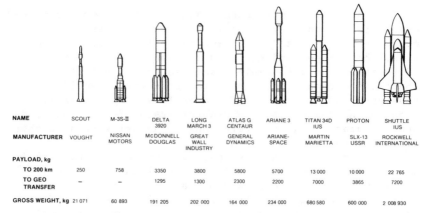

NAME	SCOUT	M-3S-II	DELTA 3920	LONG MARCH 3	ATLAS G CENTAUR	ARIANE 3	TITAN 34D IUS	PROTON	SHUTTLE IUS
MANUFACTURER	VOUGHT	NISSAN MOTORS	McDONNELL DOUGLAS	GREAT WALL INDUSTRY	GENERAL DYNAMICS	ARIANE-SPACE	MARTIN MARIETTA	SLX-13 USSR	ROCKWELL INTERNATIONAL
PAYLOAD, kg									
TO 200 km	250	758	3350	3800	5800	5700	13 000	10 000	22 765
TO GEO TRANSFER	–	–	1295	1300	2300	2200	7000	3865	7200
GROSS WEIGHT, kg	21 071	60 893	191 205	202 000	164 000	234 000	680 580	600 000	2 008 930

Fig. 1   Operational launch vehicles.

explosive or catastrophic. Some "successes" may not be
economically successful; i.e., payload delivery may be
outside the predicted 3σ state-vector condition with
consequent penalty to spacecraft life or maneuver
capability or both. A reference delivery altitude of 200
km (108 n. mi.) is selected for performance assessment.

Scout. Scout is the smallest and one of the oldest
designs in NASA use; it is a derivative of sounding rockets
used in the mid-1950's. The Scout space launch vehicle is
a four-stage, solid rocket motor (SRM) system using
inertial reference for guidance in the first three stages
and spin stabilization for the fourth stage. (Control is
effected by aerodynamic fins in the first stage.)

Of the 106 Scout launches since 1960, 92 have been
successful. Seventy-nine of the last eighty-three launches
have been successful for a recent reliability value of
0.95. The last 37 launches have been successful.

Orbital payload to 200 km is 250 kg for the G-1
configuration in current use. Alternative growth versions
have been studied for payloads of 270 and 560 kg; growth is
achieved by strap-on boosters to augment the first stage
and the use of larger second- and third-stage motors.

Launches are conducted from the Wallops Island launch
facility on the U.S. east coast, Vandenberg Air Force Base
(VAFB) on the U.S. west coast, and San Marco on the East
Coast of Africa.

Scout has had a typical growth history in that payload
mass capability has been increased by a factor of 3 and
available volume within the launch shroud by a factor of 12
from its initial configuration to the present G-1
operational configuration.

M-3S-II. The Japanese Institute for Space and
Aeronautical Science (ISAS) operates an SRM vehicle that
provides both Earth orbit and planetary injection
capabilities. Similar to the Scout, its development is
based on sounding rocket SRM, first-stage aerodynamic fin
control, spinning upper stage, and inertial guidance
technology.

The currently operational vehicle is the M-3S-II. It
has successfully deployed three payloads and is launched
approximately once per year.

Launch is from the Kagoshima Space Center.

Delta. The Delta launch vehicle was derived from the
Thor missile system developed in 1955. The Delta space
launch system has been in use since 1960, and has been
operated in more than 30 configurations over its 180
flights to date. Delta has been NASA's most extensively
used launch vehicle. There have been 12 failures for an
overall reliability of 0.93. Two-stage delivery to low
Earth orbit has had 0.95 success rate. Prior to the Delta
178 failure in 1986, there had been 43 successful launches
over 8.5 yr.

The core vehicle consists of two storable-propellant
(nitrogen tetroxide (N_2O_4), unsymmetrical dimethyl
hydrazine (UDMH)) stages inertially guided and a variety of
spin-stabilized third stages. Control is by engine vector
gimbal of the first and second stages. The first stage has
been augmented by a variety of SRM's. The 3920
configuration, currently in use, has nine Castor IV solid
rocket boosters (SRB's), which are operated in a sequence
of six ignited at lift-off and jettisoned 57 sec after
lift-off and three sustaining the first stage after the
first six have been jettisoned.

The 3920 has a 200-km payload of 3350 kg due east from
the NASA John F. Kennedy Space Center (KSC) and 1950 kg to
196° from VAFB. This payload represents a significant
growth over the life of the program, 30 times the lift of
the first Thor Delta.

The 3920 payload assist module (PAM) injects 1295 kg
into geosynchronous transfer orbit (GTO). The original
Delta launched 45 kg to geosynchronous transfer orbit in
1960.

Launches are conducted from KSC and VAFB.

N-1 and H-1. Japan has operated a comparable vehicle
built under license from McDonnell Douglas Corporation.
The N series vehicles were Japanese-manufactured versions
of the Delta design. The H-1 introduced a Japanese-
designed cryogenic second stage to enhance performance.
They have experienced a 93% success rate in 16 launches to
date.

Launch is from the Tanegashima facility south of Kyushu at latitude 30° N.

Long March 3. The Long March 3 launch vehicle uses storable propellants and has a payload capability to low Earth orbit of 3800 kg or to geosynchronous transfer orbit of 1300 kg. Only a few launches have been attempted, and available data on reliability are limited. Launch is from the Jiuquan Satellite Launch Center in Gansu Province.

Atlas Centaur. The Atlas vehicle was developed in the early 1950's as a ballistic missile. As a space launch vehicle, its first flight was in 1959. The Centaur upper stage development began in 1958 to provide a high-performance geosynchronous transfer or planetary injection capability.

Atlas Centaur has had 67 launches, of which 56 have been successful for an 84% success rate; near-term experience since 1977 is 92%. The Atlas E-F is used without Centaur for low Earth orbit payload delivery. Eighty-nine launches have been performed since 1958 with 80 successes for an overall reliability of 0.90.

The Atlas core stage is a liquid-oxygen/hydrocarbon (RP-1) system which is a stage and a half configuration using three engines at lift-off. The two booster engines are jettisoned at 153 sec and the sustainer engine continues for another 130 sec.

The Centaur is a cryogenic stage using liquid oxygen and liquid hydrogen to supply two engines. The stage has multiple start capability and a maximum burn time of 404 sec.

Guidance for Atlas Centaur is inertial and carried in the equipment section of the Centaur stage. The Atlas E-F uses a radio guidance system for low Earth orbit missions. Control is by engine gimbal.

The Atlas Centaur can deliver 5800 kg to low Earth orbit or 2300 kg to geosynchronous transfer orbit.

The Atlas Centaur is launched from KSC and the Atlas E-F from VAFB.

Ariane 3. Ariane 3 is the currently operational member of an evolutionary family of launch vehicles. Configuration of the core first stage is constant, but the second and third stages are provided in several configurations. Ariane 4 uses an assortment of strap-on boosters with solid or liquid propellants and variable in number. Delivery performance is a function of the model selected.

The Ariane provides for delivery of two spacecraft by way of an interface adapter system that deploys the two spacecraft in sequence.

Launch is from Kourou, French Guiana, at latitude 5.6° N.

In 18 launches to date, there have been 4 failures of Ariane vehicles for a reliability of 78%, which is not significantly worse than the early performance records of some other space launch systems.

Titan 34D. The Titan 34D, the current heavy lift vehicle for Department of Defense (DOD) payloads, is a derivative of the Titan ballistic missile system. The Titan 34D uses two large solid rocket boosters to augment a liquid-storable N_2O_4/UDMH two-stage core. It is the successor to the Titan III series, the development of which began in 1961. Seventy-five vehicles were launched in the Titan III series, and the total of space launches to date for the design is 135 with 5 failures for an overall reliability of 0.96.

In the Titan launch, the SRB's are used for initial lift-off and the core first-stage engine is ignited in flight. Early versions used radio guidance, but current configurations use strapped-down inertial guidance systems. Control is accomplished by liquid injection for thrust-vector control. A number of upper stages have been designed to interface with Titan - the Centaur, the inertial upper stage (IUS), the transtage, the transfer orbit stage (TOS) - and design studies are in progress for the PAM.

Delivery performance to low Earth orbit is 13 000 kg and to geosynchronous orbit, 2000 kg with the IUS.

The Titan 34D currently is launched from VAFB, and launch capability at KSC will be provided in 1989.

Proton. The Proton launch vehicle has been one of the more intensively used U.S.S.R. launch systems. For LEO missions, a three-stage configuration fueled by storable propellants (N_2O_4/UDMH) is used. For GEO missions, a fourth stage fueled by liquid oxygen and kerosene is used.

The reliability of the vehicle is reported variously as 0.91 to 0.95 depending on the inclusion of "early failures" and the failures of "experimental" fourth stages.

Launch is generally from Baykonur, Kazakhstan, at latitude 46° N.

Space Shuttle. The Space Shuttle was developed as the first reusable space transportation system to provide not only orbital delivery but recovery of payloads. The Space Shuttle flew 24 successful flights before the failure and loss of *Challenger* in January 1986 on the 25th flight.

The Space Shuttle uses two solid rocket boosters to augment initial thrust. The Space Shuttle main engines are ignited and brought to full thrust before ignition of the SRB's. The SRB's are separated at 126 sec and the system continues into an elliptical orbit under the thrust of the main engines. After separation of the external tank for ocean disposition, the orbital maneuvering engines are used to circularize and raise the orbit as required for the mission objectives.

The expected launch capability for future Space Shuttle flights is 22 765 kg, which is reduced from the original 29 010 kg by growth in system inert weight to enhance system reliability and safety and by reduced throttle settings on the main engines to minimize the risk of turbomachinery failure.

A number of upper stages are designed to be flown in the Space Shuttle; among them are the PAM, the IUS, the TOS, and several spacecraft with internal propellants (e.g., Leasat). Payload deliveries to geosynchronous orbit are consistent with the upper stage performance from 296 km; e.g., IUS is 2300 kg.

The Space Shuttle is launched from KSC. Launch pad provisions at VAFB are in caretaker status pending return to flight. The Space Shuttle no longer offers commercial satellite delivery services. Costs for sortie payloads are a function of the use of all the different resources offered by the Space Shuttle.

Under Development or Planned

Figure 2 illustrates the launch vehicles currently under development or planned for the 1990's.

Conestoga. The Conestoga launch system has multiple configurations. The Conestoga II-1 is illustrated. The

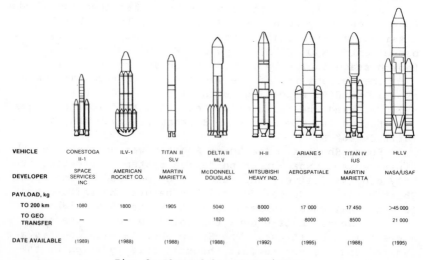

VEHICLE	CONESTOGA II-1	ILV-1	TITAN II SLV	DELTA II MLV	H-II	ARIANE 5	TITAN IV IUS	HLLV
DEVELOPER	SPACE SERVICES INC	AMERICAN ROCKET CO.	MARTIN MARIETTA	McDONNELL DOUGLAS	MITSUBISHI HEAVY IND.	AEROSPATIALE	MARTIN MARIETTA	NASA/USAF
PAYLOAD, kg								
TO 200 km	1080	1800	1905	5040	8000	17 000	17 450	>45 000
TO GEO TRANSFER	–	–	–	1820	3800	8000	8500	21 000
DATE AVAILABLE	(1989)	(1988)	(1988)	(1988)	(1992)	(1995)	(1988)	(1995)

Fig. 2 Planned launch vehicles.

Conestoga I uses no first-stage augmentation, and the
Conestoga IV uses six rather than two first-stage boosters.
All elements are solid rocket motors.

The core vehicle has two SRM stages, each with
vectorable nozzles, an inertial guidance and control
module, and a spin table for third-stage stabilization.
The Conestoga IV can be configured with six stages to place
a 180-kg spacecraft in geosynchronous orbit or to inject a
530-kg spacecraft on a planetary trajectory.

The Conestoga is in development but has flown one
successful suborbital test.

Agreements have been signed with NASA to use the
Wallops Island facility for launch of the Conestoga, and
negotiations for launch capability at VAFB are in progress.
Because its launch facility requirements are minimal,
Conestoga can be launched from other locations as well.

Launch capability is offered 18 months from authority
to proceed with launch.

Industrial Launch Vehicle. The American Rocket
Company industrial launch vehicle (ILV) design is unique in
using a hybrid propulsion system with a solid fuel and
liquid oxygen as the oxidizer. The first-stage arrangement
of the 12 fuel-motor units around the common oxidizer tank
forms a plug nozzle that can enhance high-altitude motor
performance.

The second, third, and fourth stages are, in plan
view, a hexagonal cluster of seven motor-tank units. Four
of the exterior motors are the second stage, the two
remaining outer motors constitute the third stage, and the
central unit is the fourth stage. All the motor units are
identical except for the larger expansion ratio nozzle of
the upper stage units. The seven upper stage motors each
has integral oxidizer tankage to facilitate staging.

Guidance is by inertial reference and control during
first-stage and second-stage burns and is achieved by the
injection of liquid oxygen into selected nozzles. For
third- and fourth-stage operations, head-end monopropellant
thrusters are used for pitch, yaw, and roll control.

Launch vehicle delivery capability is currently
scheduled for the end of 1988. Negotiations are in
progress for use of KSC and VAFB as launch sites.

Titan II. There are 52 Titan II ballistic missiles
that could be refurbished as space launch vehicles.
Thirteen have been placed in work to be upgraded to space
launch vehicles to launch polar Earth orbit satellites.
The first unit will be available in late 1988. At
present, this vehicle's use is planned only for U.S.
Government payloads and would replace the Atlas E-F for
such launches.

The Titan II space launch vehicle (SLV) is essentially identical to the core section of the Titan III in guidance and control systems. For launch from VAFB to Sun-synchronous orbit altitude of approximately 900 km, payload capability is approximately 1700 kg.

Delta II. The Delta II or medium launch vehicle (MLV) is another growth increment for the Delta vehicle. The growth is achieved by extending the first-stage tanks by 3.34 m to increase the storable-propellant mass and by using stretched SRM's with graphite epoxy cases. The rest of the vehicle remains basically the same.

In the Delta 7920 configuration (increased storable propellant and graphite SRM cases), the vehicle provides 4300 kg to low Earth orbit and 1600 kg to geosynchronous transfer orbit. In the enhanced Delta II configuration, performance is 5040 kg to LEO and 1820 kg to GTO.

H-II. The H-II vehicle is being developed by NASDA, the Japanese space agency. Two SRM boosters are used to augment the core vehicle, which is fueled by liquid oxygen and hydrogen for both first and second stages. The first stage uses the LE-7 engine, a staged combustion cycle engine, and the second stage uses the LE-5 engine, a gas-generator pump-fed engine.

Guidance is inertial and control is by engine gimbal to control the thrust vector.

The vehicle systems currently are undergoing testing, and the initial launch is planned for 1992.

Launch will be from the Tanegashima launch site.

Ariane 5. The Ariane 4 consists of a family of vehicles that use the core of the Ariane 3 together with a variable number of solid- and liquid-propellant boosters to augment first-stage performance and larger second- and third-stage tankage than used for Ariane 3.

Ariane 5 is a new configuration for heavy lift. Two large SRB's are used to augment the first stage, and cryogenic propellants are used in both the first and second stages. The HM60 is a pump-fed bypass flow cycle and a single engine is used in each of two stages.

The system currently is under development with first flight planned for 1995.

The design missions for Ariane 5 are to support the European Space Agency (ESA) components of the international Space Station and to provide a launch system for Hermes. Hermes is a relatively high lift-to-drag, manned spacecraft that would be boosted to orbital velocity by the Ariane 5 but have self-contained propulsion capability for performing orbital and entry maneuvers.

Titan IV. This extension of Titan capability is achieved by increasing the SRM size from 5.5 segments to 7

segments and extending the first-stage propellant tankage.
These changes increase the 200-km due-east lift to 17 450
kg. With an IUS, the system has an 8500-kg capability to
geosynchronous transfer orbit. At present, this vehicle is
planned to be used only by the U.S. Government.

    HLLV. The heavy lift launch vehicle (HLLV) being
studied by NASA and the DOD is now referred to as the
advanced launch system (ALS). The objective in these
studies is to define a system with a lift capability more
than twice that of any current system but with a lower
direct operating cost and higher reliability. The design
studies are based on analyses that indicate that advanced
technology in materials, design techniques, automation in
manufacture and operations, and performance derating can
attain such goals. The studies are oriented to first
availability of such a new system in 1995 or later.

    Concurrently with the ALS studies, NASA is examining
the performance and cost characteristics of a Space
Transportation System (STS) derived, early availability
HLLV. Such a system would use available STS components and
operating systems to provide a maximum payload capability
on the order of 100 000 kg to LEO or 10 000 kg to GTO.

    Performance. Figure 2 summarizes the low Earth orbit
performance of this family of vehicles.

    To reflect variation of launch vehicle performance as
a function of launch site and desired orbit, Fig. 3
illustrates the performance of the primary U.S. launch

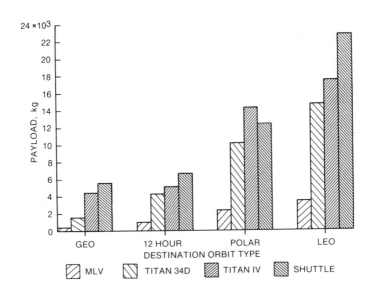

Fig. 3 Launch vehicle capabilities, payload to orbit.

vehicles of the 1990's for the most common operational orbits.

The geosynchronous orbit performance is based on the assumption of KSC launch, 3° to 5° plane change in the transfer orbit injection, and the balance of plane change, 23° to 25°, at geosynchronous altitudes. The 12-hr orbit is a semisynchronous, 20 240-km orbit at 63° inclination. Polar orbit is 920 km, Sun synchronous from VAFB, and low Earth orbit is a due-east launch from KSC to 500 km at 28.5° inclination.

Upper Stages

Existing

Figure 4 illustrates the existing U.S. stages. In this figure and the next two, details of the stages are characterized in terms of the manufacturer, primary spatial dimensions, performance parameters, mass properties, the delivery capability, the profile view, planned schedules,

CHARACTERISTICS		PAM D	PAM A	IUS (TWO STAGE)
STAGE:				
MANUFACTURER		McDONNELL DOUGLAS (MDAC)	MDAC	BAC
LENGTH	(m)	2.4	2.3	5.0
DIAMETER	(m)	1.3	1.3	2.9
ENGINE:				
MANUFACTURER		THIOKOL	THIOKOL	CSD
TYPE		(STAR 48)	(MINUTEMAN 3)	
NUMBER		1	1	SRM-1 SRM-2
FUEL		SOLID	SOLID	SOLID
COMPOSITION		TP-H-3340	AWS-3066	HTPB
TOTAL THRUST	(N)	66 300	157 000	200 000 81 200
SPECIFIC IMPULSE	(SEC)	285.6	274.3	292.9 300.9
BURN TIME	(SEC)	85.0	59.5	153 104
STAGE:				
PAD WEIGHT	(kg)	2 180	3 750	14 800
PROPELLANT WEIGHT	(kg)	2 000	3 430	9 710 2 750
BURNOUT WEIGHT	(kg)	189	318	1 160 1 140
SUPPORT EQUIP. WEIGHT	(kg)	1 140	2 080	3 350
PAYLOAD[1]:				
TO GEO - ONE WAY STAGE	(kg)	635*	998*	2 310
TO GEO TRANSFER ORBIT (GTO)	(kg)	1 250	1 950	~4 500
ILLUSTRATION: NOTES: 1. REF: 28.5° INCLINATION & 150 N.M. CIRCULAR * REQUIRES KICK STAGE		 PAM D	 PAM A	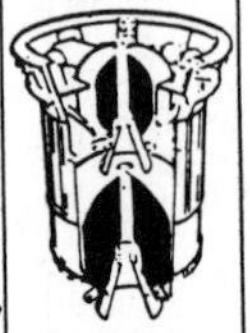 IUS (TWO STAGE)
SCHEDULE:				
START DATE		1976	1977	1978
OPERATIONAL DATE		1982	1982	1982
TYPE OF DEVELOPMENT		COMMERCIAL	COMMERCIAL	US GOVT
SPONSOR		MDAC	MDAC	USAF

Fig. 4 Upper stages, existing.

and the nature of the enterprises which sponsor and execute the development.

The PAM-D is Delta compatible, as well as STS compatible, and was developed by the McDonnell Douglas Corporation as a commercial venture under an agreement with NASA that the Government would not develop an equivalent stage. The PAM-D has been flown 18 times on the STS and 13 times on the Delta launch vehicle.

The IUS was developed to accommodate payloads to be flown on the Titan 34D or the STS. To date, it has been flown once on the Titan 34D and once on the STS. It may also be used on the Titan IV.

Under Development

Figure 5 illustrates a number of stages currently under development. These stages are in development and funded for certification, but not all have a first-use payload and a flight schedule date identified.

The Centaur-G is a wide-body modification of the Centaur stage used on the Atlas and Titan expendable vehicles.

The Centaur-G Prime (G′) is a larger version of the stage developed to support NASA planetary missions. In addition to the larger hydrogen tank, the G′ stage has a

CHARACTERISTICS		CENTAUR G	CENTAUR G	TOS	AMS	TOS/AMS	IRIS	PAM D-II	SCOTS
STAGE: MANUFACTURER		GENERAL DYNAMICS (G.D.)	G.D.	MARTIN MARIETTA (MMC)	MARTIN MARIETTA (MMC)	MMC	AERITALIA	McDONNELL DOUGLAS (MDAC)	RCA ASTRO
LENGTH	(m)	5.9	6.9	3.3	1.7	5.0	2.3	2.0	2.8
DIAMETER	(m)	4.3	4.3	3.4	3.0	3.0	1.5	1.6	1.8
ENGINE: MANUFACTURER		PRATT-WHITNEY	PRATT-WHITNEY	CSD	ROCKET-DYNE (RD)	CSD/RD	SPD DIFESA SPAZIO	THIOKOL	THIOKOL
TYPE		RL 10A-3-3B	RL 10A-3-3A	SRM-1	RS-41	SRM-1/RS-41	TBD	ISTP	63-E
NUMBER		2	2	1	1	1/1	1	1	1
FUEL		LO_2LH_2	LO_2LH_2	SOLID	N_2O_4/MMH	SOLID/N_2O_4/MMH	SOLID	SOLID	SOLID
COMPOSITION		6:1	5:1	HTPB	1.65:1	HTPB 1.65:1	HTPB	—	TPB
TOTAL THRUST	(N)	133 000/66 700	147 000/74 400	200 000	11 800	200 000/11 800	TBD	78 300	156 000
SPECIFIC IMPULSE	(SEC)	440.4	446.4	294	314.7	292.6/314.7	290.6	—	301
BURN TIME	(SEC)	436	609	150	1 050	150/1050	TBD	121	125
STAGE: PAD WEIGHT	(kg)	16 900	2a 19 200	10 800	4 230	14 900	1 830	3 490	4 350
PROPELLANT WEIGHT	(kg)	13 500	15 800	9 710	3 240	9 750/3 240	1 570	3 240	1 560
BURNOUT WEIGHT	(kg)	3 330	3 440	1 080	996	929/1 010	258	245	519
SUPPORT EQUIP. WEIGHT	(kg)	4 180	4 310	1 450	1 360	2 040	955	1 600	1 190
PAYLOAD[1]: TO GEO - ONE WAY STAGE	(kg)	4 540[2]	8 850[2] 5 990[2a]	3 080*	862*	2 690	454	907*	1 130 (REQS. KICKSTAGE)
TO GEO TRANSFER ORBIT (GTO)	(kg)	NOT APPLICABLE	NOT APPLICABLE	5 900	1 650	8 070	899	1 090	2 490
ILLUSTRATION: NOTES: 1. REF 28.5° INCLINATION & 150 N.M. CIRCULAR * REQUIRES KICK STAGE 2. STAGE PROPELLANT FULL 2a. STAGE, PROPELLANT, PAYLOAD, AND A.S.E. LIMITED TO 28 019 kg		CENTAUR G	CENTAUR G′	TOS	AMS	TOS/AMS	IRIS	PAM D-II	SCOTS
SCHEDULE: START DATE		1982	1982	1983	1984	1984	1983	1980	1984
OPERATIONAL DATE		1986	1986	1986	1987	1987	1987	1985	1986
TYPE OF DEVELOPMENT SPONSOR		US GOVT USAF	US GOVT NASA	COMMERCIAL ORBITAL SCIENCES CORP. (OSC)	COMMERCIAL OSC	COMMERCIAL OSC	GOVERNMENT ITALY	COMMERCIAL MDAC	COMMERCIAL RCA

Fig. 5 Upper stages, under development.

lower mixture ratio, a larger expansion ratio engine cone, and an increase of 6 sec in specific impulse.

The transfer orbit stage is being developed by Martin Marietta Corporation for Orbital Sciences Corporation, a new company formed to develop and market the system. The TOS uses the same first-stage motor used on the IUS and is targeted for heavy payloads beyond the capability of the PAM and not large enough to warrant the use of a Centaur-G or G' system. It offers a different avionics and data processing system than that of the IUS.

Orbital Sciences Corporation is also developing an apogee and maneuver stage (AMS) which can be used independently or as an apogee stage in conjunction with the TOS. In this configuration, the TOS/AMS has slightly better performance than the IUS because of the higher specific impulse of the storable propellants relative to the SRM of the IUS upper stage.

The Italian Research Interim Stage (IRIS) is being developed by Aeritalia under the sponsorship of the Italian Government. Its capability is a little less than that of the PAM-D. Its first use is scheduled for the LAGEOS mission in late 1987.

The PAM-D II is a larger version of the PAM-D system. The SRM is larger, as are the airborne support equipment (ASE) structure and the spin table. The avionics system is not changed in any significant detail. Its first flight was in late 1985 in support of Satcom KU-1.

RCA Astronautics has been developing the Shuttle-compatible orbit transfer stage (SCOTS) as a perigee or transfer orbit stage for future geosynchronous satellites. Its capability is slightly greater than that of the PAM-A, but not as large as that of the TOS. It will be flown first with the RCA Direct Broadcast Satellite in 1987.

Under Study

In Fig. 6, some of the stages being studied for future development by commercial firms are illustrated. Orbital Sciences Corporation is examining a larger version of the AMS and TOS/AMS combinations to provide performance comparable to that of the Centaur-G.

The satellite transfer vehicle (STV) is a storable-bipropellant stage that has been studied by Scott Science and Technology and British Aerospace. It is designed to accommodate payloads beyond the capability of the PAM-A, but not so large as to need a TOS.

The liquid propulsion module (LPM) is a storable-bipropellant stage using a pump-fed engine to get higher specific impulse and, consequently, a higher mass fraction

CHARACTERISTICS		AMS HIGH PERFORMANCE	TOS/AMS HIGH PERFORMANCE	STV	LPM	HPPM
STAGE:						
MANUFACTURER		MARTIN MARIETTA (MMC)	MMC	BRITISH AEROSPACE	AEROJET TECH SYS (ATSC)	ATSC
LENGTH	(m)	2.4	6.1	2.0	1.5	1.5
DIAMETER	(m)	3.3	3.3	4.4	4.1	3.8
ENGINE:						
MANUFACTURER		ATSC	CSD/ATC	TBD	ATSC	ATSC
TYPE		TRANSTAR 1	SRM 1/ TRANSTAR 1	TBD	TRANSTAR 1	TRANSTAR 1
NUMBER		1	1/1	1	1	1
FUEL		N_2O_4/MMH	SOLID/N_2O_4/MMH	N_2O_4 & MMH	N_2O_4 & MMH	N_2O_4 & MMH
COMPOSITION		1.8:1	HTPB/1.8:1	TBD	1.8:1	1.8:1
TOTAL THRUST	(N)	16 700	200 000/16 700	TBD	16 700	16 700
SPECIFIC IMPULSE	(SEC)	326.0	292.6/326	~310	326	326
BURN TIME	(SEC)	1 660	150/1 660	TBD	1 660	—
STAGE:						
PAD WEIGHT	(kg)	10 100	20 700	11 700	6 520	5 970
PROPELLANT WEIGHT	(kg)	8 620	9 750/8 620	9 810	6 000	5 220
BURNOUT WEIGHT	(kg)	1 450	929/1 390	1 940	522	751
SUPPORT EQUIP WEIGHT	(kg)	1 910	2 720	TBD	—	—
PAYLOAD[1]:						
TO GEO - ONE WAY STAGE	(kg)	1 610	4 540	ESTIMATED 1 100 - 1 800	1 540	1 350
TO GEO TRANSFER ORBIT (GTO)	(kg)	5 900	12 200[2] 9 710[2a]	1 800 - 3 600	4 400	2 930
ILLUSTRATION:						
NOTES: 1. REF 28.5° INCLINATION & 150 N.M. CIRCULAR * REQUIRES KICK STAGE 2. STAGE PROPELLANT FULL 2a. STAGE, PROPELLANT, PAYLOAD, AND A.S.E. LIMITED TO 29 019 kg		AMS	TOS/AMS	STV	LPM	HPPM
SCHEDULE:						
START DATE		1984	1984	1964	—	—
OPERATIONAL DATE		1988	1988	1987	—	—
TYPE OF DEVELOPMENT SPONSOR		COMMERCIAL OSC	COMMERCIAL OSC	COMMERCIAL SCOTT SCIENCE & TECHNOLOGY	COMMERCIAL ATSC	COMMERCIAL FORD AEROSPACE AND ATSC

Fig. 6 Upper stages, predevelopment.

efficiency. The Aerojet Technical Systems Company has proposed it as a stage to accommodate payloads from the PAM-D class to as high as two and a half times that mass. The Transtar 1 engine for the stage is in development, but no date for stage availability has been published.

The high-performance propulsion module (HPPM) is a joint effort by Aerojet Technical Systems Company and Ford Aerospace to examine a somewhat smaller system also based on the Transtar engine and sized to have roughly twice the PAM-D performance.

Not illustrated is a vehicle currently being studied by NASA and the DOD. It is a storable-propellant upper stage designed to place a 3350- to 4465-kg payload in geosynchronous orbit and is compatible with the STS and the Titan IV.

It is noteworthy that many of the developments and studies for upper stages identified here are commercial undertakings not sponsored by Government agencies. It also is worth noting that there is increasing interest in fluid systems rather than continued reliance upon SRM's for upper stages.

Other Significant Issues

In addition to basic lift capability, there are numerous other issues of significance to spacecraft developers and users of space systems. An extended generic discussion is not practical in a short paper, but among the more significant considerations are the following.
- Induced environments
    Acceleration
    Vibration
    Thermal
- Prelaunch access
    Physical
    Data and command
- Launch date and period of availability
- Priority in sequence for reflight in the event of launch failure
- Cost and terms of payment
- Guidance accuracy

To illustrate the significance of some of these considerations, Fig. 7, which reflects the lifetime reliability of U.S. launch systems through the end of calendar year 1985, is shown. Despite long experience and substantial efforts to enhance reliability, there are still a significant number of failures. The cost of such unreliability falls primarily on the spacecraft user who has at risk his only spacecraft or one of his few spacecraft. The effect is to limit the value that can be

LAUNCH VEHICLE	SUCCESS RATE	LAUNCH DELAYS
SCOUT	0.96	3
DELTA	.96	3
ATLAS/CENTAUR	.95	1
TITAN	.95	?
SHUTTLE	1.00	.5

1981-85 DATA FOR SCHEDULED LAUNCH DATE AS OF 60 DAYS BEFORE ACTUAL LAUNCH. TITAN DATA ARE INDETERMINATE BECAUSE OF CLASSIFICATION OF PLANNED DATES PRIOR TO LAUNCH.

Fig. 7 Launch success/launch delay equivalent days.

committed to any one spacecraft. Another unit of a launch
system in use can be ready long before the spacecraft can
be replicated.

Launch delay is another figure of merit in assessing
launch service. Do not assign significance to the delay
term as a property of the launch system. Figure 8
illustrates the sources of delay. Note that most sources
of launch delay are not the launch vehicle but other
factors. Further note that almost half of all the launch
delays assigned to the launch vehicle are due to shear
loads from winds aloft. Weather delays in the tabulation
are caused by ground fog, hurricanes, or other local
weather constraints on operation.

There are some longer term issues that will have to be
dealt with in future space launch operations. The two most
prominent are the short-term environmental effects of the
use of solid propellants and the long-term effects of
leaving spent stages and spacecraft in orbit.

Solid propellants have very good performance, but in
the near-launch-pad environment, they create a significant
source of acid rain if atmospheric conditions are not
considered carefully. Their use on orbit creates a cloud
of micrometer-sized particles that disperse or enter the
atmosphere reasonably soon but, while present, abrade
spacecraft thermal and optical surfaces. Although these
effects are minor at current levels of activity, they could
become more significant as the level of space activity
increases. When other considerations are combined with
performance evaluations, the longer term preference will be
for liquid systems.

Since space operations began, more than 17 000 objects
have been placed in low Earth orbit (i.e., between 300 and
2000 km) and more than 7000 are in orbit now. Only a small

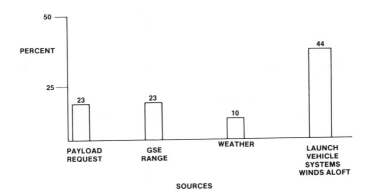

Fig. 8 Sources of changes in launch date.

fraction of these are useful operational spacecraft. These objects represent 2 000 000 kg of mass contrasted with the 10 000 kg of natural meteoritic material that passes through these same regions. Further, the manmade material is not uniformly distributed. In the not too distant future, it will no longer be wise to leave spent stages and spent spacecraft in low Earth orbit. Objects no longer useful become a hazard to those in use. There will be a performance penalty to operate placement stages for controlled entry and to enter spent spacecraft. The penalty will never be less than when it is dealt with in the initial design. Obviously, such action will require agreement and common action among all launch systems and spacecraft operators when an appropriate course of action can be defined. Studies are in progress among all the launching organizations to better understand all the issues and to define the most effective preventive and palliative measures that can be adopted.

Summary

Launch systems have improved significantly in performance and reliability over the last 30 yr. The combined effect has been to make space-based systems more cost and performance effective by more than two orders of magnitude. We may be approaching a performance asymptote as more systems use liquid oxygen and hydrogen for maximum potential chemical impulse, but there is certainly another order of magnitude or two that we can achieve in reliability. That has been the traditional path of growth for other transportation systems, and it is expected to be for space transportation. That, we believe, will be our challenge; improved reliability.

Bibliography

Scout Planning Guide, LTV Aerospace and Defense, Vought Missiles and Advanced Programs Division, P.O. Box 650003, Dallas, TX 75265-0003, May 1986.

Delta Spacecraft Design Restraints, McDonnell Douglas Astronautics Company, 5301 Bolsa Ave., Huntington Beach, CA 92647, July 1980.

Atlas Centaur Mission Planners Guide, General Dynamics Convair Division, P.O. Box 85357, San Diego, CA 92138.

Titan Space Launch Vehicles, P.O. Box 179, Martin Marietta Denver Aerospace, Denver, CO 80201, May 1986.

Ariane Launch Vehicles, rue Soljenitsyne, 91000 Evry, France.

H-II Rocket NASDA External Relations Department, World Trade Center Building, 4-1 Hamamatsu-cho 2-chome, Minato-Ku, Tokyo 105, Japan.

Industrial Launch Vehicle, American Rocket Company, 3575 B Haven, Menlo Park, CA 94025, Sept. 1986.

Conestoga User Manual, Space Services Inc. of America, Suite 140, 7015 Gulf Freeway, Houston, TX 77087, Feb. 1987.

Commercial Space, Fall 1986, McGraw Hill, P.O. Box 1523, Neptune, NJ 07754.

Space, Shearson Lehman Brothers, Aug. 1986.

Insight, The Washington Times, Oct. 20, 1986, Washington, DC, pp. 14-16.

Centaur - The NASA/AF Shuttle Upper Stage, General Dynamics Corporation, 1984.

Davis, E., Future Orbital Transfer Vehicle Technology Study, Boeing Aerospace Company, Contract NAS1-16088, 1982.

Hankins, R. A., The Liquid Propulsion Module: Low Cost Access to Space, AIAA-84-1285, AIAA-SAE-ASME 20th Joint Propulsion Conference, 1984.

IUS User's Guide, Boeing Aerospace Company, 1984.

PAM-D/DII User Requirements Documents, McDonnell Douglas Corporation, 1985.

STS PAM-A User Requirements Document, McDonnell Douglas Corporation, 1980.

Thompson, D. W., Advanced Capabilities in Space Transportation, Orbital Sciences Corporation, 1985.

Chapter II.   Material Processing and Fluid Mechanics in Microgravity

Space Processing of Metals and Alloys

Robert J. Bayuzick*

Vanderbilt University, Nashville, Tennessee

Abstract

A space environment has many useful attributes for the processing of metals and alloys. Though there are many ways of expressing these attributes, one possible listing is hydrostatic pressure tends toward zero, buoyancy convection tends toward zero, sedimentation and flotation can be eliminated, containerless processing becomes possible, and ultra-high vacuum levels at infinite pumping speed can be realized. While it is necessary for experiments to physically be located in space for the last attribute to prevail, the first four can be exploited by a number of techniques. Useful techniques involve drop tubes, drop towers, research aircraft and sounding rockets, as well as the space shuttle, space station and dedicated free-flyers.

In a general sense much of the focus for industrial applications in the U.S. may be described as being the development of high performance materials. This is true for both containerless processing and containered processing. Industry is most interested in exploring gravity as a processing parameter to gain insight for modification of and innovation in earth processing. However, actual manufacturing of metals and alloys in space for use on earth and/or use in space may have longer term possibilities.

Introduction

Nearly all of the potential benefits of the use of space for the processing of metals and alloys have their roots in one single, well-appreciated fact. Namely, the gravitational

---

*Professor, Department of Mechanical and Materials Engineering; and Director, Center for the Space Processing of Engineering Materials.

137

body forces tend toward zero. Out of this comes a number of interconnected and entwined attributes. These are:

1. Hydrostatic pressure can go to zero.

2. Buoyancy convection goes to zero.

3. Sedimentation and flotation can be eliminated.

4. Containerless processing becomes possible.

All of the above refer to attributes concerned with the fluid nature of the melts of the metals and alloys. A somewhat separate attribute of processing materials in space is the potential for ultra-high vacuum processing at essentially infinite pumping speed.

Hydrostatic Pressure

If only a gravitational field is operating, the forces acting within a fluid stem from the pressure of the surroundings and the gravitational body force. The pressure variation in an incompressible static fluid is then given by

$$p = p_o + \rho g h \qquad\qquad (1)$$

where h is in the gravity direction and its magnitude is the distance from the free surface, ρ is the density of the fluid, g is the gravity level, and p_o is the pressure of the surroundings. Hence, if g is zero, the hydrostatic component of the pressure is zero.

A direct benefit of a zero component of hydrostatic pressure in the space processing of metals and alloys is possible in float zone melting techniques. In this regard

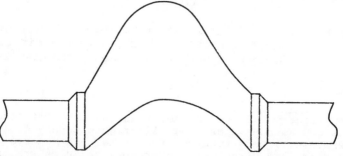

Fig. 1   Representation of a "jump-rope" instability in rotating floating zones in low gravity.

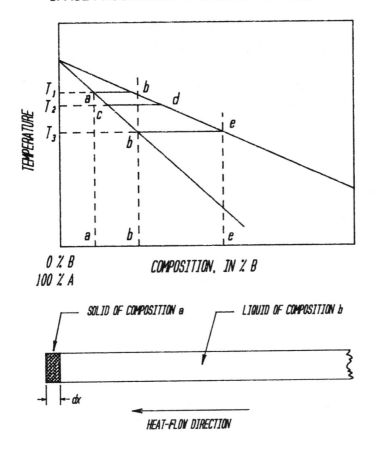

Fig. 2 Modeling of convectionless solidification: a) part of an isomorphous binary phase diagram; b) one-dimensional solidification model.

the geometry of the process can be optimized in that it is possible to extend a molten floating zone to the theoretical limit of length equal to circumference.[1] However, there are two potential problems. One is the potential for developing a "jump-rope" instability (Fig. 1) in rotating floating zones as was demonstrated as long ago as the Skylab experiments;[2] the other is the potential for Marangoni convection as discussed in the next section.

Buoyancy Convection

The buoyancy force, $\vec{F}$, on a body of volume V in a liquid can be expressed as

$$\vec{F} = -\int \nabla p \, dV \qquad (2)$$

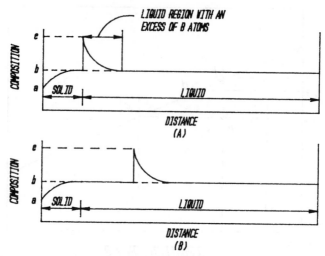

Fig. 3 Composition-distance curves corresponding to two different stages in the one-dimensional model.

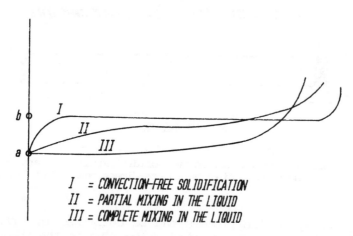

```
I  = CONVECTION-FREE SOLIDIFICATION
II = PARTIAL MIXING IN THE LIQUID
III = COMPLETE MIXING IN THE LIQUID
```

Fig. 4 Concentration profiles in one-dimensional directional solidification.

where ∇p is the gradient in pressure. If there are no body forces, ∇p is zero and there is no buoyancy force. Therefore, in a low gravity environment, buoyancy convection is for all practical purposes absent.

At first glance then, the classical textbook diffusion-dominated descriptions of solidifications[3], as represented in Figures 2 and 3, apply. In Figure 3, the first solid to freeze from liquid of composition b has composition a, rejecting solute in the process. The liquid nearby is enriched so that the next solid to freeze has a slightly greater composition than a, but again the nearby liquid is enriched.   So on and so on until the steady state is established as indicated in the figure.   The final composition profile of the solid in this one-dimensional freezing model is given by curve I in Figure 4.   This is to be compared to curves II and III for differing degrees of mixing by convection.   Thus it can be seen that improvement in homogeneity can be expected by solidification in low gravity.   However, complete homogeneity is not expected at normal freezing rates.

Improvement in uniformity of microstructure can also result by solidification under low gravity.   Figure 5 shows a schematic of a section through an earth-casting.[4]   The central equiaxed zone is caused by constitutional undercooling and/or the presence of seeds.   The origin of seeds is dendritic remelt and breakup because of convective flows as depicted in Figure 6.[5]   At low gravity this contribution to formation of the central zone would be absent.

The tolerable level of gravity in the elimination of buoyancy effects depends on other processing parameters.   For example, Coriell et al [6] have considered the critical alloy concentration and solidification rate thresholds for convective instabilities in the directional solidification of tin/lead alloys at various gravity levels.   Their results,

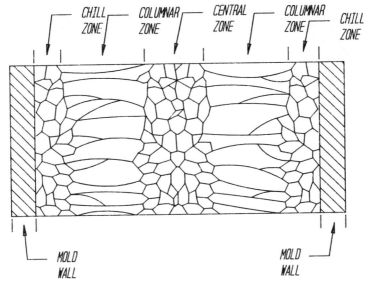

Fig. 5 Schematic of a cross section through an earth-casting.

ONE GRAVITY   LOW GRAVITY

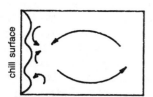

As the casting begins to
solidify, thermal and
compositional buoyancy
driven convection results.

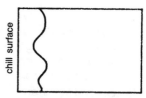

The buoyancy driven flows
are negligible.

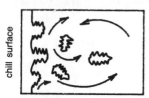

Dendritic growth occurs.
The changing interdendritic
fluid composition results
in buoyancy driven flows
leading to casting
macrosegregation.

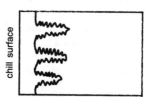

The buoyancy driven
contribution to casting
macrosegregation
is eliminated.

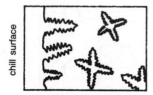

Heat pulses from
convective flows cause
dendritic remelt
and breakup.

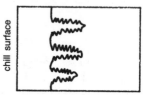

The thermal gradient
is better controlled.
Dendritic breakup
is inhibited.

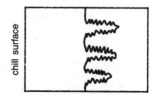

Convective flows dissipate
super heat in the melt,
allowing transported dendritic
material to grow forming
new grains.

The absence of complex
buoyancy driven phenomena
can facilitate systematic
study of important variables
in the casting process.

Fig. 6   Mechanism of dendrite multiplication due to convective
currents in an earth-gravity environment.

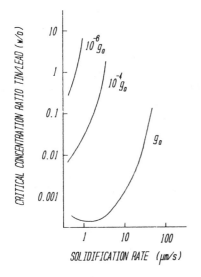

Fig. 7   The critical concentration
of tin in lead above which
convective instabilities occur.

seen in Figure 7, show that even at 10^{-6} g_0 (g_0 is the
gravity level on earth at sea level) convection is expected
to play a role at the higher alloy concentrations and slower
solidification rates, such as at a tin/lead ratio of above
about 2.4 wt.% and solidification rate of 1 μm/s.   On the
other hand, convective effects are unimportant at lower alloy
concentrations and higher solidification rates at g_0, such as
at a tin/lead ratio of 0.001% and solidification rate of 10
μm/s.

　　Furthermore, Glicksman has given   indications on the
significance of acceleration level and degree of supercooling
on the tolerable gravity levels in considerations of dendrite
velocity[7].   Langbein's formulation for convective
disturbance[8] was used so that for convection to be
inconsequential

$$a \le \frac{\delta T}{\nabla T_0} \quad \frac{\rho}{\Delta \rho_0} \sqrt{\omega^2 + (\nu/L^2)^2} \quad \sqrt{\omega^2 + (\kappa/L^2)^2} \qquad (3)$$

where a is the allowable acceleration level, δT is the
acceptable temperature change at the solid/liquid interface,
∇T_0 is the temperature gradient into the liquid, ρ is the
density of the liquid, ω is the frequency of acceleration, ν
is the kinematic viscosity, κ is the thermal conductivity and
L is the physical length scale.   In Figure 8 the calculations
show that for an undercooling of 0.3K, a gravity level of
10^{-7} g_0 is already significant if the frequency is 10^{-4} or
lower; higher frequencies at this acceleration level are of

no consequence.   As the undercooling is increased, the acceptable acceleration level increases markedly so that at a supercooling of 2.1K an acceleration level of 5×10^{-2} g_o is acceptable throughout the entire frequency range.

Another complicating factor, in considering the presence or absence of convection in the processing of metals and alloys, is the possibility of Marangoni convection.   If the process involves liquid free surfaces, or an interface between liquids, macroscopic convective flows can result if there is a gradient in surface energy.   Such a gradient can result from a thermal gradient, concentration gradient, etc. Flow occurs if the resulting shear forces exceed the viscous forces.   Marangoni convection is usually masked by buoyancy convection on earth, but it can be a very powerful effect under low gravity conditions.

Sedimentation/Flotation

In the earth processing of immiscible alloys and composites involving solid suspensions in liquids as a beginning state, homogeneity in the final product because of sedimentation or flotation can be a problem.   The separation is due to Stokes flow because of the gravitational body forces.   Under these conditions the terminal velocity of inclusions (liquid or solid) in a liquid matrix is given by

$$v_t \text{ (Stokes)} = \frac{2g\Delta\rho r^2}{3\eta_{matrix}} \quad \frac{\eta_{inclusion} + \eta_{matrix}}{3\eta_{inclusion} + 2\eta_{matrix}} \quad (4)$$

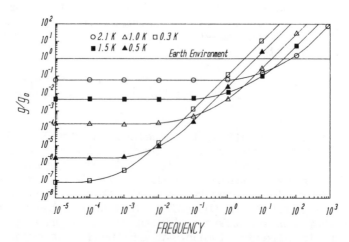

Fig. 8 Critical acceleration at various supercoolings.

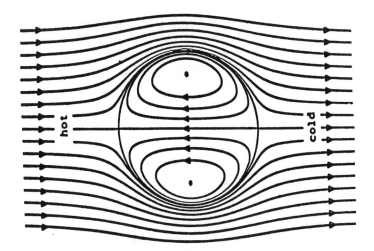

Fig. 9   Marangoni flow due to a thermal gradient.

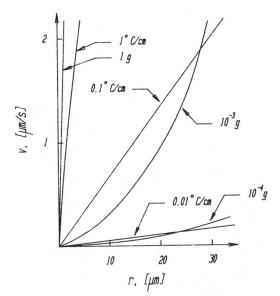

Fig. 10   Droplet migration of Bi in Zn due to Stokes flow and Marangoni flow.

Here $\Delta\rho$ is the difference in density between the inclusion and the matrix, r is the radius of the inclusion and η is the appropriate dynamic viscosity. It is clear that elimination of gravity obviates Stokes flow as a mechanism for separation.

However, as before, there is a complicating factor due to Marangoni flow when processing is done in a low gravity environment. The surface energy driving forces result in flows depicted in Figure 9 and the inclusion drives to the hot end[9]. The terminal velocity of the inclusion is given as

$$v_t \text{ (Marangoni)} = -\frac{2}{3} \frac{r}{2\eta_{matrix} + 3\eta_{inclusion}} \nabla\sigma \quad (5)$$

where $\nabla\sigma$ is the gradient in surface energy. The Marangoni effect can be very powerful and can be seen by the comparison of Bi inclusions in a Zn matrix given in Figure 10[10]. Note that a thermal gradient of 1°C/cm (resulting in a surface energy gradient) is comparable to a body force corresponding to g_o, a gradient of 0.1°C/cm is comparable to 10^{-3} g_o and 0.01°C/cm is comparable to 10^{-4} g_o. The experience in space has in fact shown separation due to the powerful Marangoni effect, but results obtained during free fall in the 105 meter drop tube at the Marshall Space Flight Center are encouraging[11]. Figure 11a shows the separation occurring by arc casting on earth in a Au-65 a/o Rh immiscible alloy and 11b shows the vast improvement in homogeneity in the free fall processed sample.

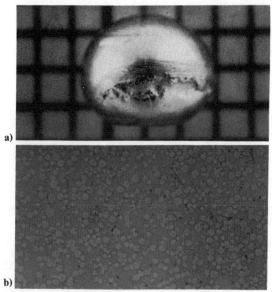

Fig. 11 Comparison of separation in Au-65 a/o Rh: a) complete separation in 4 mm button arc cast at earth gravity; b) dispersion by solidification under free fall (largest dispersoids are about 5 μm).

Containerless Processing

Obviously the same lack of body forces which precludes sedimentation in mixtures also results in the free floating of bodies.   Materials can then be melted and solidified without a container.   This has two advantages.   First, materials that react to any degree at all in the liquid state with their containers can be produced to higher purity levels than is now possible.   This is especially significant in many refractory metals and alloys.   Second, in the absence of a container liquids may freeze at temperatures significantly below their equilibrium freezing temperatures, phenomena referred to as undercooling or supercooling.   When deep undercooling occurs, it results in a bulk rapid solidification process with an accompanying possibility of the formation of metastable phases, refined microstructures, and improved homogeneity.

While the first of the advantages is straightforward and easy to understand, the second is more subtle to those not familiar with physical metallurgy.   However, it is a simple matter to provide the basis for understanding[12].

As is common for an isobaric process, the Gibbs free energy is used as an index for discerning spontaneous, equilibrium and impossible transformations.   The sign of the free energy change will be negative, zero, and positive respectively.   Assuming a spherical solid embryo forming in a liquid, the total free energy change is given by

$$\Delta G_{total} = \frac{4}{3} \, r_3 \pi (G_s - G_l) + 4\pi r_2 \, \sigma_{sl} \qquad (6)$$

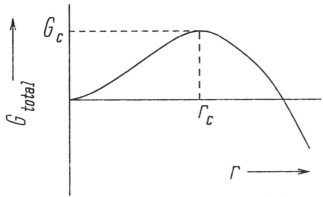

Fig. 12 Total free energy change as a function of embryo size in undercooled condition.

R. J. BAYUZICK

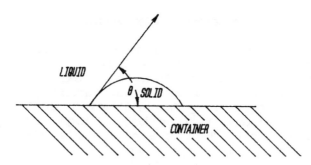

Fig. 13 Nucleation on a container.

where r is the radius of the embryo, G_s and G_1 are the free
energies of the liquid and solid respectively and σ_{sl} is the
surface energy between liquid and solid. The surface energy
term is always a positive contribution to the total while the
volume free energy term will be negative if the temperature
is below the equilibrium freezing point.   Therefore at a
temperature lower than equilibrium the behavior of ΔG_{total}
with respect to the radius has the schematic appearance of
Figure 12.   The maximum value of the total free energy
change, ΔG_c, is given by

$$\Delta G_c = \frac{16 \pi \sigma^3}{(G_1 - G_s)^2} \qquad (7)$$

or in more practical terms in relation to undercooling
approximately by

$$\Delta G_c = \frac{16\pi\sigma^3 \, T_E^2}{(H_1 - H_s)^2 \, (\Delta T)^2} \qquad (8)$$

Here T_E is the equilibrium freezing point, H_s and H_1 are
respectively the enthalpies of the solid and the liquid, and
ΔT is the amount of undercooling.   Nuclei will be of size
greater than r_c since their growth results in a decrease of
free energy and the rate of nucleation then is determined by
how fast a critical embryo goes super critical.   This
nucleation rate has the proportionality

$$I \; \alpha \; \exp(-\Delta G_c /k \, T) \qquad (9)$$

k being Boltzmann's constant. Reflection on equations 8 and
9 would lead to concluding that large degrees of undercooling
would occur.
    The above describes the expectation when there is no
container or other foreign surfaces.   It is called

homogeneous nucleation because the probability of nucleation anywhere within the volume of liquid is the same. If a container is present, the nucleation may be catalyzed at the container walls. The process is now called heterogeneous nucleation because of the existence of preferred nucleation sites. For nucleation at container walls, the total free energy change is

$$\Delta G_{total} = V(G_s - G_l) + A\sigma_{ls} + A'(\sigma_{sc} - \sigma_{lc}) \qquad (10)$$

V is the volume of the embryo adhering to the container wall as depicted in Figure 13, A is the area of the boundary between liquid and solid embryo, A' is the area between container and liquid embryo, σ_{sc} is the solid-container surface energy, and σ_{lc} is the liquid-container surface energy. As can be seen another negative contribution to the overall free energy change can be provided by the third term in equation 10. The critical free energy change for the heterogeneous process relative to that for the homogeneous process can be expressed as

$$\Delta G' = \Delta G_c [(2 + \cos\theta)(1 - \cos\theta)^2]/4 \qquad (11)$$

where θ is the wetting angle as indicated in Figure 13. It is clear that the undercooling is strongly dependent on the nature of the wetting of the container by the solid since equation 9 applies with ΔG_c replaced by $\Delta G'_c$. In the extreme of complete wetting ($\theta = 0°$) $\Delta G'_c$ is zero and no undercooling is expected.

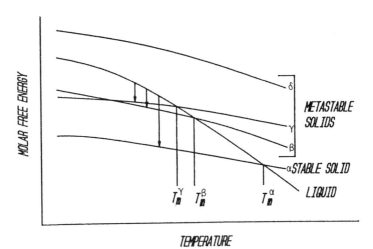

Fig. 14 Free energy as a function of temperature for hypothetical phases in a one-component system.

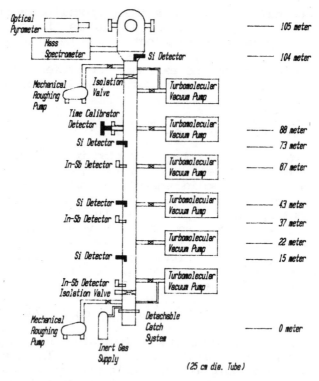

Fig. 15 Marshall Space Flight Center 105 meter drop tube.

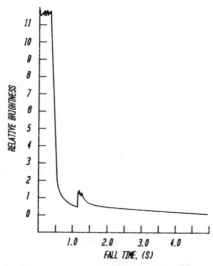

Fig. 16 Brightness curve from an undercooled niobium alloy.

Metastable phases could form by the trade-offs involving bulk free energy and surface free energy. There could be many solid phases stable with respect to the liquid at sufficiently low temperatures. In Figure 14[13], α is the stable phase at all temperatures below its freezing point, T_m. However, ß and γ are also stable with respect to liquid below their respective freezing points although both are metastable with respect to α. On the other hand, according to the previous discussion, ß or γ might be kinetically favored depending on the appropriate surface energies associated with α, ß, and γ.

One simple, efficient, inexpensive method to investigate deep undercooling in bulk metals and alloys is by containerless processing in a long drop tube. Considerable work on both metals and alloys has been done with the Marshall Space Flight Center 105 meter facility represented in Figure 15[14]. The material is melted at the top of the tube and allowed to drop in vacuum or in inert gas. Freezing is determined by monitoring the brightness as a function of free fall time and calculating the temperature corresponding to the time at which the brightness increases due to recalescence; Figure 16 gives a typical brightness curve. Some results on observed undercooling in bulk metals are given in Table 1[15]. Most of the metals listed undercooled approximately 18% of the equilibrium freezing point or greater. The two exceptions, Ru and Ir, appeared to contain impurity particles although the nominal purity given by the producer was 99.8% for Ru and 99.7% for Ir.

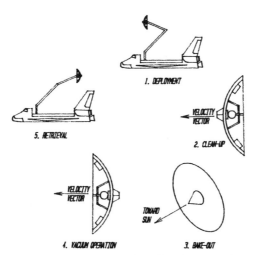

Fig. 17   Operational sequence for the Space Ultra-vacuum Research Facility.

Table 1 Summary of Pure Metals and Undercooling in Drop Tubes

Metal	Source	Nominal Purity (%)	Mass (g)	Drop Diameter (mm)	Average Time to Recalescence (sec)	Undercooling (K)	$\%T_m$ (K)
Ti	Johnson & Matthey	99.98	0.180	1.8	4.5	325	16.7
Pt	Englehard	99.99	0.175	2.6	3.0	378	18.5
Zr	Johnson & Matthey	99.8	0.39	5.0	3.5	355	16.7
	Teledyne Wah Chang Albany	98.0	0.475	5.4	3.7		
			0.80	6.4	4.5		
Rh	Englehard	99.99	0.067	2.3	3.2	457	20.4
Hf	Teledyne Wah Chang Albany	97.0 (2.7% Zr)	0.60	4.6	2.2	438	17.7
			1.10	5.6	3.1		
			1.50	6.2	3.3		
Ru	Englehard	99.8	0.73-0.88	4.8-5.1	1.5-3.8	123-377	5-15
Ir	Englehard	99.7	0.430	3.3	1.4-2.8	238-477	9-18
Nb	Materials Research Corp.	99.99	0.36	4.4	2.1	483	17.6
			0.72	5.6	2.8		
			1.0	6.3	3.2		
			1.5	7.2	3.6		
Mo	Cabot Corp.	99.9	2.3	8.3	4.0		
	GTE Type 390	99.9	0.29	3.8	1.9	590	20.4
			0.575	4.9	2.6		
			0.85	5.6	3.0		
Ta	Materials Research Corp.	99.99	0.400	3.7	1.3	737	22.6
	Cabot Corp.	99.98	1.75	6.1	2.1		
	Unknown	99.6	3.2	7.4	3.9		

Table 2 Expected vacuum levels behind the wake shield

			AT 250 km	
Species	$\bar{v}$ km/sec	n no./cm^3	wake no./cm^2/sec	P equiv Torr
H	4.60	1.2×10^5	4.2×10^7	2.1×10^{-14}
He	2.30	9.7×10^6	1.0×10^4	1.0×10^{-17}
O	1.15	1.4×10^9	3.9×10^{-14}	7.7×10^{-35}*
N$_2$	0.87	4.8×10^8	1.1×10^{-31}	2.8×10^{-52}
O$_2$	0.81	2.5×10^7	2.7×10^{-41}	7.6×10^{-62}
A	0.73	1.5×10^5	3.9×10^{-56}	1.2×10^{-76}
			AT 500 km	
Species	v km/sec	n no./cm^3	wake no./cm^2/sec	P equiv Torr
H	4.60	8.0×10^4	2.7×10^7	1.3×10^{-14}
He	2.30	3.2×10^6	6.3×10^3	6.2×10^{-18}
O	1.15	1.8×10^7	3.8×10^{-15}	1.5×10^{-39}
N$_2$	0.87	2.6×10^5	2.0×10^{-35}	5.3×10^{-56}
O$_2$	0.81	4.6×10^3	2.9×10^{-43}	7.9×10^{-64}
A	0.73	3.4×10^0	1.4×10^{-58}	4.2×10^{-79}

*Equivalent to 1 oxygen atom per cm^2 every 8,000,000 years.
Ref., U.S. Standard Atmosphere 1976.

Vacuum Processing in Space

At the typical low earth orbit, say about 500 km, the pressure is approximately 10^{-7} torr, predominantly composed of atomic oxygen. This could be improved markedly by the construction of a wake shield in the form of a plate or dish. If the velocity of the wake shield is higher, as it will be, than the resident gas molecules, then incredibly low vacuums can be realized in the region behind the wake shield. This is the basis for a concept presently under consideration, the Space Ultra-Vacuum Research Facility (SURF)[16] pictured in Figure 17. Vacuum levels to be expected behind the wake-shield at two different altitudes, 250 km and 500 km, are listed in Table 2. Not only is it possible to attain such levels, but also they can be sustained at high gas loads. Space is an infinite sink for molecules desorbed and evaporated while the convex shape of the wake shield should avoid reflection of molecules emanating from the processing area back in to the processing area. An infinite pumping speed should result.

Low-Gravity Capability

A number of techniques have been conceived for achieving low gravity. In particular, there are drop tubes, drop

Table 3 Capability of low-gravity techniques

Technique	Low Gravity Time	Gravity Level
Drop Tubes	4.5 seconds	up to 10^{-6} g_o
Drop Towers	10 seconds	up to 10^{-6} g_o
Aircraft	30 seconds	10^{-2} g_o
Sounding Rockets	6 minutes	10^{-3} g_o
Space Shuttle	several days	10^{-4} g_o *
Space Station	months	10^{-6} g_o **
Dedicated Free Flyers	months	10^{-6} g_o ***

* Number is "quiet-time" data with frequency range from 8 to 40 hz. G-jitter as high as 10^{-2} g_o with a wide frequency spectrum.[17] Predicted DC level is about 10^{-6} g_o.[18]

** Maximum DC acceleration at the center of mass due to aerodynamic drag. Will vary throughout an orbit. Superimposed on this will be g-jitter to levels of 10^{-2} g_o of higher.[19]

***Similar aerodynamic drag limit to Space Station but depends on altitude. Problem with g-jitter should be lessened.

towers, aircraft, sounding rockets, the space shuttle, the space station and dedicated free flying spacecraft. A well-rounded low gravity program would use them all. The first four in the list can be extremely useful as screening tools as well as definitive tools. Typical low gravity times and levels for each are listed in Table 3.

Drop tubes were described briefly in a previous section. Drop towers are related to drop tubes but differ in that an entire experiment package is dropped in an enclosure of some type. The enclosure must be self-contained with instrumentation, furnaces, atmosphere and/or vacuum control, power, etc. Aircraft provide low gravity by flying Keplerian trajectories as depicted in Figure 18. Sounding rockets, the space shuttle, the space station and dedicated free flyers are, of course, all long duration free fall facilities.

Some Applications of Low Gravity

The best way of illustrating the utility of low gravity is to cite examples of work being pursued. Perhaps one of the largest concentrations of projects on the applications of low gravity to the processing of metals and alloys within a single institution is at the Vanderbilt University Center for the Space Processing of Engineering Materials. The Center is a consortium of three universities and thirteen U.S. companies; they are tabulated in Table 4. The activities of the Center are summarized by the block diagram in Figure 19. Three separate objectives are given. One is eventual

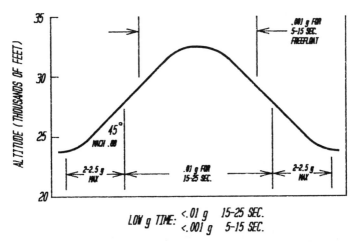

Fig. 18 Flight characteristics of the KC-135.

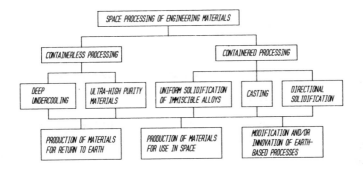

Fig. 19 Activities at the Center for the Space Processing of Engineering Materials.

production of high value materials for return to earth. A second is production of materials for use in space as mankind begins to live and work away from earth. The third is learning about gravity as a processing parameter so that earth based processes may be modified or new ones devised to optimize the properties of existing materials and/or to produce new materials.

A selection of specific projects being carried out in the center, identified by company name, is offered in Table 5. Each of these is a collaborative effort between the company and the university. A given project is dealt with by a team typically composed of an industrial investigator, a faculty investigator, and a graduate student.

Table 4   Composition of the Center for the Space Processing
of Engineering Materials

CENTER COMPOSITION

■ Vanderbilt University, Nashville, Tennessee

   -- University of Alabama, Tuscaloosa, Alabama
   -- University of Florida, Gainesville, Florida
   -- Oak Ridge National Laboratory, Oak Ridge, Tennessee

■ ALCOA, Pittsburgh, Pennsylvania

■ Allied Corporation, Morristown, New Jersey

■ Armco Inc., Middletown, Ohio

■ Boeing Aerospace Co., a division of the Boeing Company,
   Seattle, Washington; Huntsville, Alabama

■ Cabot Corp., Kokomo, Indiana

■ Engelhard Corp., Carteret, New Jersey

■ General Electric Co., Schenectady, New York

■ General Motors Corp., Anderson, Indiana; Warren, Michigan

■ GTE Corp., Towanda, Pennsylvania

■ Lockheed Missiles and Space Co., Sunnyvale, California

■ Special Metals Co., New Hartford, New York

■ Teledyne Brown Engineering, Huntsville, Alabama

■ Teledyne Wah Chang Albany, Albany, Oregon

Summary

        Characteristics of low gravity that are potentially
useful in the processing of materials all stem from one plain
fact. Simply put, this is the absence of body forces. The
accompanying lack of hydrostatic pressure results in the
absence of buoyancy convection due to thermal and chemical
gradients in a melt. Under these conditions mass transport
is controlled by diffusion and/or Marangoni convection.
Similarly, sedimentation and flotation are also eliminated
but once again Marangoni flow can cause the separation of
phases. The degree to which these attributes manifest
themselves clearly depends on the gravity level, which now
becomes a processing parameter to be added to the more usual

Table 5 Examples of research projects being pursued in the
Center for the Space Processing of Engineering Materials

ALCOA - Directional solidification to optimize the properties of
aluminum casting and wrought alloys.

CABOT - Containerless processing to improve the purity and
homogeneity of niobium, niobium-based alloys, and tantalum-
based alloys.

ENGELHARD - Containerless processing of iridium and ruthenium to
improve workability or create workability to provide
materials with excellent resistance to chemical attack and
with good high temperature properties.

GENERAL ELECTRIC - Containerless processing of titanium/rare earth
alloys.

GENERAL MOTORS (DELCO REMY) - Casting and directional
solidification of immiscible alloys systems

GTE (SYLVANIA) - Microgravity processing to produce homogeneous
composites containing fine stable precipitates.

LOCKHEED/TELEDYNE WAH CHANG ALBANY - Development of refractory
alloys for high-performance power systems.

SPECIAL METALS - Directional solidification of oxide strengthened
high temperature alloys to achieve homogeneous distribution
of particles.

---

parameters such as temperature, temperature gradient,
pressure, etc. Applications can be found in many types of
"containered processing." Some of these are the various
forms of casting and directional solidification for the
production of immiscible alloys, in situ composites,
synthetic composites of different types, and other multiphase
alloys.

Then, too, is the attribute of weightlessness which
allows solidification without a container. Deep undercooling
before freezing can then occur, terminating in bulk rapid
solidification. Refined microstructures and extended solid
solutions as well as other metastable forms may result.
Unique alloys for high performance systems become realities.
Another significant application of "containerless processing"
lies in completely bypassing any interaction between a melt
and its container. Thus, ultrahigh purity materials can be
made and very reactive materials can be produced.

Separate and apart from the fluid aspects of space processing is the potential for ultra-high vacuum materials processing in space at high gas loads. The concept of processing behind a wake shield as envisioned by SURF contributes an important added dimension to the use of space for the production of materials. Applications have been envisioned in such processes as the epitaxial growth of thin films (MBE and CBE) and the ultrapurification of metals and alloys.

Experiments involving a variety of low gravity facilities can be done. Research makes use of drop tubes, drop towers, KC-135 and other aircraft flying Keplerian trajectories, sounding rockets and earth orbit facilities such as Spacelab. Each of these offers its own unique capabilities in terms of ease of operation, turn-around time, cost of operation, sample throughput, gravity level, time at low gravity and type of information. Drop tubes and drop towers can provide 10^{-6} g_0 for seconds, aircraft 10^{-2} g_0 for tens of seconds, sounding rockets 10^{-3} g_0 for minutes and orbiters 10^{-6} g_0 for days although considerable g-jitter is to be expected. In any case, a sensible experimental program would consider all of the low gravity facilities available. Each has contributed valuable information in space processing but a much greater knowledge and data base is still necessary.

Judging by the experience in the Center for the Space Processing of Engineering Materials, industrial interest in low gravity research on metals and alloys in the U.S. has a strong thrust in high performance materials. There is a search for vastly improved behavior of materials at high temperatures. For example, much higher ceilings on high temperature strength, creep resistance, oxidation resistance, corrosion resistance, etc., and combinations of these properties at high temperature are being sought. This is certainly of no surprise since such interests are a major drive in the entire engineering materials community.

It should also be noted that there are three objectives at present for the involvement of U.S. industry in space processing. These are production in space of extremely valuable materials for return to earth, production of materials in space for use in space and innovation in earth processing as a direct result of low gravity studies. In the metals and alloys arena there is not at the present time a clear "winner" for the first category. There is presently no market for the second category, but there will be considerable applications in the future as space exploration and habitation become realities. A clear majority of the present interest lies in the third category. The unique attributes of a low gravity environment for materials processing have large potential for understanding many aspects of solidification as influenced by the presence or absence of gravity body forces. That understanding should lead to better control of earth processing with a resultant improvement of the properties of the materials being processed.

References

[1] Raleigh, J. W. S., "On the Capillary Phenomena of Jets," Proc. Roy. Soc., Vol. 29, 1879, p. 71.

[2] Carruthers, J. R., "Studies of Liquid Floating Zones on SL-IV, The Third Skylab Mission," Proceedings of the Third Space Processing Symposium, Skylab Results, NASA TM X-70252, 1974, p. 837.

[3] Reed-Hill, R. E., Physical Metallurgy Principles, 2nd Edition, Van Nostrand Company, New York, 1973, p. 583.

[4] Ibid, p. 588.

[5] Hendrix, J. C., Curreri, P.A., and Stefanescu, D. M., "Directional Solidification of Flake and Spheroidal Graphite Cast Iron in Low and Normal Gravity Environments," AFS Trans., Vol. 99, 1984, p. 435.

[6] Coriell, S. R., Cordes, M. R., Boettinger, W. J., and Sekerka, R. F., "Effect of Gravity on Coupled Convective and Interfacial Instabilities During Directional Solidification," Adv. Space Res., Vol. 1, 1981, p. 5.

[7] Glicksman, M. E., "Isothermal Dendrite Growth," Seminar at Vanderbilt University, March 25, 1987.

[8] Langbein, D., "Allowable G-levels for Microgravity Payloads," ESA Journal, to be published.

[9] Langbein, D., "Theoretische Untersuchungen zur Entwicklung nich mischbarerer Legierungen" Battelle Institute, Frankfurt, Final Report, O1QV-558-AK-SN/A-SLN 7902-9, June 1980.

[10] A. Bergman and H. Fredriksson, "A Study of the Coalescence Process inside the Miscibility Gap in Zn-Bi Alloys," in Materials Processing in the Reduced Gravity Environment of Space, edited by Guy E. Rindone, North Holland, New York, 1982, p. 574.

[11] Andrews, J. B. and Robinson, M. B., "Containerless Low Gravity Processing of Hypermonotectic Gold-Rhodium Alloys," 115th TMS Annual Meeting, New Orleans, LA, March 1986.

[12] Christian, J. W., "The Classical Theory of Nucleation," in The Theory of Transformations in Metals and Alloys, 2nd Edition, Pergamon Press, Oxford and New York, 1975, pp. 418-75.

[13] Baker, J. C. and Cahn, J. W., "Thermodynamics of Solidification," in Solidification, ASM, Metals Park, OH, 1971, p. 27.

160   R. J. BAYUZICK

[14]Bayuzick, R. J., Hofmeister, W. H. and Robinson, M. B., "Review on Drop Towers and Long Drop tubes," Undercooled Alloy Phases, edited by E. W. Collings and C. C. Koch, The Metallurgical Society, Inc., Warrendale, PA, 1987, p. 207.

[15]Hofmeister, W. H., Roberson, M. B. and Bayuzick, R. J., "Undercooling of Bulk High Temperature Metals in the 100 Meter Drop Tube," Proceedings of the Symposium on Materials Processing in the Reduced Gravity Environment of Space, Materials Research Society, Pittsburgh, PA, 1987, p. 149.

[16]McDonnel Douglas Astronautics Co., Space Ultra-High Vacuum Research Facility User Requirements Briefing Document, February 12, 1987.

[17]Chassay, R. P. and Schwaniger, A., "Low-g Measurements by NASA," Proceedings of the Workshop on Measurement and Characterization of the Acceleration Environment on Board the Space Station, Teledyne Brown Engineering, Huntsville, AL, 1986, p. 9-1.

[18]Hamacher, H., Merbold, V. and Jilg, R., "The Microgravity Environment of the D1 Mission," Ibid, p. 8-1.

[19] Teledyne Brown Engineering, Final Report on Low Acceleration Characterization of Space Station Environment, SP85-MSFC-2928 Revision B, Contract No. NAS8-36122 Modification 6, Huntsville, AL, 1985.

Economics and Rationale for Material Processing Using Free-Flying Platforms

Richard Boudreault*

Canadian Astronautics Ltd., Ottawa, Canada

ABSTRACT

Material processing in space (MPS) consists of the production of materials and goods in a microgravity environment. These goods can be used for research and development purposes or for industrial and individual consumption. The interest in MPS has increased steadily during the last 15 years which demonstrates the staying power of this new human endeavour. Furthermore, it is likely that a large portion of the space capable countries will invest into microgravity research in the next 20 years.

This paper reviews the utilization of diverse space platforms for the production of material in a microgravity environment. The platforms reviewed are the shuttle, the space station, and the existing unmanned free-flying platforms.

The economics of each type of platform is reviewed within the constraints of the different MPS needs (for example pharmaceutical and semiconductor production). The minimum breakeven price for products is described in terms of production volume, initial investment, and space platforms. This cost is then compared to the price of different materials.

It is concluded after a comparison of the platforms that the free-flying platform provides the best microgravity environment and is most economically viable.

*Senior Staff Scientist, Space Systems Group.

161

A costing model for the free-flying platform
is then demonstrated. It showed that custom designed
freeflying platforms could be optimized and made available
to experimentors and entrepreneurs.

1. INTRODUCTION

Material Processing in Space (MPS) has been
performed since the early 1970s by Apollo astronauts.
Further experiments were performed during Skylab missions
in 1973 and 1975 and during the Apollo-Soyouz mission
in 1975.

Today MPS is a well organized research and development area supported by companies such as McDonnell Douglas,
3M, General Motors, John Deere, etc. The first commercial
application of an MPS product involved the sale of mono-
disperse latex spheres as calibration standards by the
U.S. National Bureau of Standards. Another commercial
application is the 1.5 kg semiconductor crystal of the
cadmium selenide class produced on Saluut 7 by the U.S.S.R.
Additionally, improvements in ground based processes
such as the John Deere's carbon iron foundry have been
achieved.

The market for MPS products in North America
and the U.S.S.R. is expected to be over $40 billion
by the year 2000 as shown in Figure 1. The breakdown
of MPS products into three categories with their expected
sales growth is illustrated in Figures 2, 3, and 4.
The product categories are listed as pharmaceuticals,
semiconductors, and optical materials. The market potential

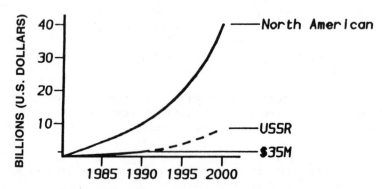

Fig. 1   USSR and North American MPS markets.

of the product lines have been estimated by McDonnell Douglas Aircraft Company (MDAC), the Center for Space Policy (CSP), NASA, and Rockwell.

Canadian Astronautics Limited (CAL) has delivered a study report to the National Research Council of Canada concerning Canadian industrial opportunities in materials processing in space (ref. 3). Three of the opportunities underlined in the study report are being investigated.

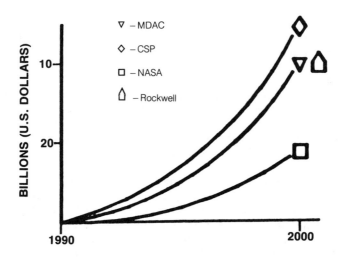

Fig. 2   Pharmaceutical market.

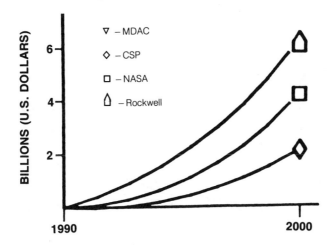

Fig. 3   Semiconductor market.

2. MICROGRAVITY PROPERTIES

Most MPS research and development is focusing on the utilization of the microgravity properties in a space environment as provided by Low Earth Orbits (LEOs). The advantages of material processing in a microgravity environment are described in the ensuing paragraphs.

Gravitational accelerations are greatly reduced in a Low Earth orbit and may vary from 10^{-4} to 10^{-6}g. This reduction in gravitational acceleration can be beneficial for various processes, however, its absence can accentuate the importance of secondary acceleration forces such as crew operations. A review of published data of known secondary forces are listed in Table 1.

A reduction in natural convection (buoyancy driven) in the order of 400% to 500% can be accomplished in a microgravity environment (ref. 1). This property provides great benefits for the production of pharmaceuticals, metals, glass, and crystals.

When producing semiconductor crystals on earth, the container is the limiting factor in the resulting purity of the crystals. This is caused by hydrostatic pressure on the container which results in residual stress, misalignment, and nucleation points within the crystals. The walls of the container also produce radial

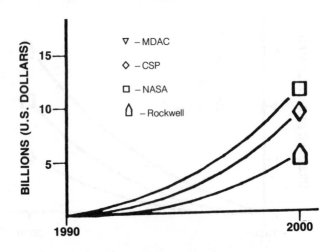

Fig. 4 Optical material market.

Table 1   Secondary acceleration forces important in Microgravity.

Factor	Influence
• Remnant gravitational acceleration	$\leq 5 \times 10\text{-}4$ g
• Coulomb Forces	$\leq 1 \times 10\text{-}5$ g
• Coriolis Forces	$\leq 5 \times 10\text{-}4$ g
• Aerodynamic Forces	$\leq 4 \times 10\text{-}4$ g
• Solar wind forces	$\leq 4 \times 10\text{-}4$ g
• Radiation pressure	$\leq 1 \times 10\text{-}5$ g
• On-orbit maneuvers	$\leq 1 \times 10\text{-}1$ g
• Crew operations	
Crew operation	$\leq 5 \times 10\text{-}3$ g
Breathing	$\leq 1 \times 10\text{-}3$ g
Sneezing	$\leq 2 \times 10\text{-}3$ g
Translating	$\leq 3 \times 10\text{-}3$ g

temperature gradients which cause undesired mixing to occur. Microgravity processing can be performed without a container, thus eliminating the effects mentioned above and producing a higher crystal purity.

In a microgravity environment, a fluid's entropy is greatly reduced and enables perfect spheres to be formed. This property is used to manufacture perfectly round latex spheres for use as calibration standards. The nuclear industry can also benefit from this property since perfectly spherical pellets of hydrogen can be manufactured for use in nuclear fusion.

3. SUPPORT SYSTEMS

Support systems for the microgravity environment include the Space Shuttle (STS), the Space Station (SS), and Long Duration Exposure Facility (LDEF). The operating characteristics for these support systems are listed in Table 2. The Space Shuttle can provide Getaway Special Canisters and Hitchhiker facilities to provide inexpensive support systems for payloads of up to 1,300 kg for approximately 10 days. The MPS breakeven price as a function of initial investment for various transportation methods and production mass is shown in Figure 5. All costs are given in 1984 Canadian dollars.

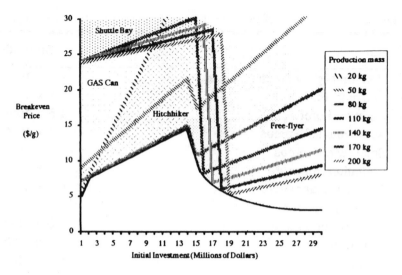

Fig. 5 MPS breakeven price vs. initial investment.

Table 2 Microgravity support systems.

	Payload (kg)	Power (kW)	Altitude (km)	Duration
Space Shuttle	27,272	5-7	276	10 days
Space Station	large	150	497	years
EURECA	1,000	2	500	6 months
Leasecraft	145,000	1-7	460	years
LDEF	5,500	none	270	years

Prior to performing full scale space flights
using one of the above-mentioned support systems, it
is desirable to test critical components and processes
using a drop tower, aircraft flights, or sounding rockets.
The cost of achieving microgravity for manned and unmanned
missions are listed in Figure 6. The cost is shown
as a function of the time spent in microgravity. The
cost of microgravity can also be expressed in dollars
per kilogram-hour as shown in Figure 7. The longer
the production period the smaller the unit cost.

It is found that the cost is lowest for the
free-flying platform. The free-flying platform is also
free of the crew-related secondary acceleration forces

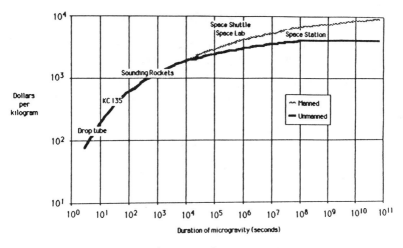

Fig. 6   Cost of microgravity.

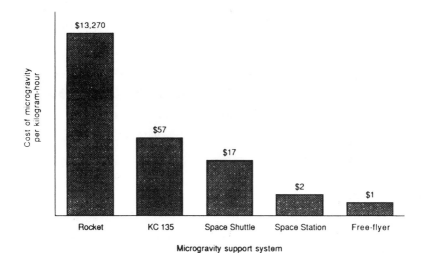

Fig. 7   Cost per kilogram-hour of microgravity support system.

listed in Table 1.   Other advantages of the free-flying
platform are:
-   Large payload capacity
-   Ability to achieve higher orbit
-   Long exposure to the microgravity environment
-   Lower microgravity accelerations
        From the above observations, the free-flying
platform is the obvious selection for a microgravity
support system.

4. MPS PLATFORM COST MODEL

A costing model for a MPS free-flying model is described in the following paragraphs.

The volume, mass, and cost characteristics of small size free-flying platforms for Low Earth Orbits

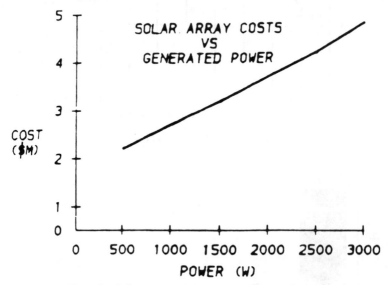

Fig. 8 Solar array cost vs. generated power.

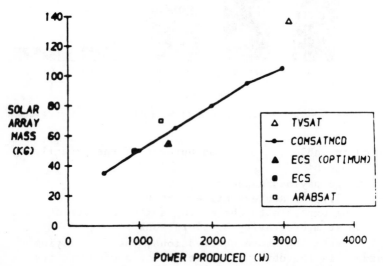

Fig. 9 Solar array mass vs. power produced.

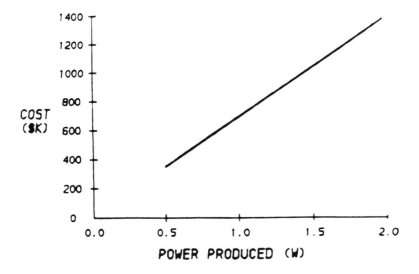

Fig. 10   Battery cost vs. power produced.

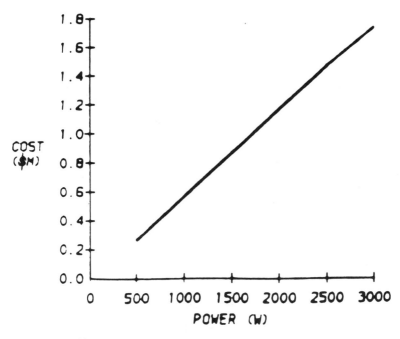

Fig. 11   Power conditioning costs vs. generated power.

(LEO) were modelled. Single and multi-user platforms
with payloads ranging from 500 to 5,000 kg were considered.

The model is an adaptation of the ComSatMod
communications satellite system produced by CAL. The
technology assumed by the model is proven and generally
considered as "off the shelf". The model includes project
management and administration costs as well as the recurring costs for hardware and services such as:

- Power System (generation and distribution)
- Structure and Integration
- Reaction Control
- Telemetry Tracking and Command (TT&C) and
    Attitude Control
- Thermal Control
- Launch and launch services

Subsystems

The subsystems are discussed below and any relevant
assumptions made in the model are described. The costs
are projected in 1984 Canadian dollars.

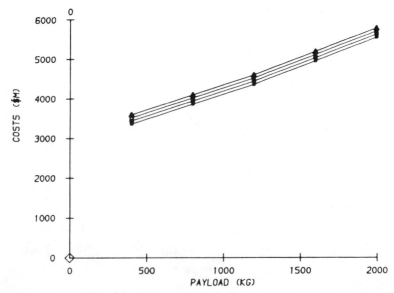

Fig. 12   Structure cost vs. payload mass.

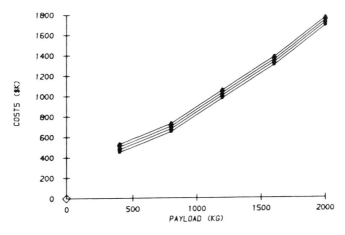

Fig. 13 Mechanical intergration cost vs. payload mass.

Power Subsystems

The power subsystem consists of solar arrays,
batteries, and power conditioning equipment. Power
requirements affect the system cost in two ways. For
higher power requirements the solar arrays and batteries
must be larger and more expensive thus increasing the
system cost. These larger components also have greater
mass which increases the structure, the integration
and particularly the launch costs. Figure 8 shows the
solar array cost as a function of generated power.
Figure 9 shows solar array mass as a function of power
produced.

It was estimated that the platform would spend
45 minutes of each 90 minute orbit in the earth's shadow.
For the duration of the shadow period, power is supplied
by 24 Ni H_2 cells of 1.2 Volts each.

Figure 10 illustrates battery cost as a function
of produced power. Power conditioning costs as a function
of generated power are shown in Figure 11.

Thermal Control

Heat produced by carrier operation will be dissipated by the thermal control subsystem. The overall
carrier efficiency is assumed to be 30% and hence 70%

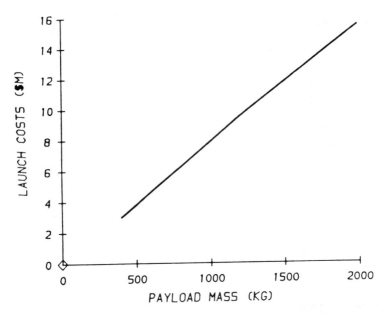

Fig. 14 Launch cost vs. payload mass.

of the power produced must be dissipated. The cost
of this subsystem depends on power requirements.

Attitude Control and Telemetry
Tracking and Command TT&C

        The mass of the attitude control subsystem is
a function of the number of fine pointing mechanisms.
The model assumed that five fine pointing sensors would
be used as listed below:

-   Two startracker sensors
-   One fine sun sensor
-   One mission specified sensor
-   One infrared unit

        It should be noted that no earth sensors would
be used. The cost for this item is fixed. The TT&C
subsystem costs were derived from a Geostationary Earth
Orbit (GEO) satellite which may be considered to be
conservative since a LEO TT&C subsystem requires less
power.

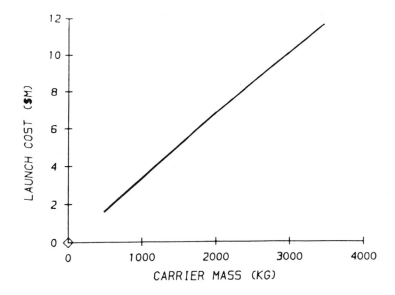

Fig. 15 Launch cost vs. carrier mass.

Reaction Control

The reaction control subsystem (RCS) is propelled with hydrazine and the cost of the propellant is negligible as compared to the cost of the RCS motors, even for a ten year design life. The model used a fixed cost for this item.

Structure and Integration

The structure must provide a platform for mounting all subsystems and the payload. Figure 12 shows structure cost as a function of payload mass. Figure 13 shows mechanical integration costs as a function of payload mass.

Launch and Launch Services

Launch costs were obtained from NASA customer service, Arienespace, and Rockwell International STS User Service Center. These costs are determined by the percentage of either Shuttle bay length (60 feet) or payload mass capacity (65,000 pounds) which is occupied

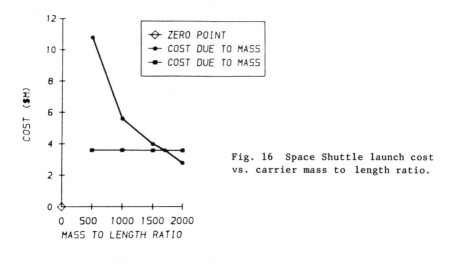

Fig. 16 Space Shuttle launch cost vs. carrier mass to length ratio.

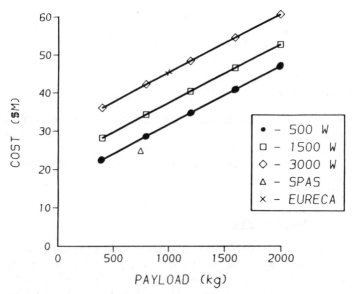

Fig. 17 Total system cost vs. payload mass.

by the carrier. Figure 14 shows launch costs as a function
of total carrier mass. Figure 15 shows launch costs
as a function of carrier payload mass. Clearly, any
effort to reduce the carrier mass can reduce the launch
cost by large amounts. Figure 16 shows the launch costs
as a function of the mass to length ratio of the carrier
and also the optimum (lowest cost) mass to length ratio.

The model determines total system cost as a function of payload mass. This information is shown in Figure 17 for various power requirements. These results were obtained after making the subsystem assumptions described earlier.

The total power/mass cost determined by the model agrees with the actual cost of the Eureca and SPAS platforms. This demonstrates the model's high degree of accuracy in the prediciton of MPS platform costs. The model will enable platform designers to assess the economic impact changes in the configuration of platform subsystems.

5. CONCLUSIONS

The microgravity environment characteristics have been reviewed along with means to manufacture MPS products in space.

It has been estimated that potential markets of up to $40 billion will be available by the year 2000 in the areas of pharmaceuticals, semiconductors, and optical materials. Economic and technical considerations suggest free-flying platforms are the optimum vehicle for space manufacturing. The design of such platforms can be optimized using the ComSatMod model discussed since predictions made using the model were consistent with existing spacecraft.

ACKNOWLEDGEMENTS

I would like to thank my supervisors, Dr. Joseph Gore and Mr. John Graham, and Canadian Astronautics Limited for encouraging and financing this paper. Special thanks should also go to Chris Caterall, Paul Barry, Rick Escher, and Randall Budd whose invaluable help transformed this paper from a draft to its present state.

REFERENCES

[1]Boudreault, R., "Numerical Simulation of Convections in the Microgravity Environment", Proceedings of the 5th European Symposium on Materials Science Under Microgravity; Schloss Elmau, Germany, November 1984.

[2]Boudreault, R., "Microgravity Environment Quality Aboard a Low Earth Orbit Space Station", 36th Congress of the International Astronautical Federation, Stockholm, Sweden, October 1985.

[3]Canadian Astronautics Limited, "Space Station Industry Joint Endeavour Program Microgravity Final Report", National Research Council of Canada, June 1985.

[4]Canadian Astronautics Limited, "Study to Expand on Algorithm and Associated ComSatMod Computer Program", 1980.

[5]Koelle, D.E., "Personal Communications", June 1985.

[6]Napolitano, L.G., "Microgravity Sciences and Applications - A Critical Overview", Proceedings of the Symposium on Industrial Activities in Space, Stress, Italy, May 1984.

[7]Regan, F.J., "Re-entry Vehicle Dynamics", AIAA, 1985.

[8]Vinh, Busemann, Culp, "Hypersonic and Planetary Entry Flight Mechanics", University of Michigan Press, 1980.

[9]Covault, C., "Evolving Government Policy Eases Way for Space Ventures", Commercial Space, Spring 1985, page 16.

[10]ESA BR-16, "Eureca, An Introduction to Europe's Free-Flying Retrievable Carrier", July 1983.

[11]Deskevich, J., "Leasecraft: An Innovative Space Vehicle", IEEE Transactions on Aerospace and Electronics Systems, Vol. AES-20, No. 1, Jan. 1984.

Commercial Prospects for Bioprocessing in Space

D. W. Clifford*

*McDonnell Douglas Astronautics Company,
St. Louis, Missouri*

Abstract

The microgravity environment of space holds interest for both scientific and commercial bioprocessing applications. The physical processes most affected by microgravity are those involving fluid systems where buoyancy driven convection, sedimentation and the resulting turbulence and disruption of quiescent conditions are absent. Since biological systems are inherently aqueous and usually function in a fluid environment, fluid dynamic effects are naturally of interest to scientists interested in understanding operational mechanisms and functions of living cells and their products.

Three areas of biological processing are strong candidates for space study and applications. They are: a) protein crystallization in space for crystallographic study of the three-dimensional structure of selected protein molecules, b) electrophoretic separation of biological materials (cells, proteins, nucleic acids, etc.), and c) tissue culture and the associated study of cell physiology and behavior in microgravity. Several Space Shuttle missions have already been flown where protein crystallization and electrophoroetic separation experiments were conducted. Both the EOS and PCG programs have produced interesting and encouraging results. In the case of cell culturing in space, plans are being developed by NASA to conduct bioreactor experiments in microgravity. The Space Bioreactor will be an automated tissue culture system which will grow living cells in a fluid media, controlling the constituents of the media which affect cell growth. The commercial potential of each of these bioprocessing applications will be discussed

as will the results and status of the respective flight
research programs.

Introduction

One of the exciting new possibilities resulting from
the space program is the opportunity to develop new
materials in an environment free from gravity-induced
buoyancy, convection flows, and sedimentation. The
turbulent mixing which accompanies these terrestrial
phenomena, though hardly discernible at times, has already
been shown to have a profound effect on a wide range of
materials. From work done to date, it is clear that the
microgravity environment opens a new frontier for
materials scientists and engineers by providing new
understanding of the effects of gravity on materials and
how materials properties may beneficially be altered or
processed in the absence of gravity. By eliminating some
of the disruptive influences of gravity, it is anticipated
that materials produced in space may provide exciting new
possibilities in several different technical disciplines.

The microgravity environment is especially exciting to
many investigators working in the biological sciences.
The elimination of gravity-driven convection in fluid
materials can preclude undesirable stirring and mixing
encountered during the growth of crystals, or during the
growth or separation of biological materials. In fact,
the physical processes most affected by microgravity are
those involving fluid systems where gravitational forces
are able to produce uncontrolled movement of some sort
within the fluid. Since biological systems are inherently
aqueous and usually function in a fluid environment, fluid
dynamic effects can be expected to affect in some way the
functioning of biological systems, or at least the
processing of the biological materials. These
environmental differences, and their impact on biological
systems are naturally of concern to scientists intereted
in understanding operational mechanisms and functions of
living cells and commercial organizations and independent
entrepreneurs who forsee the possibility of obtaining
useful materials unobtainable on Earth.

This paper will discuss the commercial prospects for
the processing of biological materials in space,
contrasting biotechnology projects to other space
processing candidate technologies. It will then describe
the most promising candidates for space bioprocessing and
the status of their respective flight research programs.

Space Processing Commercial Prospects

Candidate Materials

Materials which are candidates for space processing are those having high commercial value per unit mass, or those which cannot be obtained on Earth at all. Basic economics limits severely the number of candidates for space processing to those which can justify high production costs, which include an estimated transportation cost of $3,000-10,000 per pound. The only materials likely to justify space production initially are semiconductor materials and pharmaceuticals. For example, a semiconductor detector material, mercury cadmium telluride, has a projected market value of 1 Million Dollars per kilogram. A substance such as interferon can command a market price of $250,000 per gram, or $250 million per kilogram. A Japanese Journal[1] estimates a 'hot therapeutic epidermal cell life extension drug' is valued at 120 million yen per gram. At today's rates, that value is approaching $1 billion per kilogram. It has been estimated that if a product's annual market value is at least 10 to 100 million dollars, and it can be produced from some reasonable starting volume, then money can be saved by processing it in space.

Commercial Projections For Space Bioprocessing

Worldwide, the total annual market for biological products with human health or pharmaceutical applications was estimated by the Center or Space Policy (CSP) to be on the order of $76 billion (U.S.) in 1980[2]. The U.S. and Japan together accounted for about 45 percent of this total. Assuming a modest annual growth rate of eight percent (the average annual rate for 1975-1980 was 15.2 percent), this market would reach over $350 billion by the turn of the century, and the American market alone would then be $92.3 billion.

In the report, issued before the Challenger accident, the Center for Space Policy estimated the total space processing market in the year 2000 at 2.6 to 17.9 billion dollars. Although the time scale can no longer be considered valid, the projection that approximately 75 percent of the total, or $2.0 to $14.9 billion, will eventually be generated by bioprocessing, is still reasonable. By way of comparison, the previously referenced Japanese journal estimates that the projected

Japanese space manufacturing market after the space
station is in routine operation will be well over
200 billion yen (over 1.3 billion dollars U.S.). The
pharmaceutical market in space will amount to about
80 billion yen (500 million dollars U.S.), or about
4 percent of the total Japanese pharmaceutical market.
The four percent number is in agreement with the high end
of the CSP projection of $14.9 billion for space processed
phamaceuticals in a total world market of $350 billion.

Of course, many of the pharmaceuticals included in the
total market aggregate are unsuitable candidates for space
processing. Many are simple and cheap to produce on
Earth. However, the fastest growing categories of
pharmaceuticals are those for cancer therapy, diuretics,
and anti-arthritics; categories for which demand will
increase as the general population ages. Treatments for
diseases such as cancer, requiring even more sophisticated
drugs, are prime candidates for space processing, because
the cost of production increases dramatically as the
complexity of the molecular structures increases.

The CSP space bioprocessing projections mentioned
earlier were based in part on estimates made by the
McDonnell Douglas Astronautics Company (MDAC)
Electrophoresis Operations in Space (EOS) project[3].
Table 1 lists a number of candidate pharmaceutical
products which MDAC feels could be produced in space by
separation and purification of biological materials. MDAC
projected a potential $2 billion per year revenue for a
space processing business which markets four to six of
these high value pharmaceuticals on a world wide basis.

One of the related biotechnologies which is a candidate
for space processing is protein crystal growth. The
commerical potential of protein crystallography in
general, and of protein crystal growth specifically, has
been addressed in a report from a group at the Harvard
Business School that was sponsored by NASA[4]. This
report included a detailed analysis of the market
potential for protein crystal growth in space. The market
survey indicated that U.S. pharmaceutical companies alone
spent $3.5 billion in 1984 on research and development
efforts, amounting to 8.5 percent of their sales of
29.9 billion. R&D expenditures increased 15 percent over
the 1983 level while sales increased only 3.9 percent. In
the field of biotechnology, another 250 firms, from large
chemical components to small start-ups, spent $200 million
in 1984 on R&D. Since protein crystallography is an R&D

Table 1 Typical candidate biologicals for processing in space.

TYPICAL PRODUCTS	MEDICAL USE	ESTIMATED NUMBER OF ANNUAL PATIENTS (USA)
IMMUNOGLOBULINS	EMPHYSEMA	100,000
ANTIHEMOPHILIC FACTORS VIII & IX	HEMOPHILIA	20,000
BETA CELLS	DIABETES	600,000
EPIDERMAL GROWTH FACTORS	BURNS	150,000
ERYTHROPOIETIN	ANEMIA	1,600,000
IMMUNE SERUM	VIRAL INFECTIONS	185,000
INTERFERON	VIRAL INFECTIONS	10,000,000
GRANULOCYTE STIMULATING FACTOR	WOUNDS	2,000,000
LYMPHOCYTES	ANTIBODY PRODUCTION	600,000
PITUITARY CELLS	DWARFISM	850,000
TRANSFER FACTOR	LEPROSY/MULTIPLE SCLEROSIS	550,000
UROKINASE	BLOOD CLOTS	1,000,000

activity, the market for this service will be firms with R&D programs in protein crystallography or rational drug design. Pharmaceutical firms who are involved in protein crystallography indicated to the Harvard group that they would be willing to pay $100,000-200,000 per protein sample that is crystallized in space.

Space Bioprocessing Technologies

Bioprocessing in space was recognized as the first space processing technology to have potential for commercial production. Three areas of biological processing are strong candidates for space study and applications, and each will be discussed in turn. They are: 1) crystallization of biological molecules for structural information, 2) biological separation for purification purposes; and 3) culturing of living cells in space for both scientific and commercial applications.

Macromolecular Crystallography

Crystallography Applications. Crystallography is a
powerful method for determining the three-dimensional
structures of complicated molecules. Crystallographic
studies of biological macromolecules including nucleic
acids and proteins, are currently the only approaches that
reveal the complete three-dimensional arrangements of
atoms in these molecules. Such crystallographic studies
have played key roles in establishing the structural
foundations of biochemistry and molecular biology. They
have also revealed structure and function relationships
that are of great importance in our understanding of how
enzymes, nucleic acids, and other macromolecules operate
in biological systems. More recently, crystallographic
studies of biological macromolecules have become of
considerable practical interest to the pharmaceutical and
chemical industries, as promising tools in drug design,
protein engineering, and other applications to
biotechnology.

Most of the drugs that are now available have been made
by tedious trial-and-error methods. Hundreds or thousands
of compounds are synthesized and tested in an effort to
obtain inhibitors of key proteins within target cells. A
more logical approach, which is the foundation of
"rational drug design," is to obtain the complete
three-dimensional structure of a target protein using the
techniques of crystallography, and to then specifically
design compounds that have the proper shape and
distribution of chemical groups to bind tightly within key
sites of the protein, and thereby block its biological
action.

Unfortunately, each of the major steps that are
involved in determining a protein structure are subject to
a number of experimental difficulties and uncertainties.
Many of the proteins that have been studied during the
past three decades have required many man years of effort
before the complete three-dimensional structure was
known. Fig. 1 illustrates a computer generated graphical
representation of the protein purine nucleoside
phosphorylase, developed at the University of Alabama in
Birmingham after over five years of work. Because of the
large amount of effort involved, there has been limited
interest in using protein crystallography as a tool in
pharmaceutical research until recently.

Several recent advances in the technologies required
for protein crystallographic studies have now made it much

easier to determine the atomic structure of a protein or other macromolecule, once good crystals are available. These advances have generated new interest in protein crystallography among pharmaceutical companies because of the commercial possibilities in rational drug design. However, there are problems that need to be overcome before crystallographic analysis of macromolecules can become a general tool of use to the pharmaecutical, biotechnology, and chemical industries.

Protein Crystal Growth. Nearly any protein crystallographer would immediately point to protein crystal growth as being a major bottleneck in the widespread development of this field. Most macromolecules are extremely difficult to crystallize, and many otherwise exciting and promising projects have terminated at the crystal growth stage. Proteins often yield small microcrystals readily, but it can then take several years of trial-and-error experimentation before these microcrystals can be induced to grow large enough for a complete structural analysis. Many examples can be cited of proteins that have never been obtained as suitable crystals, despite tremendous effort.

Fortunately, there is good evidence that much larger and higher quality crystals can be grown in space, under microgravity conditions. Enhanced crystal growth under microgravity conditions has now been demonstrated for several simple inorganic and organic compounds. The experiments performed in space so far have indicated that

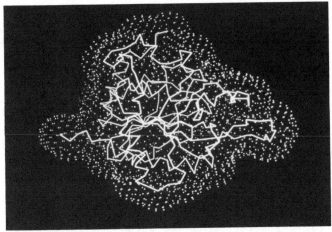

Fig. 1   One unit of the portein purine nucleoside phosphorylase (PNP).

microgravity can improve crystal homogeneity and decrease
the number of defects in crystals[5-11]. Since protein
crystals are extremely fragile, being stabilized by
relatively weak crystalline interactions, one might expect
that protein crystal formation would be especially
affected by fluctuations in the growth environment,
including those caused by sedimentation or convection in
gravitational fields.

Several laboratories around the world are involved in
efforts to investigate gravitational effects on protein
crystal growth. The first reported space experiments are
those of Littke and John[7] which describe the growth of
lysozyme and beta-galactosidase crystals in Spacelab 1.
These preliminary studies indicated that space-grown
protein crystals are considerably larger than crystals of
these proteins obtained under the same experimental
conditions on Earth. A U.S. team headed by Dr. Charles
Bugg of the University of Alabama at Birmingham has
conducted preliminary protein crystal growth experiments
on four shuttle flights (STS-51D in April 1985; STS-51F in
July 1985; STS-61B in November 1985; and STS-61C in
January 1986). All experiments have been performed in the
middeck area of the shuttle, which provides a controlled
environment with pressure and temperature typical of
laboratory conditions on the ground. All of the
preliminary experiments have been performed using small
amounts of space available in the lockers that occupy one
wall of the middeck area. Two of the flights were flown
as part of the MDAC EOS experiment package, which will be
described later in this paper.

Fig. 2 depicts the hand-held apparatus that has been
developed for protein crystal growth by vapor diffusion
techniques. Each experiment is performed within one of
24 closed chambers, each having a volume of approximately
2 cm^3. Clear plastic windows cover the chambers so that
crystal growth can be monitored visually and
photographically. The syringes containing the proteins
and syringe capping mechanisms protrude into opposite
sides of the growth chambers. Prior to activation of the
experiment, the protein sample is held within its syringe,
which is stoppered during launch and landing. The chamber
contains a wicking material that can be saturated with a
equilibration reservoir prior to loading the experiment
aboard the shuttle. The syringes contain screw-operated
pistons. Growth is activated by extruding a droplet of
the protein solution onto the syringe tip where it
equilibrates with the reservoir solution within the closed

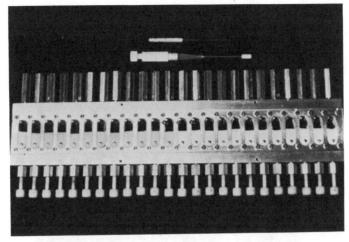

Fig. 2 Handheld vapor diffusion apparatus shown with activation tool and seeding syringe.

chamber. After the crystallization experiment is complete, the protein solution along with any crystals is withdrawn back into the syringe. The crystals produced by this method can then be examined by reextruding the samples from the syringes in the ground based laboratory after landing.

During the most recent shuttle experiments on STS-61C crystals were grown of all proteins that were tested, including hen egg white lysozyme, human serum albumin, human C-reactive protein, bacterial purine nucleoside phosphorylase, canavalin, and concanavalin B. The flight results are discussed in detail in the report by DeLucas, et. al.[12]. That particular shuttle mission was prematurely shortened, and the protein crystal growth experiments were deactivated during the third day of the flight. Although many of the protein solutions had not completely equilibrated during that period of time, relatively large x-ray quality crystals were obtained for all of the proteins except lysozyme. In addition, photographic records of the crystallization solutions in the vapor diffusion apparatus were obtained while in orbit. Fig. 3 shows a photograph of crystal growth in one of the solution droplets during the flight. Fig. 4 shows photographs of some of the space-grown crystals, which were taken within 72 hours of the time that the shuttle landed.

Although quantitative conclusions about protein crystal growth cannot be made at this stage, there are several

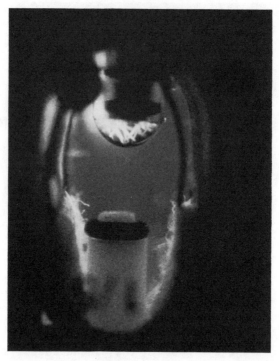

Fig. 3   Crystals of concanavelin protein growth on earth.

interesting qualitative observations that can be made as a
result of these preliminary shuttle experiments. It can
be stated with some certainty that the elimination of
density-driven sedimentation can affect crystal
morphology. The best example of this is canavalin, which
grew crystals in space that were dispersed through the
volume of the droplet (see Fig. 4a). Nearly all the
space-grown crystals of canavalin appear to have formed
from separate nucleation sites, resulting in uniform
morpologies. On the other hand, canavalin crystals grown
by this method on Earth generally form as fused aggregates
at the bottom of the droplets (Fig. 5). In the case of
human C-reactive protein, an entirely new crystal form,
which had not previously been identified in ground-based
crystal growth experiments, was obtained from shuttle
experiments. Crystallization of C-reactive protein has
been studied extensively over the past eight years in
Birmingham and only one crystal form, with space group
different from that of the space crystals grown in
microgravity, has been grown. The new crystal form was
first observed for C-reactive protein from experiments on

(A) CANAVELIN

(B) C - REACTIVE PROTEIN

Fig. 4   Space grown
crystals.

(C) HUMAN SERUM ALBUMIND

(D) CONCANAVELIN B

STS-61B, and quantities of this crystal form were obtained
on STS-61C. The new crystal form diffracts to an
appreciably higher resolution than the original crystal
form.

Despite these encouraging results, it is not yet clear
if the internal order or diffraction resolutions of
space-grown protein crystals are, in general,
significantly different from those of crystals grown on
Earth. It will be necessary to do detailed comparisons
involving large numbers of crystals grown under well
controlled conditions on earth, and in space, before the
potential effects of microgravity on protein crystal
quality can be conclusively evaluated.

Biological Separations

The Electrophoresis Process. The area of space
processing receiving the earliest commercial attention has
been the separation of protein materials by free flow

electrophoresis. Electrophoresis is, in general, the separation of charged particles in an electric field[13-15]. The technique is based on the fact that dissimilar molecules acquire a unique charge/size ratio when exposed to an electrical field in an electrolytic buffer fluid. The technique used routinely to separate samples of biological materials in laboratory research and diagnostic analyses is gel electrophoresis. However, this technique is restricted to sample sizes no larger than a single droplet. Space electrophoresis is of interest because of the possibility of separating very large quantities of biological materials, leading to the commercial production of therapeutic products. For this application, a technique is needed which can handle much larger volumes of material than gel processes can accommodate.

Continuous Flow Electrophoresis. The McDonnell Douglas Astronautics Company (MDAC) began a program in 1977 specifically designed to utilize the microgravity environment of space to enhance electrophoretic separations. This program, called Electrophoresis Operations in Space (EOS), included the development of new continuous flow electrophoresis apparatus, designed specifically for operation in space[3]. The R&D system, known as CFES (Continuous Flow Electrophoresis System), was built to fly in the shuttle middeck and study the effects of weightlessness of the separation process. The

Fig. 5 Protein crystals of canavelin B protein growth on Earth.

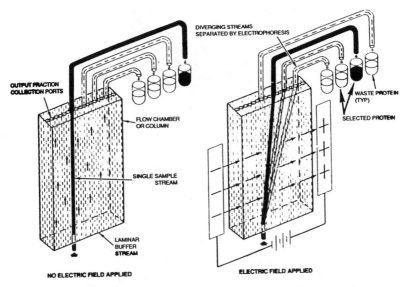

Fig. 6 Seperation of biological materials by continuous flow electrophoresis.

CFES unit employs a long flow channel between two closely spaced parallel plates. A carrier fluid (the buffer) flows continuously through this channel from one end to the other as shown in Fig. 5. The substance to be separated is then injected continuously as a thin stream at one end of the channel. The sample is carried along with the flow and exits at the other end of the channel through one or more of a large number of output (fraction collection) ports. A thin piece of plastic tubing connects each output port to a separate collection chamber.

When an electric field is applied across the chamber, ions begin moving across the electrolytic carrier fluid. The proteins in the sample stream also acquire an electric charge and begin to move laterally across the channel at a velocity which is proportional to their acquired charge. Each protein will have a unique velocity and will therefore separate from the other dissimilar molecules. As a result a number of separate, discrete streams of proteins will be formed, diverging continuously as they proceed along the length of the channel. Each stream then exits the column at a unique set of output ports and is collected in separate collection vials as shown in Fig. 6.

Microgravity Effects. Fig. 7 illustrates the CFES unit installed in the middeck of the shuttle. Figs. 8 and 9

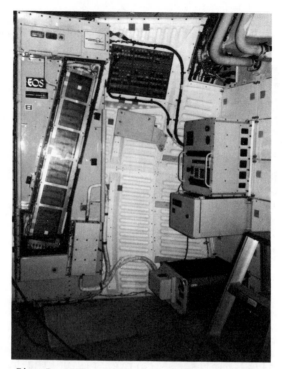

Fig. 7 CFES installed in Shuttle middeck.

illustrate the separation capability and microgravity
performance of the CFES unit. CFES has flown seven times
on the Space Shuttle, beginning with four flights in the
research mode where the unit was operated by NASA mission
specialists or flight personnel. Beginning with
flight STS-41D in August of 1984, the CFES unit flew with
MDAC's payload specialist, Charles Walker, operating the
system. Table 2 summarizes the performance objectives and
results of these seven flights.

   MDAC has also developed a number of these systems for
ground operation, and therefore has had an opportunity to
compare the effects of operation in space directly with
the ground systems. A number of differences has been
observed and attempts have been made to quantify the
magnitude of the gravitational effects, both analytically
and experimentally. It was found that on the ground,
thermal convective currents were generated by electrical
heating in the column. These heated areas become less
dense and thus more bouyant than the surrounding areas,
causing turbulence which upsets the delicate flow

Table 2   EOS results

STS	DATE	OBJECTIVE	RESULTS
4	6/82	≥ 400X THROUGHPUT	463X
6	4/83	≥ 556X THROUGHPUT	718X
		≥ 4X SEPARATION TWO NASA SAMPLES	4.28X
7	6/83	PRODUCT NO. 1 DATA BASE TWO NASA SAMPLES	POSITIVE
8	8/83	CELL SEPARATIONS	POSITIVE
41-D	8/84	PRODUCT NO. 1– MATERIAL FOR CLINICAL TESTING	PRODUCT CONTAMINATED
51-D	4/85	ELIMINATE CONTAMINATION	POSITIVE
61-B	8/85	REPEAT TRY FOR PRODUCT NO. 1 CLINICAL TESTING	POSITIVE

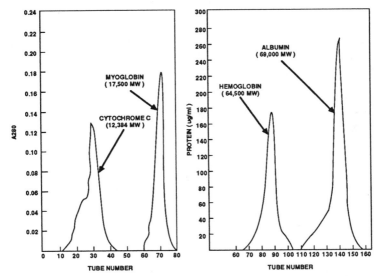

Fig. 8   Continuous flow electrophoresis seperation of proteins with similar molecular weights.

balance. Two steps were necessary to reduce this convective buoyancy on the ground to an acceptable level; first, the column faces had to be placed very close together, and secondly, the electric field across the column had to be reduced. In space, there is no buoyancy factor, so the plates can be separated at least twice as far as they can on the ground. The wider spacing allows the sample stream to be doubled in diameter, giving four times the ground-based sample stream volume.

The related factor which serves to improve the performance of the CFES in microgravity is the ability to utilize higher electrical fields than on the Earth. The higher fields produce greater joule heating because of higher ionic current flow through the resistive buffer. The heated areas again produce turbulent flow due to buoyancy driven convection, limiting the separation capability on the ground to very minimal levels. On board the shuttle, the field strength was increased a factor of four with no disruption of flow conditions. The higher field strengths allow either greater throughput (flow velocity) at the same separation, or greater separation (purity) at the same throughput.

The most significant factor, however, involves the density of the protein sample in comparison with the carrier fluid. Because proteins are so dense, ground operation requires that the sample stream be diluted to a fraction of a percent by volume in order to allow it to be carried smoothly by the buffer flow. In microgravity, the protein sample can be as dense as solubility limits allow and still be carried uniformly by the buffer flow. This factor alone can produce a factor of up to 100 times the throughput of a comparable ground column.

Multiplying these three factors together, it was predicted analytically that a space throughput advantage of 556 times the ground throughput rate should be observed. The improvement actually observed was a factor of 728 over the comparable ground system. This means that the work of one ground system for an entire year could be accomplished on orbit in one afternoon.

Commercial Objectives. The ultimate commercialization goal of the CFES flights has been to develop the technology for a production factory to be flown in the payload bay of the shuttle, and eventually on the space station or on a free flying satellite. The initial product is a hormone to be used in the treatment of patients suffering from anemia. This hormone, erythropoietin, is produced on the ground in genetically engineered cell cultures. A liquid substance is harvested from the supernatant fluid media in which the cells grow.

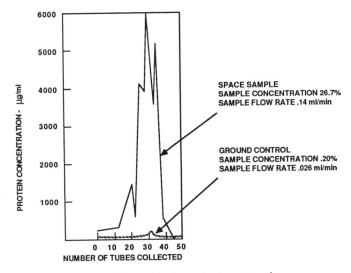

Fig. 9 Space/ground throughput comparison.

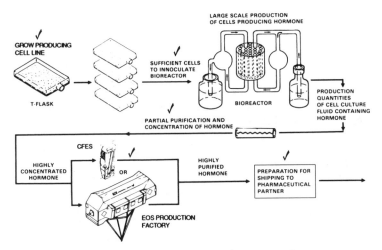

Fig. 10 EOS hormone production scenario.

Generally, the liquid contains a large fraction of
unwanted materials, most of which are proteins and salts
used in the tissue culture process. The function of CFES
is to separate the pure erythropoietin from all the other
impurities still remaining after the best ground
preprocessing is completed. Fig. 10 illustrates the
overall process beginning with the initial cell culture to
the final purification of the product in the CFES unit on
the shuttle.

Dr. Paul Todd of the Bioprocessing and Pharmaceutical Research Center in Philadelphia noted recently that "It is a common misconception that the current ground-based technologies are sufficient for producing and purifying in economically viable quantities, synthetic peptides, monoclonal antibodies, transplantable cells, and products of genetic engineering. These products are normally present in heavily contaminated extracts containing up to 5,000 different unwanted substances. Free-fluid two-phase extraction and free-fluid electrokinetic separation methods have great potential for achieving rapid high purity separations of industrial quantities of modern biologicals, especially those that are difficult to purify by high-volume column or batch chromotography or high performance liquid chromotography. It is somewhat paradoxical that, although analytical electyrophoresis offers the greatest resolution of any separative technique, no form of electrophoresis or isoelectric focusing is currently used for industrial-scale purification of biological compounds[16]."

Cell Culturing in Space

Biological Pharmaceutics. It should be noted that erythropoietin is representative of a new type of pharmaceutical drug which is beginning to make its appearance in the marketplace. These drugs, referred to as "biologicals," are actual biological materials extracted from living systems, or synthesized from biological material taken from living systems. They are not simple chemical formulations, nor even complex formulations developed in a pharmaceutical laboratory. They are generally very complex proteinous substances derived directly from living cells, either in vivo, that is directly from an animal, or in vitro. e.g. in a test tube or laboratory system. The general area of biotechnology has, of course, grown rapidly in recent years with the advent of genetic engineering. The ability to clone genes and modify cellular functions has allowed researchers to enhance the cellular production of natural substances such as hormones or enzymes from cell cultures in the laboratory. The materials so obtained may be produced in such quantity as to be of interest for therapeutic research, or even commercial pharmaceutical purposes.

Cell Culture Technology. The biotechnology explosion has produced a demand for related processing technology and services to recover and purify the target substances. Therefore, there is a close relationship between the separation sciences employed in electrophoresis and the science of cell culturing. Cell culturing, or tissue culturing as it is sometimes called, is the branch of biology concerned with growing and maintaining living cells. Cell biology and biochemistry, on the other hand, deal mainly with the internal structure and functions of a cell.

If the cells are mammalian cells, they must be handled very carefully, and maintained under conditions which simulate the environment in the body as closely as possible. That generally means very closely controlled temperature, a good supply of oxygen, the proper nutrients, and a means to remove waste products produced by the cells. Above all, sterile conditions must be maintained in the system. The cells are extremely susceptible to bacterial or viral infections.

A simple yet efficient system for maintaining the proper growth environment is a suspension system where the cells are freely suspended in a liquid medium, maintained at the proper temperature and chemical composition. Other approaches for large-scale cell culturing include growing the cells on the outer surfaces of hollow fiber membranes with nutrients flowing through the hollow fiber cores, and microcarrier perfusion systems which will be described later.

Microgravity Effects. The question which is of interest to this discussion is whether there might be an advantage in growing cells in the microgravity environment of space. One of the apparent attractions of cell growth in microgravity is the potential for controlling and maintaining a uniform and constant environment around every cell, while eliminating cell damage produced by stirring action. In the Earth's gravity, cells tend to sediment and become concentrated, and aeration with its accompanying foaming is necessary to keep them suspended. The replacement of nutrients and growth factors is difficult to control uniformly throughout the growth volume. In the microgravity environment of space these problems might be reduced significantly. The lack of bubble buoyancy means that oxygenation can be done by better means than sparging air through the culture, a method that often causes foaming and damage to the cells. The overall effect of microgravity might be higher cell viability and less stress on the cells. This in turn may

allow a higher cell concentration, longer survival times,
and consequently larger quantities of cellular products.

Coupled to this attraction is the notion that secreting
cells might behave differently in microgravity, spending
more energy multiplying and secreting and less energy
developing and maintaining viability against gravity.
Finally, if the electrophoresis purification process
proves feasible and advantageous, then an on-board
bioreactor could be used to provide raw material directly
to the separation column, thereby reducing the requirement
for many resupply missions to replenish the
electrophoresis system.

Unfortunately, only a handful of cell biology and cell
culture experiments has been performed in space.
Consequently, there is a paucity of information concerning
the behavior of cells and their biological functions in
space. A knowledge of gravitational effects on cell
biology is essential for efficient development of space
biotechnologies as well as for understanding the origin of
physiological changes in astronauts (e.g. loss of bone
calcium and decrease in the immuno-defense system). The
few space biology experiments with living cells have
generally produced little enlightenment in either area.
The early experiments showed no important differences in
growth rate, cell morphology, karyotype, or cell migration.

However, encouraging new data are now becoming
available from the most recent U.S. and European space
missions and there is optimism that more emphasis will be
placed on cell research in space as a result. For
example, Hymer has shown dramatic reductions in growth
hormone production by rat pituitary cells[17]. In
addition, the Spacelab D-1 mission included a Biorack with
a 1-G centrifuge which allowed controlled experiments on a
number of cell types. The direct comparison of
microgravity and 1-G experiments in a controlled flight
situation eliminated many of the uncertainties of effects
from other launch and flight factors. The results of the
D-1 mission showed significant effects of microgravity on
cell growth, development and differentiation[18]. A
renewed interest in cellular effects of microgravity has
been generated, therefore, which may lead to new insights
into human physiologic responses to prolonged
spaceflight. The results also indicate that space process
studies should be continued until the effects of
microgravity on biological systems are assessed well

enough to determine the potential advantages and
trade-offs for a particular commercial application.

  Space Bioreactors. From the commercial point of view,
the interest in cell growth in space hinges around the
possibility that cellular systems in microgravity will
outperform ground-based systems significantly, either in
quantity or quality of product produced. There is
commercial potential for large quantities of cell products
that can be produced and separated on orbit if the
technical obstacles to space operation can be overcome.
In the U.S., the NASA has been working on the development
of a space bioreactor for some time[19]. A small space
bioreactor is being constructed at NASA Johnson Space
Center for flight experiments aboard the shuttle. The
experimental bioreactor will verify systems operation
under microgravity conditions and measure the efficiencies
of mass transport, gas transfer, oxygen consumption and
control of low shear stress on cells. An early schematic
of the experimental bioreactor concept is shown in
Fig. 11. The flow diagram of the complete process shown

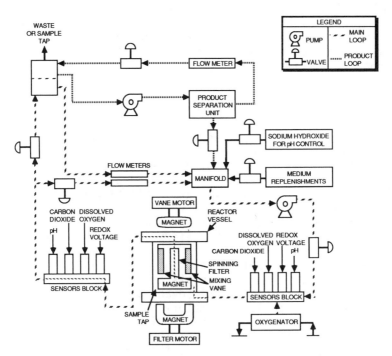

Fig. 11 NASA space bioreactor flow chart (courtesy of
D. Morrison, NASA, JSC).

in Fig. 10, for the combined ground and space operation
for CFES, could all be conducted in space if the
bioreactor were located in space also.

The short term objectives for the bioreactor are to
develop the capability to accomplish bulk cell culture
experiments in a research quality continuous perfusion
bioreactor, designed for operation on the space shuttle.
It is also necessary to gain metabolic, biomechanical, and
physiologic knowledge of bulk cell culture properties
unique to microgravity. NASA would like to demonstrate
the feasibility for pharmaceutical biosynthesis and
product harvesting in microgravity by optimizing and
validating cell culture vessels for microgravity and
enhancing the "state of the art" for earth based
bioreactors.

The longer term objectives of the bioreactor program
are to determine which cell lines and products are most
enhanced by the unique properties of microgravity bulk
cell culture and to enhance product extraction,
concentration, and detection technology. To achieve these
goals, it will be necessary to determine culture
parameters which optimize product biosyntheses. After the
optimal system configuration for space operations is
determined, commercial systems can be scaled up for
economic efficiency.

Conclusion

The prospects for commercial bioprocessing in space can
be characterized as guardedly optimistic. The whole field
of biotechnology is undergoing a period of explosive
growth which can be expected to continue as the general
population ages and as new pharmaceutical products are
developed. Presently, access to space is severely limited
because of the Challenger accident and the measures being
taken by NASA to ensure the safety and reliability of
future missions. However, the prospects for a space
station which will allow a continuous manned presence in
space are good. Those prospects are enhanced by the
promise of new technological advances in bioprocessing,
once the space station laboratory is a reality. More data
are needed before anyone can say with certainty just what
advantages may be realized from the microgravity
environment of space. However, the very fact that the
possibilities are there, ensures that adventurous
researchers will continue to seek the rewards that await

new discoveries, and hopefully new cures for the diseases
that plague mankind.

References

[1]Nihon Keizai Shimbum, 5 December 1984; p. 9.

[2]Commercial Space Industry in the Year 2000: A Market
Forecast, The Center for Space Policy, Cambridge, Massachusetts,
June 1985.

[3]Rose, J. T., "Role of A Space Station In Pharmaceutical
Manufacturing," in Space Station: Policy, Planning and
Utilization, AIAA Aerospace Assessment Series, Vol. 10,
October 1983.

[4]"Commercializing MPS: Protein Crystal Growth. The
Opportunity, A Business Plan, and Recommendations," prepared for
NASA Marshall Space Flight Center by The Harvard Business School;
May 1985.

[5]Rindone, G. E. (editor), Materials Processing in the Reduced
Gravity Environment of Space, Materials Research Society Symposia
Proceedings, Vol. 9, North-Holland, New York (1982).

[6]Wiedemeier, H., "Results of Vapor Crystal Growth Experiment
on STS-7," presented at OSTA-2 Preliminary Science Review,
NASA Headquarters March 1984.

[7]Littke, W. and John, C., "Protein Single Crystal Growth Under
Microgravity," Science, 225, 1984, pp. 203-204.

[8]Littke, W. and John, C., "Protein – Einkristallzuchtung in
Mikrogravitationsfeld," A. Fluqwiss. Weltraumfrosch, 6, 1982,
p. 325.

[9]Wilcox, W.R., "Morphological Stability of a Cube Growing from
Solution Withot Convection," J. Crystal Growth, 38, 1977, p. 73.

[10]Yee, J. F., Lin, M. C., Sarma, K., and Wilcox, W. R., "The
Influence of Gravity on Crystal Defect Formation in Sb-GaSb
Alloys," J. Crystal Growth, 30, 1975, pp. 185-192.

[11]Wilcox, W. R., "Influence of Convection on the Growth of
Crystals from Solution," J. Crystal Growth, 65, 1983, pp. 133-142.

[12]DeLucas, L. J., et. al., "Preliminary Investigations of
Protein Crystal Growth Using The Space Shuttle," J. Crystal
Growth, 76, 1986, pp. 681-693.

[13]Gaal, O., Medgyesi, G. A., Vereczkey, L., Electrophoresis in
the Separation of Biological Macromolecules, John Wiley & Sons,
1980.

[14]Heftmann, Erich, Chromatography, Second Edition, Chapter 10;
Von Nostrand, Reinhold, 1967.

[15]Morrison, D. R., et. al., "Electrophoresis Separation of Kidney and Pituitary Cells on STS-8," Advances in Space Research, 4, 1984, pp. 67-76.

[16]Todd, P., "Space Bioprocessing" in Biotechnology, Vol. 3, September 1985, pp. 786-789

[17]Hymer, W.C., Grindeland, R., Farrington, M., Fast, T., Hayes, C., Motter, K., and Patil, L., "Microgravity associated changes in pituitary growth hormone (GH) cells prepared from rats flown on Spacelab 3: Preliminary results." The Physiologist, 28, 1985a, p. 377.

[18]Naturwissenschalten, 73, A series of reports on preliminary science results obtained on the German Spacelab D-1 Mission, 1986, pp. 404-446.

[19]Morrison, D. R., et. al., Editor, Bioprocessing In Space, NASA TM X-58191, Johnson Space Center, January 1977.

Electronic Materials Processing and the Microgravity Environment

A. F. Witt*

Massachusetts Institute of Technology, Cambridge, Massachusetts

Abstract

The nature and origin of deficiencies in bulk electronic materials for device fabrication are analyzed. It is found that gravity generated perturbations during their formation account largely for the introduction of critical chemical and crystalline defects and, moreover, are responsible for the still existing gap between theory and experiment and thus for excessive reliance on proprietary empiricism in processing technology. Exploration of the potential of reduced gravity environment for electronic materials processing is found to be not only desirable but mandatory.

Introduction

Progress in solid state technology during the past two decades was dominated by innovative device design and engineering. Projected further advances, according to steadily mounting evidence, are contingent on the availability of electronic materials with a degree of crystalline and chemical perfection which exceeds significantly the capabilities of established processing technology.[1] When considering the background and nature of the existing technological deficiencies, the exploration of the potential of reduced gravity environment for electronic materials processing and for related R&D appears not only attractive, but also logical.[2] Manufacturing technology centers undergo global redistribution and trends indicate clearly that developing countries assume at a steadily increasing rate a primary supplier position in the advanced materials sector. While this shift in manufacturing activities is

---

*Professor, Department of Materials Science and Engineering.

proceeding, the related R&D activities are almost exclusively conducted in the traditional centers of high technology. By definition, the developing countries become developed countries and are as such expected to also provide for timely advances of their adopted technologies. The broadbased involvement of developing countries in the exploration and anticipated exploitation of microgravity environment for electronic materials processing must thus not only be considered as desirable, but rather as mandatory.

Status

The fundamental semiconductor property requirements dictated by device technology are matrices with controllable, uniformly distributed charge carriers of maximized mobility and lifetime. This requirement translates into the need for semiconductor single crystals of uniform composition and a maximized degree of crystalline perfection.[3] The complexity and changing nature of property requirements, dictated by device engineering, is reflected in some of the emerging GaAs device technology where a primary property requirement extends to the controlled introduction of uniformly distributed interactive point type defects.[4]

Semiconductors of primary industrial concern, Si and Ge originally, now include in addition GaAs, GaP, InP, CdTe, and related ternary and quarternary compounds. All of these materials in bulk form are used either directly or indirectly (as substrates) for device fabrication.

The fundamental electrical properties are conventionally achieved through the incorporation of appropriate electrically active dopant elements into the single crystal semiconductor matrix while simultaneously keeping any contamination by impurities at an absolute minimum.

In excess of 90% of the semiconductor crystals used for device fabrication are produced by the Czochralski[5] process: A single crystal seed of a particular orientation attached to a pulling rod is contacted with the semiconductor melt which is maintained in an inert atmosphere at a temperature somewhat above its melting point. After thermal equilibration, the seed is pulled (at a rate ranging from 0.1 to about 5"/hr) under simultaneous seed and crucible rotation (Fig. 1). Through appropriate temperature adjustments of the melt, the diameter of the growing crystal

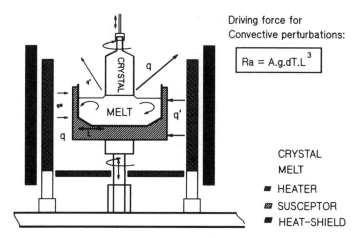

Driving force for
Convective perturbations:

$$Ra = A.g.dT.L^3$$

CRYSTAL
MELT
■ HEATER
▨ SUSCEPTOR
■ HEAT-SHIELD

Fig. 1   Conventional Czochralski configuration used for pulling of semiconductor single crystals from the melt.   Notice lateral, asymmetric heat input (q) and heat loss giving rise to three-dimensional, turbulent convective melt flows which affect adversely the crystal properties.

(ranging in weight from 3 to in excess of 50 kg) can be increased or decreased as desired.   Inherent to this crystal growth configuration is radial heat input to the melt and the withdrawal of the crystal along the rotational axis of the system.   Accordingly, the melt experiences radial and vertical thermal gradients which are destabilizing in the gravitational field on earth.   The thermal field distribution thus gives rise to convective melt flows which affect the growth behavior[6] as well as the incorporation of the dopant elements (Fig. 2).   The driving force for convection in commercial-scale systems is of a magnitude which renders the resulting melt flows turbulent; they are characterized by significant time dependent velocity as well as temperature fluctuations.

These gravity induced melt flows affect the heat and mass transport in the bulk melt, control the nature of all associated boundary layers, and thus by-and-large dictate the crystalline and chemical perfection of the forming crystal.   Effects of concern that can be directly related to convection are summarized in Fig. 3.   It should be noted also that inherent to processing on earth is the need for charge confinement and, related to it, the potential of melt contamination.   Gravity induced melt flows accelerate mass transport and for that reason are also responsible for increased impurity levels in crystals grown from or in containers.

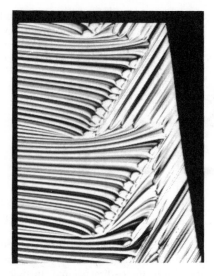

Fig. 2 Interference contrast micrograph (x402) of etched longitudinal section of doped silicon. The distinct intensity patterns reflect charge carrier fluctuations in the matrix caused by non—uniform dopant incorporation due to gravity induced convection.

Containers, moreover, modify the thermal field distribution in growth systems and adversely affect heat transport control, a critical element in efforts to optimize crystal properties for particular device applications.

Semiconductors for advanced device applications, produced by the established growth technology, fail at a steadily increasing rate to meet property requirements and it is acknowledged that property

> Macro—segregation of dopants and impurities
>
> Micro—segregation of dopants and impurities
>
> Deviations from stoichiometry
>
> Formation and distribution of non—equilibrium point—defects
>
> Formation of dislocations and other extended crystal—defects
>
> Constitutional supercooling effects
>
> Growth interface morphology controlled by convective heat—transport
>
> Deterioration of radial vertical thermal gradients in the melt

Fig. 3 Summary of the most conspicuous property deficiencies in melt grown semiconductors which can be attributed to gravity induced convective melt flows, prevailing during their formation.

requirements for projected device technology will likely not be met by presently practiced procedures.

Research aimed at the development of novel approaches to semiconductor growth are severely impeded by the fact that the theoretical framework for solidification and segregation is not quantitatively applicable to crystal growth systems, primarily because of convective interference and the existence of non-quantifiable boundary conditions.

Gravitational effects are clearly recognized as the primary cause of our inability to bridge the existing gap between theory and practice of crystal growth and are therefore the direct cause for our failure to achieve theoretical property limits in materials, the prerequisite for optimization of device performance.[7] They are thus also responsible for industry's heavy reliance on proprietary empiricism in processing technology.

Crystal Growth in a Reduced Gravity Environment

The driving force for convective melt flows in semiconductor growth systems is primarily controlled by the magnitude of the gravitational constant. Accordingly, thermal convection can be expected to be virtually absent in a microgravity environment where the g value is reduced by four to five orders of magnitude. This effect provided the primary motivation for the first semiconductor

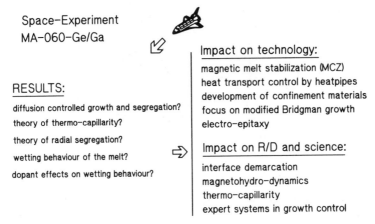

Space-Experiment
MA-060-Ge/Ga

RESULTS:

diffusion controlled growth and segregation?
theory of thermo-capillarity?
theory of radial segregation?
wetting behaviour of the melt?
dopant effects on wetting behaviour?

Impact on technology:

magnetic melt stabilization (MCZ)
heat transport control by heatpipes
development of confinement materials
focus on modified Bridgman growth
electro-epitaxy

Impact on R/D and science:

interface demarcation
magnetohydro-dynamics
thermo-capillarity
expert systems in growth control

Fig. 4 Summary of results obtained from semiconductor growth experiments conducted in space and their impact on technology developments.

growth experiments in space.[8,9] These early space activities, conducted in relatively unsophisticated hardware, provided more questionmarks than answers to existing issues (Fig. 4). Most of all they exposed serious deficiencies in the database of these materials and processes. They also demonstrated that attempts to apply theory to growth experiments are severely limited by the non–quantifiable nature of prevailing boundary conditions.

Shortcomings in the design and execution of the space experiments notwithstanding, their impact on growth technology on earth and their stimulation of growth related R&D is conspicuous. A direct outgrowth of efforts aimed at emulating, to the extent possible, conditions prevailing in reduced gravity environment led to the evolution of magnetic melt stabilization and to the development of magnetic Czochralski growth as practiced in growth technology (Fig. 5). Heat transfer control, virtually neglected prior to these experiments, emerged as a primary issue and resulted in major efforts aimed at the development of high temperature heat pipes, now considered essential for the establishment of quantifiable boundary conditions in crystal growth systems.

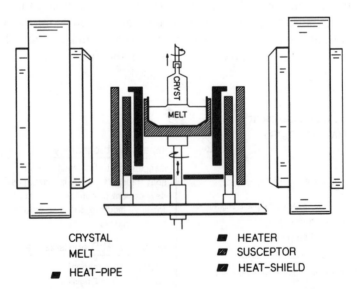

CRYSTAL
MELT
HEAT-PIPE

HEATER
SUSCEPTOR
HEAT-SHIELD

Fig. 5 Modified Czochralski configuration for advanced semiconductor growth. The modifications (magnetic melt stabilization, coaxial heat pipe between heater and crucible) constitute significant advances, which are based on results from experiments conducted in reduced gravity environment.

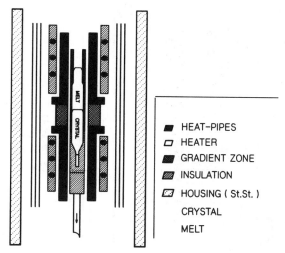

Fig. 6   Modified vertical Bridgman–Stockbarger configuration for growth of semiconductor single crystals. This heat pipe based system provides for unprecidented heat transfer control and permits meaningful mathematical modelling.

Knowledge, reproducibility, stability, and controllability of boundary conditions are realized as the fundamental prerequisites for meaningful mathematical modelling activities of growth processes and ultimately for efforts aimed at model–based growth control schemes.

Outlook

Experiments in reduced gravity environment have focused attention on Bridgman type growth geometries, primarily because of the fundamental incompatiblity of the conventional Czochralski geometry with space environment. These studies stimulated the application of heat pipes for heat transport control (Fig. 6) and resulted in the first applied modelling approach in which the theoretically predicted thermal field distribution was quantitatively confirmed by the experiment.[10,11]   More recently this growth geometry (in gradient freeze configuration) has been shown to permit single crystal growth of InP and GaAs with unprecedented low dislocation density.[12]   Of interest in context is the finding that vertical Bridgman growth, although characterized by stable axial thermal gradients, does exhibit pronounced thermally driven

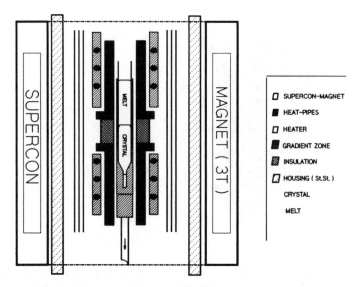

Fig. 7 Vertical magnetic Bridgman configuration for growth of semiconductor crystals with diffusion controlled dopant segregation. (Using a 3 T axial field it has thus been possible to achieve a k_{eff} of 1 for segregation of Ga in Ge.)

convective melt flows because of unavoidable radial temperature gradients. These convective melt flows which, as in Czochralski growth, adversely influence the growth and segregation behavior, remain virtually unaffected by axial or transverse magnetic fields of up to 0.3 T (Fig. 7). Increasing the strength of the applied magnetic field by one order of magnitude, to 3 T, it is most recently found that convective melt flows in doped Germanium are reduced to a point where they no longer interfere with dopant incorporation (Fig. 8); the segregation behavior exhibits characteristics which are very similar to those observed during crystal growth in a reduced gravity environment.

The results of growth experiments with magnetic melt stabilization are of technological interest, primarily because they indicate that convective flows in stabilized melts are laminar and the melt temperature is no longer subject to uncontrollable fluctuations. Such experiments, however, are still subject to other gravity induced side effects, such as melt contamination through interaction with the confinement material and thermal field distortion due to heat conduction along the crucible walls. Finally, it must be realized that melt stabilization through magnetic fields is

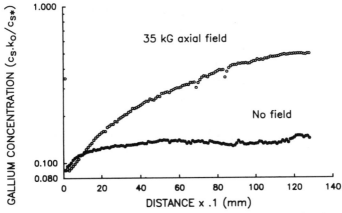

Fig. 8 Macrosegregation behavior (of Ga in Ge) in a heat pipe based vertical Bridgman system with and without magnetic field stabilization. The transient segregation in the presence of the magnetic field indicates the absence of convective interference with dopant accumulation at the growth interface.

only possible in highly conductive melts which severely limits the range of its applicability.

Melt stabilization of growth systems by magnetic fields, in spite of shortcomings, clearly demonstrates elements of the potential of reduced gravity environment for electronic materials processing. The results suggest that growth of crystals (in space) with compositional homogeneity on both a micro- and macro-scale appears achievable. Also achievable appears to be the suppression of constitutional supercooling effects and heat transport control to a point where thermal stresses can be maintained at levels of less than the critical resolved shear stress. The experiments as well as the underlying theory indicate unambiguously that realization of the full potential of space environment is contingent on the achievement of a high degree of heat transfer control, the establishment of quantifiable boundary conditions, and on the availability of a viable theoretical framework for solidification.

Considering the advances of processing technology on earth which resulted from the rather limited number of space experiments, it is apparent that direct science and technology transfer from space experiments is an important element. Equally important for significant improvements of ground based operations is the broadening of the database which is expected to accompany experimentation in space.

The fundamental drawbacks of space experimentation relate to costs, limitations in access to space, and uncertainties in man–experiment interaction. Although costs for experiments in space will always be high, the costs for exploring the potential of reduced gravity environment, an element of concern for developing nations, will become less as the scientific framework for processing activities is complemented by results obtained in space. A further cost reduction can be expected from process automation which will make it possible to reduce man–tended experiments to a minimum and allow experiment monitoring and control from the ground.

Conclusions

The capabilities of established semiconductor growth technology are considered inadequate to meet property requirements for projected advances of device technology. Encountered materials deficiencies, inadequate crystalline and chemical perfection, can primarily be related to gravitational interference with the melt growth process. It has been shown that semiconductors grown in a reduced gravity environment do not exhibit these deficiencies and that results of growth experiments in space can provide substantive input to the advancement of processing technology on earth.

Considering the fact that semiconductor manufacturing technology is at a steadily increasing rate being relocated to developing nations, their participation in the exploration and possible exploitation of the potential of space environment for semiconductor growth appears not only desirable but mandatory.

Acknowledgment

Support by the National Aeronautics and Space Administration (Contract No. NSG–7645) for this work is gratefully acknowledged.

References

[1]Nowogrodski, M., Advanced III–V Semiconductor Materials Technology Assessment, Noyes Publications, Park Ridge, NJ, 1984.

[2]Naumann, R.J., and Herring, H.W., Materials Processing in Space: Early Experiments, NASA–SP–443, Washington, D.C., 1980.

[3]Bylander, E.G., Materials for Semiconductor Functions, Hayden, Inc., New York, 1971.

[4]Ferry, D.K., Gallium Arsenide Technology, Howard W. Sams, Inc., Indianapolis, IN, 1985.

[5]Brice, J.C., Crystal Growth Processes, Wiley, New York, 1986.

[6]Witt, A.F., Lichtensteiger, M., and Gatos, H.C., "Experimental Approach to the Quantitative Determination of Dopant Segregation During Crystal Growth on a Microscale: Ga−Doped Ge", Journal of The Electrochemical Society, Vol. 120, August 1973, pp.1119−1123.

[7]Avduyevsky, V.S., Scientific Foundations of Space Manufacturing, MIR Publishers, Moscow, 1984.

[8]Witt, A.F., Gatos, H.C., Lichtensteiger, M., Lavine, M., and Herman, C.J., "Crystal Growth and Steady State Segregation Under Zero Gravity: InSb", Journal of The Electrochemical Society, Vol. 122, February 1975, pp. 276−283.

[9]Witt, A.F., Gatos, H.C., Lichtensteiger, M., Lavine, M., and Herman, C.J., "Crystal Growth and Segregation Under Zero Gravity: Ge", Journal of The Electrochemical Society, Vol. 125, November 1978, pp. 1832−1840.

[10]Jasinski, T., Rohsenow, W.M., and Witt, A.F., "Heat Transfer Analysis of the Bridgman−Stockbarger Configuration for Crystal Growth. Part I: Analytical treatment of the axial temperature distribution", Journal of Crystal Growth, Vol. 61, March 1983, pp. 339−354.

[11]Jasinski, T., and Witt, A.F., " On Control of the Interface Shape During Growth in Vertical Bridgman Configuration", Journal of Crystal Growth, Vol. 71, March/April 1985, pp. 295−304.

[12]Gault, W.A., Monberg, E.M., and Clemans, J.E., "A Novel Application of the Vertical Gradient Freeze Method to the Growth of High Quality III−V Crystals",, Journal of Crystal Growth, Vol. 74, April 1986, pp. 491−506.

Containerless Science for Materials Processing

T. G. Wang*

*Jet Propulsion Laboratory, California Institute of Technology
Pasadena, California*

Abstract

One of the unique features of space which cannot be reproduced on earth is long-duration zero gravity. Experiments that take the greatest advantage that space can offer are containerless processing experiments. This is precisely the reason why NASA's Microgravity and Science Applications Program has a long-term interest in the containerless processing of materials, i.e., the ability to melt, solidify, or otherwise process a specimen without physical contact with walls or other holding devices. This technique eliminates container-induced contamination and heterogeneous nucleation as well as restricts the specimen deformation either self-induced from hydrostatic pressure or due to physical contact with the container. However, when conducting experiments on space flights, we cannot escape from residual gravitational effects. Therefore, many of the materials research experiments to be conducted in space require the manipulation and control of weightless molten materials in a noncontaminating environment. In these experiments, the melt is positioned and formed within a container without physically contacting the container's wall.

Containerless Science Experiments

Materials processing in space[8] allows the study of earth-based processes with one parameter, the gravity forces, eliminated. This type of work could lead to improvements of the future processing on Earth. A general area in this category is the increase of basic scientific understanding and its ultimate application to general

---

*Program Manager, Microgravity Science and Applications.

processing. The last category is, of course, the new material processes never before tried on earth or in space.

(1) Drop Dynamics - The dynamics of free liquid drops has long been a subject of interest, both for the sake of basic scientific understanding and for various applications in meteorology, the chemical industry and material science.[6] The problem of liquid drops oscillating and rotating in a gravitational field is complicated by the fact that, for most droplets of practical size, surface tension forces and gravity are two competing factors influencing the dynamics of such behavior. This is the limitation inherent to all laboratory work. In immiscible liquid systems, the effects of gravity can be made negligible, although other difficulties arise owing to the mass loading and boundary-layer dissipation from the outer host liquid. The containerless study of dynamics of drops will be the first rigorous test of the classical theory.

(2) Nucleation Studies - Nucleation studies of the condensed phases are notoriously difficult to carry out in the laboratory, and their results are rarely definitive because complete control over the experimental environment is ever elusive. Containerless processing conditions ease some of these experimental constraints by eliminating the more obvious sources of heterogeneous nucleation by removing all physical contact with the foreign matter of containers. Microgravity also alleviates the adverse effects caused by large levitating forces found under 1g by reducing the need for high intensity levitation sources. In addition, experimentation in space will extend the accessible range to higher temperatures (by removing the need for an external supporting liquid), thus making it possible to study a larger number of high melting materials. The highest density materials available will be accessible since gravitational effects will be reduced by a factor of 10^{-3} to 10^{-4}. According to the current nucleation theory, the liquid-solid interfacial energy may be estimated from known homogeneous nucleation temperatures.[7] More reliable experimental data for this latter parameter will doubtless result in more exact interfacial energy values entering in the analysis of many aspects of the growth process of the solid phase, thus impacting the details of materials processing.

(3) Metastable Solid Phases - Metastable solid phases have been of practical use in metallurgy for a long time. Some steels, for example, are strengthened by the incorporation of martensites, a metastable phase derived from austenite. The obtention of these metastable products has

generally required only moderate quenching rates ($\sim 10^3$ °C/s). Many more metastable solid alloys have been discovered through the technique of ultra fast splat quenching (cooling rates between 10^6 °C/s and 10^9 °C/s). It is also possible, however, to arrive at nonequilibrium solid structures by solidifying highly undercooled melts. It is, therefore, quite likely that the combination of the capability for undercooling melts together with the availability of moderately fast quenching rates through containerless manipulation will yield new metastable solid phases not obtainable on earth.

(4) <u>Glass Research</u> - The traditional method of glass formation consists of reacting and fining mixtures of crystalline materials at high temperatures, and then undercooling the molten material to the glassy state. A variety of difficulties have been encountered in the attempt to prepare certain nonstandard glass compositions by such traditional techniques.

A variety of fluoride compounds form glasses in binary and ternary mixtures. These glasses are based on ZrF_4, HfF_4, PbF_2, and BeF_2, with additions of alkaline earth, rare earth, aluminum and transition metal fluorides.

Several glasses in the ZrF_4, HfF_4, and PbF_2 systems have good optical transmission in the middle infrared region (5-8 µm) where silicate glasses absorb, and are being intensively investigated for optical applications. (Glasses based on BeF_2 are of particular interest for laser optics because of their low refractive index and optical losses.) Some glass compositions are difficult to form because of nucleation at container walls. It is possible that these glasses could be formed only in a containerless environment.

(5) <u>Biological Materials</u> - One of the more ingenious approaches to overcome the organ transplant immune rejection problem is the technology so called "cell microencapsulation." Cell microencapsulation is to immuno-isolate living cells inside biocompatible, semipermeable, spherical shells. This allows the nutrients and hormones to pass through, but not antibodies and lymphocytes. The microgravity environment would offer the opportunity to study the behavior of the fluid motion and progress of the reaction and centering mechanism influenced by gravity.

(6) <u>Protein Crystal Growth</u> - X-ray crystallography is an important tool to understand the molecular structure and the functions of complex proteins. This understanding is essential for the bioengineering industries to design and manufacture drugs that are much more effective and with less harmful side effects. Containerless, together

with the buoyancy-free environment of space, might provide a unique opportunity to grow small quantities of large-sized protein crystals for study.

Containerless Technology

In order to perform these experiments mentioned above and others, NASA has been developing a set of container-less processing technologies for materials research in space - namely: Acoustics, Electrostatic, and Electromagnetic.

(1) Acoustics - The acoustic radiation force on a sphere in a standing wave field was first calculated by L. V. King[3] in 1934. King approached the problem by solving the linear wave equation with scattering corrections, and found F to be

$$F = \frac{5\pi}{6} \frac{p_1^2}{\rho c^2} ka^3 \sin 2kx, \qquad (1)$$

where p_1 is the pressure amplitude of the fundamental frequency, k is the wave number, a is the sphere radius, x is its center position, ρ is the gas density, and c is the speed of sound. Placed in an acoustic resonance chamber of length L, and driven by a standing acoustic wave of length $\lambda = 2L$, the sphere will experience a maximum acoustic restoring force at $x = L/4$, $3L/4$, and will experience no force at the center.

Fig. 1   A laboratory tri-axial acoustic levitation chamber.

An acoustic containerless system utilizing the acoustic radiation force has been developed. The laboratory model of the rectangular resonant chamber is shown in fig. 1 and consists of a $4\frac{1}{2}$ inch x $4\frac{1}{2}$ inch x 5 inch plexiglass rectangular box and three commercially made speaker drivers. These drivers are mounted at the center of the three orthogonal sides of the box. In order to maximize the efficiency of the system, three appropriately dimensioned aluminum spacers were inserted. The resonant chamber and the speaker driver units were acoustically coupled through the spacers which are located at the center of the three orthogonal walls. When the chamber is driven at one of its resonant modes by acoustic compression drivers, the ambient pressure is minimum at the pressure nodes of the wave and is maximum at the antinodes. Consequently, there is a tendency for liquids and particles introduced into such enclosures to be driven toward the nodes, where the materials collect and remain until excitation ceases.

The chamber has been designed so that two sides have the same dimension. When the acoustic drivers for these directions are driven 90° out of phase, a maximum torque is applied on the sample. This torque is in the (x,y)-plane and is proportional to the sine of the phase difference.[1] This torque is a result of a viscous effect rather than of the Bernoulli effects as proposed by Lord Rayleigh in 1882. Consider a thin disk with a radius (r_o) very small compared to wavelength and its axis in the z-axis. When the disk is subjected to the influence of two orthogonal acoustic waves with 1/4-wavelength phase lag, the disk is surrounded by a gas in which the particles move circularly. Thus the disk will experience a torque due to viscosity. The torque T can be computed

$$T=\left(\frac{r_o k}{2}\right)^2 l_\eta A \frac{P_x P_y}{2\rho c^2} \sin \omega t_o \left[1 - \frac{\left(r_o k\right)^2}{2} + ... \right], \qquad (2)$$

where l_η is the viscous length defined as $(2\nu/\omega)^{\frac{1}{2}}$, ν is the kinematic viscosity, A is the total surface of the disk, P_x and P_y are pressure amplitudes of the orthogonal waves, and ωt_o is the phase angle between the orthogonal waves.

To generate oscillations of a liquid drop, it is necessary to excite the drop surface at its normal-mode oscillation frequencies. For a liquid sphere of surface tension σ, density ρ, and radius a, the normal-mode angular frequencies are given by

$$\omega_n^2 = n(n-1)(n+2)\frac{\sigma}{\rho a^3}.$$ (3)

The lowest order nonzero frequency corresponds to $n=2$ ($n=0$ corresponds to radial modes which cannot exist in an incompressible fluid, and $n=1$ describes translational motion of the whole drop). The surface deformation for the n-th mode is described by

$$r_n = a + a_n P_n(\cos\theta)$$ (4)

in the axisymmetric case. P_n is the Lengendre polynomial of order n. The modulation of the acoustic force is obtained electronically through a balanced modulator. In the simplest case, a sinusoidal signal (frequency f_c) is multiplied by a second sine wave of much lower frequency (f_m). The voltage across the transducer terminals is then

$$V_T = V_c \sin(2\pi f_c t) \cos(2\pi f_m t).$$ (5)

The acoustic pressure is proportional to this voltage[4] for linear operation of the transducer. The radiation pressure force is proportional to the time average of the acoustic pressure squared, and can be described as

$$P_r \sim <P^2_{acoustic}> \sim \cos^2(2\pi f_m t).$$ (6)

In the case of a standing wave, the periodically oscillating component drives the drop into the oblate shape, while the interfacial tension provides the restoring action.

(2) <u>Electrostatic</u> - In electrostatic containerless[2,5] processing, the material is electrically charged and the force is produced by electric fields between the electrodes. Since the static electric field cannot provide a stable position for a charged object (Earnshaw's theorem), the positioning had to be accomplished by a feedback system which constantly monitors the position of the object.

The force F on a charged sphere can be expressed in terms of force F_0 on an uncharged sample

$$F \simeq F_0\left(1 - K\frac{Q_s}{Q_e}\right)$$ (7)

where K is a constant, Q_s is the sample charge, and Q_e is the electrode charge. As predicted, opposite-sign charges increase the attractive force, and same-sign charges will decrease the force and eventually overcome the attractive force due to polarization, resulting in repulsion.

The object within the levitation chamber is monitored by a charge-coupled device (CCD) camera (or two cameras for the three dimensional control) with ~120 Hz frame rate. From this, the position and the velocity of the object are calculated by a minicomputer. For the present system, the computer-generated signal which goes into the high voltage units is composed of (1) the position error signal, (2) the velocity damping signal, and (3) the integral signal which is used during the launching process.

There are basically three different types of electrostatic levitators: the dish levitator, the ring levitator, and the tetrahedral levitator. The first two levitators have axially symmetric shapes, and have drawbacks of only single axis (symmetry axis) damping. The tetrahedral levitator having closer resemblance to a sphere is not only capable for three-dimensional positioning and damping, it is particularly suitable to the reduced gravity environment. A stably levitated glass shell of ~1.5 cm in diameter and ~0.1 gm in weight in the ground-based laboratory has been demonstrated, as shown in Fig. 2.

Performance tests of these levitators in extreme physical conditions, such as in high temperature or high vacuum environments, are underway.

Fig. 2   The tetrahedral electrostatic levitation system.

Fig. 3   Electromagnetic containerless
processing system.

(3) Electromagnetic - The electromagnetic containerless processing technology originally developed at
Westinghouse,[9,10] is to melt and resolidify electrical conducting materials in a contactless manner. In this
approach the levitation force is provided by the eddy
current in an electrical conductor, induced by an
alternating electromagnetic field. In the case where the
skin depth, δ, is much smaller than sample dimension, a,

$$\delta = (2/\omega\sigma\mu_o)^{\frac{1}{2}}$$ (8)

where ω is the frequency, σ is the electrical conductance,
and μ_o is the permeability of free space, the eddy current
at the surface can completely shield the interior from
magnetic flux. Under that condition, the electrical
magnetic levitation force on a spherical sample can be
written as

$$\overline{F} = K\mu_o N^2 I^2$$ (9)

where K is geometric constance, N is number of turns in a
coil, and I is the rms of the current. To demonstrate the
capability of this technique in a zero-gravity environment, General Electric Space Science Laboratory has flown
an electromagnetic containerless processing system (See
Fig. 3) on a Space Processing Applications Rocket to levi-
tate and melt a beryllium sample containing beryllia.
They found that the low-gravity grown sample had a more
uniform dispersion of BeO particles.

Conclusion

Materials processing in space has received a great deal of academic and congressional support for its scientific content and for its thrust on international competition. Containerless science and processing is leading its charge. Currently, NASA is planning a new 2000°C Modular Containerless Processing Facility for Spacelab and Space Station. The work on high temperature and high field containerless systems reported in this paper has established a solid foundation for this new initiative.

Acknowledgments

The author is grateful to Dr. W. K. Rhim and Dr. G. Wouch for providing the photos of their experimental hardware (Figures 2 and 3 respectively). The research described in this paper was carried out by the Jet Propulsion Laboratory, California Institute of Technology, under contract with the National Aeronautics and Space Administration. It was funded through NASA's Materials Processing in Space Program Office.

References

[1]Busse, F. H. and Wang, T. G., Journal of the Acoustical Society of America, Vol. 69, 1981, p. 1634.

[2]Clancy, P. and Lierke, E. G., "Electrostatic and Acoustic Instrumentation for Material Science Processing in Space," IAF Vol. 79, 1979, F-62.

[3]King, L. V., Proceedings of the Royal Society of London, Series A: Vol. 147, 1934, p. 212.

[4]Leung, E., Jacobi, N., and Wang, T. G., Journal of the Acoustical Society of America, Vol. 70, 1981, p. 1762.

[5]Rhim, W. K., Collender, M., Hyson, M. T., Simms, W. T., and Elleman, D. D., "Development of an Electrostatic Positioner for Space Materials Processing," Review of Scientific Instruments, Vol. 2, 1985, p. 56.

[6]Trinh, E. and Wang, T. G., Journal of Fluid Mechanics, Vol. 122, 1982, p. 315.

[7]Wang, T. G., "Applications of Acoustics in Space," Nouve Tendenze dell'Acoustica Fisica, "Frontiers in Physical Acoustics," 1986.

[8]Wang, T. G., Saffren, M. M., and Elleman, D. D., "Acoustic Chamber for Space Processing," AIAA paper No. 74-155, 1974.

[9]Wouch, G., Frost, R. T., Pinto, N. P., Keith, G. H., and Lord, A. E., Jr., Nature (London), Vol. 235, 1978, p. 274.

[10]Wouch, G., and Lord, A. E., Jr., "Eddy Current: Levitation, Metal Detectors, and Induction Heating, "American Journal of Physics, Vol. 46, (5), 1978.

Kinetic of the Soret Effect and Its Measurement Under Microgravity Conditions

S. R. Van Vaerenbergh* and J. C. Legros†
Free University of Brussels, Brussels, Belgium

Abstract

The Soret coefficient will be measured on twenty different systems under microgravity conditions during the automatic platform EURECA 1 mission.
A description of the hardware is given. The kinetic of the Soret separation is described, showing that the variation of the concentration is much faster near the solid boundaries than in the bulk of the liquid phase.
The Soret separation influence on the hydrodynamic stability of liquid system is analysed.

Nomenclature

D	= Isothermal diffusion coefficient
D'	= Thermodiffusion coefficient
D'/D	= Soret coefficent
N_1	= Mass fraction of the denser component
ΔT	= Temperature difference (>0 if a layer is heated from below)
ν	= Kinematic viscosity
$\varkappa$	= Thermal diffusivity
Ra	= Rayleigh number
R_{th}	= Thermodiffusion Rayleigh number
$Pr = \nu/\varkappa$	= Prandt 1 number
$Sc = \nu/D$	= Schmidt number
k^2	$= (k_x^2 + k_y^2)^{1/2}$

---

*Scientist, Chef de Travaux, Service de Chimie-Physique.
†Fellow Engineer, Service de Chimie-Physique.

a $= A(z) \exp (ik_x + ik_y) \exp (-\sigma t)$ general
           expression for a perturbation

s' $= (D'/D) |\Delta T| (1-N_1^{in})$ Soret number

Introduction

The hydrodynamic stability of the rest state of a horizontal layer of liquid submitted to a vertical temperature
gradient is controlled by the dynamic of the fluctuations.
The motions appear for a finite value of the thermal constraint, when the balance between fluctuations and their
dissipations by the viscosity cannot be maintained. This
critical value of the temperature difference is given under
dimensionless form by the critical Rayleigh number, its numerical value is dependent only on the boundary conditions.

The concentration gradients and their fluctuations must
be taken into account when analysing the stability of non-
isothermal binary liquid system. It is now established
that when the contribution of the concentration gradients
to the density distribution is opposite to the contribution
arising from the thermal expansion, a number of surprising
things can happen .

In such a system the hydrodynamic stability of the rest
state is not guaranteed by the existence of a total density
distribution decreasing upwards. The difference of magnitude between the heat diffusion and the mass diffusion of
the components has to be considered in the stability analysis.

A borderline case to this phenomena is the so-called salt
fountain. The mass diffusion is artificially reduced to
zero by the use of an impervious, long, narrow tube of good
heat conducting materials and which is inserted in a layer
of salty water. This layer is warm, salty and less dense
near the surface than near the bottom where it is cold and
fresh. Hydrostatically this layer is thus stable. If water
is pumped upward in the pipe, the temperature distribution
of water in the tube is nearly the same as in the layer due
to heat conduction through the wall of the tube, while it
remains fresh and thus lighter than its surrounding. A flow
of liquid is standing and will continue as long as the salt
gradient is existing and supply evaporation increases the
salt concentration in upper warm layer.

Not only the concentration gradient could be initially
established, maintained by evaporation, established by the
rejection of a component near the solidification front of a

mixture, it can also be induced by the thermal migration of
components due to the Soret effect. In this paper we shall
emphasize the study of the establishment of the concentration gradient induced by a heat flux and particularly
we shall look at the behaviour of the system near its solid
boundaries just after the establishment of the temperature
difference.

We shall recall briefly in the following paragraphs the
linear stability analysis results, showing the existence of
oscillatory behaviours when the system is heated from below
with the Soret effect inducing downwards migration of the
denser component. This time dependent regime has been experimentally confirmed. Furthermore, in a layer heated from
the top with upwards migration of the denser component,con-
vection can appear even if the total density gradient remains directed downwards.
This behaviour is due to the fact that a concentration fluctuation has a timelife a hundred times larger than a temperature fluctuation. In usual binary mixtures, Prandtl number
($Pr = \nu/ \varkappa$) is around 10 and Schmidt number ($Sc = \nu/ D$) is
around 1000.

Keeping in mind that the convection during materials
processings (e.g. crystal growths) is deeply related to the
mixing of the different components and thus to the concentration distribution, we are studying the time dependent
convection which could happen at low Rayleigh number under
normal gravity conditions. In the case of directional
solidification, this could also contribute to back melting
at the level of the solidification interface and to bands
of non-uniform solute concentration.

The determination of the Soret coefficient in one step
in the study of the influence of the Soret effect on the
hydrodynamic of a system. No reliable values of these coefficients exist because it has to be measured in a
non-isothermal fluid system and because of the long relaxation time of the phenomenon, any residual convection
deeply influence the obtained results. We have proposed to
ESA to measure this coefficient in twenty different systems
during the first mission of the automatic platform EURECA.
The description of this experiment is given herebelow.

Kinetic of the Soret effect

Molecular fluxes appear in mixtures submitted to a temperature gradient. If the temperature gradient is maintained externally at the boundaries of the system, a steady
state can be reached where the thermal diffusion flux is

balanced by the diffusion. A chemical separation is thus induced, which is called Soret separation.

In a binary mixture at mechanical equilibrium and not too far from the thermodynamical equilibrium, the mass barycentric fluxes $\rho \, \bar{J_i}$ of component i are related to the existing gradients by linear relations [1-2]. For thermodiffusion, those relations are :

$$\rho \, \bar{J}_1 = - \rho D_{12} \, \nabla N_1 - \rho D'_{12} \, N_1 \, N_2 \, \nabla T \qquad (1)$$

$$\rho \, \bar{J}_2 = - \rho D_{21} \, \nabla N_2 - \rho D'_{21} \, N_1 \, N_2 \, \nabla T \qquad (2)$$

where appear the mass fraction and the temperature gradients ∇N_i and ∇T describing the (small) thermochemical non-equilibrium state.

The phenomenological coefficients are : the mass diffusion coefficients (D_{12} and D_{21}) arising from Fick's low and the thermodiffusion ones, which are multiplied by the $N_1 . N_2$ factor so that they have no effect in a pure compound . It can be shown that D must be positive and D' can be of both signs. We have chosen the convention that D' is positive when the denser component migrates towards the colder boundary.
In usual liquids, the magnitude of D is of the order of 10^{-5} cm^2/s and the ration D'/D, called the Soret coefficient, is generally estimated between 10^{-2} and $10^{-3} K^{-1}$ Their dependence with respect to state variables must be specifically discussed. The distribution of these state variables evolves according to conservations laws. With the mechanical equilibrium assumption, the barycentric velocity vanishes in a closed cell or remains constant in a flow cell. In this last case we shall suppose that the viscous effects can be neglected, so that the energy conservation law reduces to the Fourier equation describing the evolution of the temperature distribution. The establishment of a steady temperature profile through a liquid layer is characterized by a relaxation time about three orders of magnitude smaller than the relaxation time corresponding to a concentration profile establishment. In our approach to the study of the establishment of the Soret seperation we shall suppose that ∇T is stationary Thus only N_1 will be time dependent. This behaviour in a non-reacting system is described by the relation :

$$\frac{\partial \rho N_1}{\partial t} = - \operatorname{div}(\rho N_1 \overline{V}_1) = - \operatorname{div}(\rho \overline{J}_1 + \rho N_1 \overline{V}) \qquad (3)$$

where $\overline{V}_1 - \overline{V}$ is the barycentric velocity of component 1 and ρ the specific mass of the mixture.

Eq. (3) yields the continuity equation :

$$-\frac{\partial \rho}{\partial t} = - \operatorname{div}(\rho \overline{V}) \qquad (4)$$

We do not consider in this paragraph the possible existence of gravity or double diffusive instabilities. A stable solution Eq. (3) exists describing the evolution towards a steady state, taking into account the signs and magnitudes of the different phenomenological coefficients. The so-defined phenomenon is called Soret effect; a steady state of non-chemical equilibrium is reached in a non-reacting mixture submitted to a temperature difference imposed at the boundaries of the system.

The final Soret separation is small, proportinal to D'/D and to the temperature difference that we assume to be of the order of 10 K. The phenomenological coefficients and in particular D'/D and the thermal diffusivity are considered constant.

This yields the temperature gradient and the specific mass of the mixture are also constant all over the system.

The behaviour of the system is described by

$$\frac{\partial N_1}{\partial t} + \nabla \cdot \overline{J}_1 = 0 \qquad (5)$$

$$\overline{J}_1 = - D \nabla N_1 - D' N_1 (1 - N_1) \nabla T \qquad (6)$$

in which only the flux and the mass fraction are dependent on time and on space variable x.

This system of equations is non-linear. Because of the choice of the phenomenological coefficient of ∇T, their dependence on N_1 remains an open question, but a linearization is justified by the small mass fraction difference which is reached. If necessary, D' will depend on the initial concentration.

Let us write $\beta \ \delta N_1$ the linearized form for the variation $\delta N_1(1-N_1)$

Substitution of Eq. (6) into Eq. (5) yields :

$$\frac{\partial N_1}{D\partial t} = \nabla^2 N_1 + \overline{S} \cdot \nabla N_1 \qquad (7)$$

where
$$\overline{S} = \beta \frac{D'}{D} \nabla T$$

One can also substitute N into Eq. (6). This leads to :

$$\frac{\partial \overline{J}_1}{D \partial t} = (\nabla + \overline{S}) \nabla \cdot \overline{J}_1$$

$$= \nabla \cdot (\nabla + \overline{S}) \overline{J}_1 + \nabla x \left[(\nabla + \overline{S}) x \overline{J}_1 \right]$$

(8)

The normal flux is vanishing at impervious boundaries; the mechanical equilibrium assumption yields the following boundary conditions :

$$\overline{J}_1 (\overline{x}_B, t) \cdot \overline{1}_B = 0$$

(9)

where the subscript B stands for "boundary".
If the initial mass fraction N is homogeneous, the flux remains parallel to s and then one gets for the components of the flux the same evolution equation than for N :

$$\frac{\partial \overline{J}_1}{D \partial t} = \nabla^2 \overline{J}_1 + \overline{S} \cdot \nabla \overline{J}_1$$

(10)

with the initial conditions

$$\overline{J}_1 (\overline{x}, 0) = - \alpha N_1^\circ D \overline{S}$$

(11)

where
$$\alpha = (1 - N_1^\circ)/\beta$$

In order to simplify the notations, let us consider the one dimensional problem.

Let us make the preliminary transformation :

$$J_1 (x,t) = \exp (- \frac{S^2}{4} Dt - \frac{Sx}{2}) \cdot U (x,t)$$

(12)

The solution of the eigenvalue problem for U leads a complete set $\{ U_p \}$ in the L_2 space with the corresponding discrete set of negative eigenvalue $\{ \lambda_p \}$.

Thus :
$$U (x,t) = \sum_{p=1}^{\infty} A_p \exp (\lambda_p Dt) U_p (x)$$

(13)

The completeness property can be used to compute the Ap;
Using the boundary conditions and the fact that λp are negative, one founds :

$$A_p = \alpha\, N_1^{\circ}\, DS\, \frac{\Delta[\exp\,(\frac{Sx}{2})\,\frac{dU_p}{dx}]}{a_p} \tag{14}$$

where $a_p = |\lambda_p| + \frac{S^2}{4} > 0$

and $\Delta[\,y\,]$ stands for $y\,(x_{sup}) - y\,(x_{inf})$.

The expression for the flux is given by Eq. (12)   and
for the concentration by

$$N_1\,(x,t) - N_1^{\circ} = -\int_0^t \nabla\cdot\overline{J}_1\,(\overline{x},t)\,dt' \tag{15}$$

The steady concentration distribution is then   :

$$N_1(x,\infty) - N_1^{\circ} = -N_1^{\circ}\,\alpha\,S\,\exp\,(-\frac{Sx}{2})\,.$$

$$\sum_{p=1}^{\infty}\frac{\Delta\,[\exp\,\frac{Sx}{2}\,\frac{dU_p}{dx}]}{a_p^2}\cdot(\frac{d}{dx} - \frac{S}{2})\,U_p\,(x) \tag{16}$$

The could have been found easily form Eq. (6) where $\overline{J}_1$ vanishes and with the condition

$$\int_V \rho N_1\,(x,t)dV = \rho N_1^{\circ}\,V \tag{17}$$

where V is the seperate volume of the experimental cell in
which the Soret seperation is performed.
This steady problem can be solved exactly and the solution
compared to Eq (16). This can be used to adjust the parameter α .

In the case of diluted binary solutions, the mass fraction
distribution is rather well approximated by

$$N_1\,(x,\infty) = -N_1^{\circ}\,\alpha\,S\,\exp\,(\alpha\,Sx)/\Delta[\exp{-\alpha Sx}] \tag{18}$$

and thus one has to take $\alpha = 1$, corresponding to the approximation

$$\delta N_1 (1-N_1) \approx (1-N_1^\circ) \delta N_1 \qquad (19)$$

For higher concentrations, a way has been proposed[3] to linearise the parabola $N_1(1-N_1)$ around the initial concentration. Then the final obtained separation is

$$\Delta N_1 = - N_1^\circ S \Delta x$$
$$= - \frac{D'}{D} N_1^\circ (1-N_1^\circ) \Delta T \qquad (20)$$

In this way, the Soret coefficient measurement can be performed by measuring at steady state, the mass fraction difference between two points at known temperatures.

Measurements of the diffusion coefficient have also been performed in the same kind of experiment by analyzing the evolution with respect to time of ΔN.
This can be done using the fact that the time dependence could be governed by the factor

$$\exp -(|\lambda_1| + \frac{S^2}{4}) Dt$$

This leads a relaxation time of the phenomena expressed by

$$\frac{a^2}{\pi^2 D}$$

Near the impervious boundaries of the system, some authors (see Refs. 1, 3, and 5) suspect a more complicated time behaviour.
Let us write the flux, along space coordinate x, varying from 0 to a, as:

$$J_1 (x,t) = - N_1^\circ DS' \exp(- \frac{S'x}{2} - \frac{S'^2}{4}\tau) \sum_{p=1}^{\infty} [1-(-1)^p \qquad (21)$$
$$\exp \frac{S'}{2}] \cdot \frac{\pi p}{(\pi p)^2 + \frac{S'^2}{4}}$$
$$\sin (\pi px) \exp (- \pi^2 p^2 \tau)$$

where $X = \frac{x}{a}$, $S' = a S$ and $\tau = \frac{Dt}{a^2}$

Fig 1 shows the computed values of the flux with respect to time. For t $\downarrow$ 0, one gets the step distribution, corresponding to initial values.

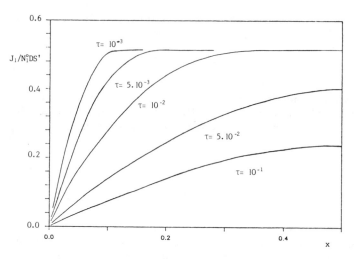

Fig. 1 Evolution of normalised flux (J_1/ N_1^o D S') distribution
for S'= 0.01 (space coordinate x varying from 0 to 1) .

As a consequence, using Dirac delta distribution, we may
write symbolically that

$$\lim_{t \downarrow 0} \frac{\partial N}{\partial \tau} = \lim_{t \downarrow 0} \frac{\partial J}{\partial x} = - \delta(x) + \delta(x-a) \qquad (22)$$

Physically, this means that for short times (but longer
than to establish the temperature distribution) concentration varies mainly at the boundaries as illustrated on
Fig.2 where the following expression is plotted as a function of X :

$$N_1(x,t)-N_1^o = N_1^o \text{ S'}\exp \left(\frac{-S'x}{2}\right) \sum_{p=1}^{\infty} \left[1-(-)^p \exp \frac{S'}{2}\right]$$

$$\frac{(\pi p)^2}{\left[(\pi p)^2 + (S'/2)^2\right]^2}\left[\cos(\pi px)-S'/2 \pi p \quad \sin \pi px\right]$$

$$\left[1-\exp(-\pi^2 p^2\tau- S'^2 /2\tau)\right]$$

$$(23)$$

This behaviour prevails at the first instants. For S'=0.01
and τ= 0.05, 10 % of the final concentration is already
obtained at the boundaries, at the same time, only 1 % of
the final concentration is reached at x = 0.1. This shows

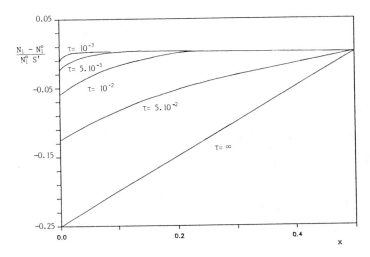

Fig. 2 Evolution of normalised concentration ($N_1 - N_1^o$)/ N_1^o S' for
S'= 0.01 .

clearly that the separation rate is highly stressed at
boundaries in the first moments.

This was already experimentally observed by Tanner[5] .
On the other hand, Thomaes[3] established by an approximate
method that $\partial^2 N/\partial\tau\partial x$ was infinite in τ = 0 and x = 0,
due to the fact that the separation propagates progressi-
vely in the bulk of the cell from the boundaries.
This interesting behaviour is presently being investigated
in detail.

Thermal migration and stability of a horizontal liquid
layer.

    The linear stability analysis of the rest state of a
horizontal binary liquid layer submitted to a vertical temperature gradient can be performed analytically for boundary conditions corresponding to impervious, free surfaces[6].
The classical Boussinesq approximation is maintained, the
Dufour effect is neglected and the mass fraction variation
by the Soret effect is supposed sufficiently small to
replace in the expression of the thermodiffusion flux
N_1 . N_2 . by a constant N_1^{in} . N_2^{in}

    A 3-dimensional perturbation analysis can be performed
by linearisation of mass, momentum and energy conservation
laws. The resulting five differential equations depends
on Schmidt, Prandt, Rayleigh, thermodiffusion Rayleigh and

Soret numbers (respectively Sc, Pr, Ra, Rth, and s') and
on the sign of $\Delta T/|\Delta T|$ taken to be positive when the layer is
heated from below (see Nomenclature).

A classical normal mode analysis with exponential time
dependence (exp σ t) allows the elimination of pressure,
concentration and temperature perturbation amplitudes.
The resulting dispersion equation for the vertical
component of the velocity amplitude is associated with
homogeneous boundary conditions.
To study stability the fundamental mode can be choosen.
Conclusions are discussed in the $\Delta T/|\Delta T|$ Ra-s' plane.
Stable and unstable domains are separated in this plane
by the lines of marginal stability (given by $\sigma = 0$)

$$\frac{\Delta T}{|\Delta T|} \; Ra^{cr.ex.} = 657,5 - Rth \; s'(1+\frac{Sc}{Pr}) \frac{\Delta T}{|\Delta T|} \qquad (24)$$

For R_{th} s' vanishing this expression gives the value of
critical Rayleigh number for a pure compound.
The domain of overstability is given by Real $\sigma = 0$ and
Im $\sigma \neq 0$ wich leads to

$$\frac{\Delta T}{|\Delta T|} \; Ra^{cr.ov.} = 657,5(1+\frac{1}{Sc})(1+\frac{Pr}{Sc})-s'\frac{Pr}{Pr+1} \frac{\Delta T}{|\Delta T|}Rth \qquad (25)$$

and by $s' < 0$

$$|s'| > \frac{[1+(k/\pi)^2]^3}{(k/\pi)^2} \frac{Pr \; (1+Pr)\pi^4}{Rth \; Sc^2} \qquad (26)$$

On Fig. 3 $Ra^{cr.ex}$ and $Ra^{cr.ov.}$ are given as function of
the Soret number (for a given mass fraction, Rth is only
depending on physical constants related to the system and
not to any imposed constraint).
On this diagram, the section BC, DE and FG have no physical
meanings because they do not correspond to Ra > 0.

The transition through an oscillatory state can occur only
in quadrant I for sufficiently large values of $|$ s' $|$ (see
condition 26). For a layer heated from the top ($\Delta T/ |\Delta T|$
= - 1) the only possibility for convection is through
exchange of stability (for details see Ref. 6).
How must the diagram be interpreted, eg. for a system with
D'/D > 0; the representative point for ΔT increasing is
moving from the origin by following the segment OA. The
convection appears at the intersection of OA with the line
S defined by Eqs. (24) or (26), for a critical temperature
difference, ΔT^{cr}.

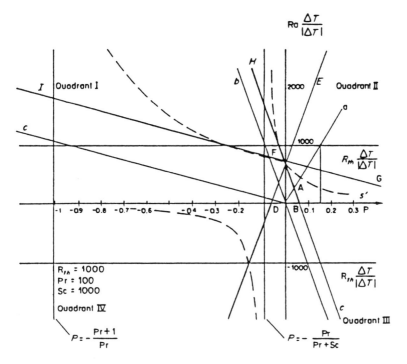

Fig. 3 Stability diagram in the Ra/s' and Ra/P planes .

$$tg\ \beta = (\frac{g\alpha d^3}{\varkappa\nu}\ \frac{\Delta T}{|\Delta T|})\ /\ \ D'/D\ (1-N_1^{in}) \tag{27}$$

From this diagram, $Ra^{cr.ex.}$ and $Ra^{cr.ov.}$ can be evaluated as function of P, which is the ratio of the two contributions to the total density gradient

$$P = \frac{\gamma}{\alpha}\ D'N_1^{in}\ N_2^{in} = \frac{Rth\ S'}{Ra}\ \frac{\Delta T}{|\Delta T|} \tag{28}$$

Construction of similar triangles allows to pass from the first representation to the second one (in dotted line on Fig. 3) $Ra^{cr.ex.}$ is a function of P presenting infinite asymptotic value for $P = -\frac{Pr}{Pr + Sc}$, depending on the numerical value of the mass and of the heat diffusion. This leads to the fact that motions can appear in quadrant IV (heating from the top) with a downward total density gradient. $P < 0$ means that the two contributions (concentration induced by Soret effect and thermal expansion) are opposite.

P = -1 means that these two effects are opposite and equal, i.e. the total density gradient is vanishing.

Instabilities can be observed in the range

$$-1 < P < -\frac{Pr}{Pr + Sc} \qquad (29)$$

i.e. when the total density gradient is "stably" stratified, the density is decreasing with elevation.

Eq.(29) can be rewritten as

$$-1 < P < -\frac{1}{1 + \varkappa/D} \qquad (30)$$

indicating clearly that the stability is closely related to the lifetime of concentration and temperature fluctuations.

If D→ 0 (as in the example of salt fountain) this unusual instability can appear in the range -1< P< 0, on the contrary for D→ ∞ , this instability disappears.

For system with Pr = 10 and Sc = 1000 (as in most of usual organic mixtures) instabilities can appear for P <-1 i.e., when the destabilizing contributions of the mass fraction gradient is only 1 % of the total downwards density gradient.

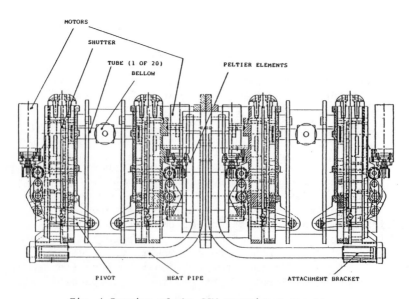

Fig. 4 Drawing of the SCM experiment reactor .

This unexpected convection could affect most of the experimental values of D'/D < 0 reported in the literature (see ref 7) .

In the case where $\sigma_I \neq 0$, in quadrant 1, the equation for the vertical asymptot is

$$P = - \frac{Pr + 1}{Pr}$$

Oscillatory convection could appear in the range

$$- \frac{Pr + 1}{Pr} < P < -1 \qquad (31)$$

where the total density gradient is "stably oriented" this range can be extended if Pr is small (e.g. liquid metals). The onset of this convection is only Pr dependent.

Soret coefficient measurement under microgravity conditions

Hardware description of the Soret coefficient measurement (SCM) cell

The available volume and the weight budget are the more restrictive parameters which were taken into account in the definition of the hardware during design and production by Pedeo Techniek (Oudenaarde, Belgium). It is planned to integrate this cell into one of the four reactors of the Solution growth Facility (SGF) produced by Laben (Milano). This reactor is a cylindrical aluminium container whose temperature is maintained constant at 40°C by the use of electical heaters monitored by the SGF system.

The experimental cell is sketched on Fig. 4. It is constituted by two sets of ten tubes which are symmetrically distributed with respect to the central plane. These 8 cm long tubes are in titanium. The choice of this material arises from weight and corrosion resistance considerations.
At each end of these tubes, are located sampling volumes of 1 ml; they can be insulated from the tube contents by the action of shutter valves. The shutters can be displaced at the end of the experiment by strong loaded springs triggered by small electrical motors. These sliders are in titanium coated with FEP in order to decrease the friction. Special o-rings for the tightness are also in FEP and tita-

nium. These sampling volumes are inserted in massive aluminium parts to be as isothermal as possible.
The hot parts of the cell are in good thermal contact with
the SGF container and are thus maintained by the SGF system at $40°C \pm 0.1°C$. The thermal gradient is created by
using Peltier elements. A carefull study showed that for
the expected heat flux of 1.5 W, the more efficient configuration for each half cell is two layers of four Peltier
elements each, the eight Peltier elements being electrically connected in Series (MARLOW MI 1902-02 AC). The
heat rejected by the Peltier elements will be flown back
to the hot side by the use of ammoniac heat pipes in stainless steel produced be SABCA (Brussels). It is planned
that the cold part of the tubes will be maintained at $30°C \pm 0.1°C$
by the use of an electronic regulator thermistors
(FENWALL GB 34 JM 86) distributed on the tubes in order
to measure as accurately as possible the temperature distribution.

Rotating valves, at the ends of all the tubes, will
allow to fill the cell without any bubble to avoid Marangoni convection and to sample when the experiment will be
recovered in our laboratory after the accomplishment of
the EURECA 1 mission.

In a tube containing aqueous nitrate solution, the rotating parts of the valves will be in silver. The e.m.f.
taken by these electrodes will be measured six times per
minute by the use of a preamplifier integrated to the SCM
cell.
This preamplifier is produced by the Belgian Space Aero-
nomy Institue (Brussels). These data will be dumped regularly in quasi real time. An electronic mail box will
react if some malfuctionning is detected (e.g. save results
by closing the valves before the end of the mission if the
thermal gradient is not maintained).
Only one volume compensation per tube are used. They are
constituted by titanium bellows connected at the central
parts of the tubes (where the concentration remains
constant). Their design allows to compensate volume
variations of 1 ml with a minimum initial volume content.

After the mission, the content of the different sampling
volumes of the SCM cell will be analysed in our laboratory
by at least two different methods.

Systems to be studied

We have listed in Ref. 9 the different binary systems
that it is planned to study in this first SCM experiment.

We have choosen systems which are reported in the literature
as having a Soret coefficient negative (the denser component is migrating towards the hot side). It is suspected
that most of the reported values are wrong[10] due to the
unavoidable existence of horizontal temperature gradients
leading to convection in ground experiments, and in some
extent to possible double diffusive instabilities. (cf.
quadrant IV, Fig. 1).
Reference ground tests with the SCM hardware cannot lead to
reliable results, because the apparatus is not designed in
order to minimize horizontal temperature gradients.
In our laboratory, we are performing two types of different
ground tests. The Soret coefficients of the reported systems are measured, using a flow cell, with a very thin
liquid layer in order to minimize the convection[11].

These reference ground tests are needed in order to
compare with the values obtained under microgravity conditions with a temperature difference of 40 - 30°C.
The Soret coefficient is generally not determined on earth
for such thermal conditions, which have been chosen for
the EURECA experiments due to the technical constraint of
the SGF.
The sign of the Soret coefficient is also determined by the
Schmidt Milverton plot outlook obtained during Bénard experiments.[9]

Conclusions

Generally the Soret effect is neglected as well in the
density distribution as in the hydrodinamic stability analysis because it is considered as a small effect. We have
shown that it can produce instabilities even in quadrant IV.
Furthermore its constribution to a density distribution is
generally of the same order of magnitude than the thermal
expansion as it can be seen on the following examples

Let us consider a binary system with:

$$D'/D = 10^{-2} K^{-1} , N_1 = N_2 = 0.5 , \rho_1 = 1 , \rho_2 = 0.8$$

submitted to a temperature difference of 10 K, the mass
fraction difference induced by the Soret effect is 0.025
corresponding to a density difference of 0.0005.
The density difference induced in water by the thermal expansion between 20 and 30°C is equal to only 0.0025, a factor of two smaller than the Soret contribution.
One can object that the Soret effect differs by the long
time needed to reach a steady concentration distribution

but we have shown the concentration varies rapidly
at the limit of the system due to the squashing of the
thermodiffusion mass flux on the rigid walls.
By this set of experiments to be performed in the very good
microgravity conditions of the automatic platform EURECA,
it is expected to obtain reliable values of the Soret coefficient allowing to compare them with some theoretical evaluations and with ground obtained experimental values.

Acknowledgments

We are deeply indebted to Professor A. Jaumotte for his
constant support and interest into our work.
This research is financially supported by an "Action
de Recherche Concertée" contract with SPPS and by a "Fonds
de la Recherche Fondamentale Collective" contract.
We have also received funding from "La Loterie Nationale."

References

[1] De Groot, S. R. and Mazur, P., Non-equilibrium Thermodynamics,
Dover, New York, 1983.

[2] Prigogine, I., Introduction a la Thermodynamique des Processus
Irreversibles, Dunod, Paris, 1968.

[3] Thomaes, G., "Recherches sur la Thermodiffusion en Phase Liquide,"
Physica, Vol 17, No. 10, Oct. 1951.

[4] Richtmeyer, R. D., Principles of Advanced Mathematical Physics,
Vol. 1, Springer-Verlag, New York, 1978.

[5] Tanner, C. C., "The Soret Effect. Part I," Transactions of the
Faraday Society, Vol. 23, 1927. p. 75.

[6] Platten, J. K. and Legros, J. C., Convection in Liquids, Springer-
Verlag, New York, 1984.

[7] Verlarde, M. G. and Schetchter, R. S., "Thermal Diffusion and
Convective Stability, III. Critical Survey of Soret Coefficient
Measurements," Chemical Physics Letters, Vol. 12, No 2, 1971.

[8] Solution Growth Facility, Sample Interface Agreement, Experiment
No. 123, ESTEC, Doc. No. G.P./SGF/123/DF/JP, 1986.

[9] Legros, J. C., "Double Diffusive Instabilities and the Soret
Coefficient Measurement under Microgravity Conditions," Acta
Astronautica, Vol. 15, No. 6/7, 1987, p. 455.

[10] Legros, J. C., Goemeare, P., and Platten J. K., "Soret
Coefficient and the Two-Component Bénard Convection in the

Benzene-Methanol System," Physical Review A, Vol 32, No. 1, July 1985.

[11]Legros, J. C., Rasse, D., and Thomaes, G., "Thermal Diffusion in a Flowing Layer with Normal and Adverse Temparature Gradients in the CCl_4-C_6H_6 System," Physica , Vol. 57, 1972, p. 583.

[12]Turner, J. C., Butler, B. D., and Story, M. J., "Flow-Cell Studies of Thermal Diffusion in Liquids. III," Transactions of the Faraday Society, Vol. 63, No. 8, 1967.

Chapter III. Satellite Communication

Communication Satellite Technolgy
Development Survey

E. W. Ashford *

European Space Agency, ESTEC, Noordwijk, The Netherlands

ABSTRACT

The use of satellites in the geostationary orbit has shown
itself to be a cost-effective and highly flexible means of
providing global, regional, and national telecommunications
and broadcasting services. The rapid growth in the demand
for such services in the only slightly more than two decades
since the first such satellite was launched, combined with
the commercial nature of the provision of such services, has
necessitated the development of increasingly advanced technology in a wide range of areas. This has enabled communications by satellite to be one of the very few areas in our
global society that has become, in real terms, progressively
cheaper with time, reversing the international inflationary
trends that have been experienced in most other areas in the
same time frame.

The development of the requisite technology has itself been
expensive, however, and a number of national and international space agencies have arisen whose yearly budgets
are, to a significant extent, devoted to defining the
anticipated needs of the future and making technology available in the areas and time frames where and when it will be
needed.

The forecasting of the communication satellite technology
needs of the future is not an exact science. Care must be
exercised when using past trends to draw conclusions about
future needs. A number of factors and constraints dictate
the need for technology advances, which must be considered
along with past technology trends. Summaries of the tech-

*Head of Satellite Systems Division, Communication Satellite
Department.

nology programmes of several organisations show, however,
that similar conclusions about future needs have been drawn.
A number of areas where technology developments are taking
place will have a large potential future impact.

INTRODUCTION

Since the first civilian experimental communication satellite, TELSTAR, was launched into an inclined low orbit in
1962, there has been an amazing expansion in the demand for
the services that communication satellites can provide. The
first commercial geostationary orbit communication satellite, Intelsat's "Early Bird", followed its experimental
precursor by only two years, and bigger and more performant
satellites have been developed and put into service since
then at an expanding pace.

The types of services provided by communication satellites
have also increased with time. The first satellites were
used primarily for telephone traffic and occasional relay of
trans-oceanic television programmes. Since then, increasing
use has been made of such satellites for television distribution and business communications, the organisation
INMARSAT has arisen to provide satellite communications to
ships and, soon, to planes in flight, and satellites have
and are increasingly being put into service to broadcast
television programmes directly to viewers in their homes
over wide geographical areas.

The increase in the number of types of services being provided by satellites, and the increase in demand for traffic
capacity in each of these service areas, have together
pushed satellite designers to develop more and more performant satellites to meet the demands in the most cost
effective manner. That this latter point in particular has
been emphasised and achieved may be inferred from the fact
that, using Intelsat as an example, the cost of leasing a
trans-oceanic telephone circuit dropped by a factor of five
in their first 15 years of satellite operation, despite the
fact that this was a period of relatively high inflation.

In order to meet such increasing service demands, satellite
designers have had to incorporate increasingly more sophisticated and complex technology into their satellites. The
development of this technology has been an expensive evolutionary, and sometimes revolutionary undertaking. How the
needs for the specific technological developments that will
be required are identified, and how the requisite technology

is developed and made available for operational satellite usage, is the subject of following sections of this paper.

ASSESSMENT OF FUTURE TECHNOLOGY NEEDS

Technology Requirements Definition

Identification of the technology that will be needed for use in future communication satellites is not always a straight-forward affair. Improvements in satellite performance and cost effectiveness can come about both through evolutionary improvements in the equipment embarked on the satellite, the trends for which are relatively easy to determine and extrapolate some way into the future to determine technology goals, and through revolutionary changes in basic satellite system or subsystem design, where prediction of future needs is both more difficult and significantly less accurate. For example, the early satellite designs were all spin stabilized. These then developed with time into dual spin designs, which can be considered partly as an evolutionary change (as far as the satellite bus is concerned), and partly as an entirely new concept. The next stage was the development of three axis stabilized satellite designs, which were revolutionary as far as the bus is concerned, but could employ payloads that were an evolutionary outgrowth of their predecessors.

The development of new technology for use in space is, in general, a lengthy process when compared to technology for usage on earth, due to a number of factors. These include, in particular, the need to develop equipment that will be highly reliable for many years, without maintenance, in what is in many respects a relatively harsh environment. Figure 1 shows a hypothetical project life cycle, from technology requirements definition to launch of a satellite, with typical times scales shown. In practice, many satellite programs are developed in a considerably shorter time scale than indicated, but if brand new technology is involved, such durations are not generally bettered to a great extent.

The consequence of long development times is that it is necessary to plan for and initiate the development of the technology to meet the anticipated needs of communication satellites of the future, at least in some cases, as far ahead as 8-10 years before it is expected that the technology will be needed in space. This same technology must then, of course, continue to be suitable throughout the 7-10 year mission life of the satellites once they are in orbit.

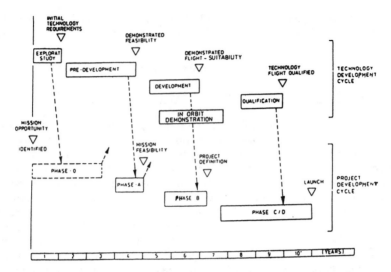

Fig. 1 Hypothetical satellite project development cycle.

Deciding what particular technologies should be developed to meet needs so far in the future is difficult, for in most cases the specific requirements that the technology will have to satisfy cannot be predicted with any accuracy that far ahead. Nevertheless, there are certain motivating factors that can at least indicate the direction that improvements in technology should take, and analysis of past trends can often help to identify targets to aim for in the developments.

The provision of satellite communication satellite services is a commercial business which is also, in most instances, a highly competitive one. It is thus driven by the "Cardinal Rule" that applies to all service providers in a competitive environment, "to succeed, one must provide improved services, at lower cost, to more customers".

To comply with this rule, technology developments are continuously being initiated, which may be either evolutionary or revolutionary in nature. The term "evolutionary" in this context can be understood as meaning technology that improves an existing method of providing a service, while "revolutionary" implies either finding another and better way to provide a service, or providing totally new services.

The way the Cardinal rule influences communication satellite technology development can perhaps best be understood by giving some illustrative (but by no means comprehensive) examples associated with its three basic tenets: Better, Cheaper and More.

o The drive towards better service has caused satellites to progressively tend toward:
- higher radiated output powers and higher gain "spot beam" and/or contoured beam antennas to increase EIRP,
- designs with automatic on-board fault detection and correction to minimize outages,
- lower noise receivers.

o The drive towards cheaper service has pushed designers to provide:
- longer lifetime, more reliable satellites so that investment costs can be amortized over a longer period,
- lighter and more efficient designs to reduce the mass dependent costs of launching satellites,
- increased satellite EIRP and G/T so that user earth terminals can be cheaper.

o To cope with the increase in customers that the better and cheaper services attract, satellites have had to be developed that:
- make maximum use of the limited RF spectrum available in the satellite communication bands by reusing the spectrum a number of times on different polarizations or in different spot beams,
- operate in higher allocated frequency bands when congestion fills the lower bands.

To obtain more customers, satellites originally intended for one purpose are increasingly being used to provide new services once they are in orbit. This in turn has often required technology improvements in earth terminals.

The types of technology improvements described under each aspect of the Cardinal Rule are not, of course, restricted to a single aspect. Many possible improvements could effect two or all three areas, and potential improvements under one aspect could adversely affect one or both of the others (e.g., an improvement in performance may only be possible with an attendant significant increase in cost which would have to be passed onto the users in the form of higher tariffs, resulting in a decrease in the number of users who are willing to use a service). It is thus the task of those who are responsible for deciding upon which technologies should be developed, given their limited budgets, to try and determine areas which could maximise the improvement in two or three areas simultaneously and which can have the greatest financial benefits for the lowest investment costs.

Technology Development Trend Analysis

As mentioned above, historical and projected trends in satellite subsystem and/or equipment performance can provide a useful guideline to establish future technology development goals. Before giving some examples of such trends, it is important to stress that trying to predict future needs on the basis of past history is only even remotely accurate if one understands the reasons why there has been a certain trend in the past, and can be reasonably assured that these same reasons will continue to be valid in the future. There are many examples that can be given of technology developments being initiated on the basis of trend analyses that have subsequently been found not to be needed due to the reasons for the past trend having disappeared. The classic academic example often given is the "demonstrated need" for an improved whip to be used on horse-drawn carriages, which disappeared with the advent of the age of automobiles.

Nevertheless, trend analysis, if used carefully, can be a useful tool to help decide upon future development needs. Thus it is worth giving some examples of the results of such trend analyses in certain satellite subsystem areas.

Figure 2 shows a trend chart for satellite eclipse power requirements, as a function of the actual or planned time of launch of a number of satellites (not all of which are, however, communication satellites). It illustrates two things about such analyses. First, that to interpret the significance of a trend, one must analyse what types of technology are applicable for certain regions along the vertical axis. It is apparent from the figure, for example, that it would make little sense to invest money to develop 10 Kw nickel-cadmium batteries, since this is in a regime where metal-hydrogen batteries are most suited. Second, it shows that trend analyses must take into account the fact that different applications of a technology may show different trends (i.e., the separation between LEO and GEO trends on the right side of the chart).

Figure 3.a shows a scatter plot of some 73 different communication satellite designs, where beginning of life

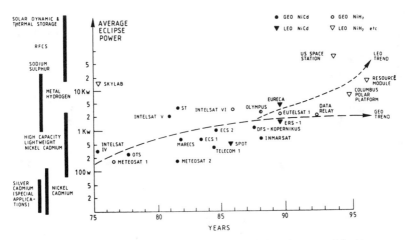

Fig. 2 Evolution of eclipse power requirements and battery
technologies.

solar array power is related to the year of launch of the
satellite (or the year of first launch, where a series of
satellites is involved). Figure 3.b shows the results of an
analysis of what types of array technologies are best suited
to various power and efficiency requirements.

   Plots such as on Figures 2 and 3, taken individually,
are of use to, respectively, battery and solar array designers to point the way in which their technologies should
evolve. Taken together, and properly interpreted, they can
also provide useful guideline to power subsystem electronics
designers (to estimate future power handling requirements)
and to thermal subsystem engineers (to estimate future thermal dissipation requirements).

   Figure 4 shows, for another subsystem, another way of
showing the same type of relations. It takes into account
the different types of technology by showing discrete curves
for each.

   Figure 5 shows yet another way of showing trends, this
time for communication satellite repeaters. This time, the
trend is not shown versus time, except as an indicated parameter at points on the various curves. It can, however, be
equally as useful as the previous figures in indicating
goals towards which future developments should aim.

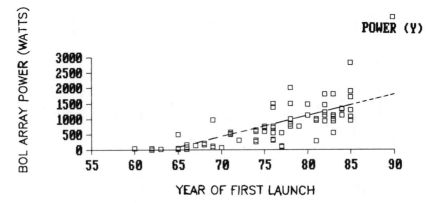

a) Evolution of solar array output power with time.

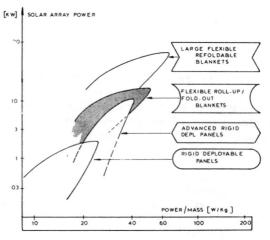

b) Solar array technology areas of emphasis.

Fig. 3   Solar array power and efficiency.

ONGOING DEVELOPMENTS

In light of the above, what then are the ongoing developments in the world to satisfy the future needs for communication satellite technology ?   These are being undertaken by
a large number of countries and agencies throughout the
world, and it would be impossible to cover all of them in a
paper of this sort. What follows then will be a summary of
the activities being undertaken by the European Space
Agency, about which the author is most familiar, and a less
comprehensive summary of activities in the United States and

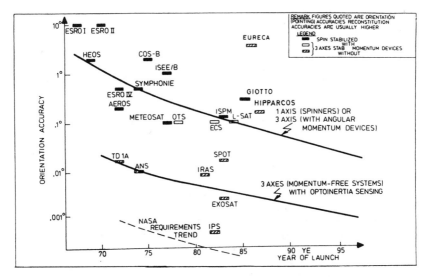

Fig. 4   Orientation accuracies of European spacecraft.

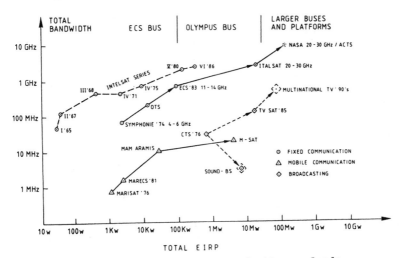

Fig. 5   Trend chart for communication payloads.

Japan, on whose programmes there is considerable available literature for the author to draw upon.

ESA Communication Satellite Technology Programmes

History

The European Space Agency (ESA), and its predecessor (ESRO) became involved in communication satellites in a major way in the beginning of the 1970's with the development of first the technology for, and then an experimental satellite (OTS) to demonstrate, the feasibility of implementing a European Communication Satellite (ECS) System. On the basis of projections that showed a future congestion in the C-Band frequencies that were being used at that time for satellite communications, the technology developments were oriented toward the next available frequencies at Ku Band. Also, because of the projected needs in solar array power, emphasis was given to three-axis stabilized bus designs, rather than to the dual-spin satellites then being deployed.

Prior to the launch of OTS, however, ESA had expanded its originally conceived programme to include the initiation of the development of the technology, and later of an entire payload concept for maritime communications via satellite at L Band frequencies. This led to the development of the MARECS satellite series, two of which were put into service for INMARSAT starting in early 1982.

After OTS had demonstrated the viability of its design concepts in orbit, the first of the five satellite ECS series was launched in mid-1983 and a second in mid-1984. These, and later ECS satellites, once successfully in orbit, are used by EUTELSAT to provide telephony, television distribution, and specialized business services throughout western Europe.

The Present and Future

Further communication satellite development programmes in ESA include the OLYMPUS satellite, an experimental multi-payload satellite in the 2-to-3 metric ton class that will be launched in 1988, two further multi-payload experimental satellites foreseen to be launched in the 1991-1993 time frame, and an operational data relay satellite system, due to begin operation in 1994/1995.

Such satellite programmes can only be successful if they are supported by an on-going programme of system studies and technology development at subsystem and equipment level. In ESA, a comprehensive programme is made up of several separately funded programmes, differing in their particular orientations and content. These are as follows:

o The Technological Research Programme (TRP). This programme, financed by all the 13 member countries of ESA in direct proportion to their gross national products, aims at the development of technology in a wide range of areas to meet the potential needs of satellite programmes that could arise a decade or more in the future.In 1987, approximately 20 % of the 23 M$ budget of this programme will be directly oriented toward communications and/or data relay satellite needs.

o   The System Studies Programme. This programme, financed in a similar manner to the TRP, performs preliminary feasibility and definition studies on potential future satellite programmes. In 1986, approximately 18 percent of the 7.5 M$ budget of this programme was in the field of communication satellites.

o   The Advanced Systems and Technology Programme (ASTP). This programme, financed by participating member countries in amounts depending upon their own wishes and the aspirations of their space industries, is devoted completely to the field of communication satellites. Some 85 percent of its budget is devoted to specific technology developments, directed toward usage on anticipated new ESA, national, or competitive international satellite projects which are expected to materialise within the next 5-to-10 years. The presently approved ASTP budget for the period from mid-1986 through mid 1990 stands at present at some 96 M$, and is anticipated to increase to 125 M$ before the end of that period.

o   The Telecommunications Preparatory Programme (TPP). This programme was funded by participating member countries essentially to carry out system studies and satellite project (or major subsystem) Phase A feasibility studies. Its purpose was to supplement the amount of effort that the Agency could expend on such topics, since the overall budget of the System Studies Programme is a fixed one, and only a relatively small share of that budget could be devoted to communications satellites. The TPP budget over 83/86 was 11.5 M$. The programme, which was of a fixed duration, will come to an end in the middle of 1987, and will be supplanted by the first line of the PSDE programme described below.

o   The Payload and Spacecraft Development Programme (PSDE).This is a major new ESA telecommunications programme which initially extends from mid-1986 through mid-1995. It is an omnibus programme that combines, under one basic framework: studies, the development on the ground of a relatively large number (ten defined at

present, with more expected in the future) highly
advanced satellite payloads, the flight of a number of
these on two or more experimental satellites,
experimentation programmes with satellites in orbit, and
selected technology developments to maintain and improve
the competitiveness of European industry. The programme,
with an estimated cost over the first ten year period of
1091 M$, was approved in principal by the ESA member
states in mid 1986, but the funding for its various
phases will be approved in a sequential fashion over the
next six years. The first phases to be initiated, whose
budgets were approved in 1986, are:

- The Basic Support Line - a continuous programme of
  system studies to prepare for later developments of
  satellites or technology. It is budgeted at an annual
  expenditure of 4.8 M$

- The Development of Payloads Line - a phase, budgeted
  at a total of 65.4 M$ over the years 1986-1988, to
  initiate the development of a number of communication
  satellite payloads, including:

  . Advanced Repeater for Aeronautical and Maritime
    Integrated Services (ARAMIS)
  . Land and Aeronautical Mobile Experiment (LAMEX)
  . High Gain Single-Access S-Band Data Relay
  . S-Band Phased Array Multiple Access Data Relay
  . Optical Inter-Satellite Links
  . Optical Inter-Orbit Links
  . Reconfigurable Television Broadcasting
  . Millimetre-wave Propagation Measurement
  . Navigation
  . On-board Processing Repeater

- The Configuration Definition Line, budgeted at 14.4 M$
  over the years 1986-1988, to assess the feasibility
  and cost versus benefits of embarking various sub-sets
  of the payloads being developed in the line above on
  modifications of one or two of the satellite busses
  existing in Europe, and in defining in detail the preferred configurations of two experimental satellites
  to be launched in the 1991-1993 time frame.
  Subsequent phases of the PSDE programme will be approved to develop and fly the two satellites mentioned
  above, and to continue the development of the payloads
  not selected for flight on these through the engineering model test phase. These, as well as other payloads
  which may be developed in the meantime, will become

candidates for flight on a much further advanced experimental satellite system (the Advanced Orbital Test System - AOTS) foreseen to be flown in the latter half of the next decade under an extension of the PSDE programme.

o   The Data Relay Satellite Preparatory Programme (TRPP). At a meeting held at Ministerial level in Rome in 1985, the ESA member countries approved a long term European Space Programme to be conducted by the Agency that included, as one major element, the development of a Data Relay Satellite (DRS) System to be part of the European In-Orbit Infrastructure foreseen for the mid-1990s and beyond. In 1986, the member States approved the detailed content and funding for the DRPP, the first phase in the development of the DRS System.

The DRPP, budgeted at 31.3 M$ over a $2\frac{1}{2}$ year period beginning in mid-1986, will carry out the major System Definition and supporting studies necessary to establish detailed schedule and cost information for the subsequent development, implementation, and operation of the DRS System. In addition, time critical items of technology that must be available for the DRS will be identified and their development will be begun within the DRPP or PSDE.

In this latter regard, it should be noted that an apparent duplication of DRS related technology development in the PSDE and DRPP does not, in fact, exist. While both programmes will carry out technology developments in this area, there is a firm demarcation between them. DRS technology developed within the PSDE will be that which it is felt will require in-orbit demonstration and experimentation before being used operationally in the DRS System, while that developed within the DRPP and later DRS programmes is that which can be adequately qualified and tested on the ground to allow its use thereafter directly on the DRS System, without requiring intermediate in-orbit testing.

NASA Communication Satellite Technology Programmes

History

NASA was an early leader in the development of communication satellite technology, with a vigorous Research and Development programme and with its Applications Technology Satellite (ATS) series. Besides testing various types of stabilisation and demonstrating various types of bus subsystems in

orbit, the ATS satellites were designed to be at the forefront of technology in experimenting with the provision of a
number of different types of satellite communication services in the VHF, L, C, Ku and Ka frequency bands, as well
as providing data on propagation through the atmosphere of
Millimeter Waves. Six ATS satellites were launched between
1966 and 1974, although two of these unfortunately were
unable to satisfy their original mission objectives due to
launch vehicle malfunctions. Nevertheless, the experience
and data obtained with the successful satellites was highly
important in leading the way to the later developement of
commercial communication satellites exploiting what had been
learned from the ATS series.

Following the launch of ATS-6 in early 1974, NASA shifted
its priorities away from communication satellites. The prevailing opinion in the U.S. Administration at that time was
apparently that commercial firms were capable of carrying on
developments in this field without the need for any significant government R&D impetus. Although communications technology continued to be developed by NASA thereafter, it was
at a considerably slower pace than in previous years. Figure 6 illustrates this situation graphically.

In the ensuing years, continued advances made in Europe and
in Japan eroded, and in many cases overtook the lead in communication satellites previously held by the U.S. This led
finally to a re-appraisal within NASA and the U.S. Government, and in 1982 a major new programme, the Advanced Communications Technology Satellite (ACTS) was initiated.
Associated with this was a major emphasis in the development
of technologies which NASA felt would be the drivers in the
satellite systems of the future.

ACTS was an outgrowth of a series of activities begun by
NASA in 1978 to develop the technology to use the 30/20 GHz

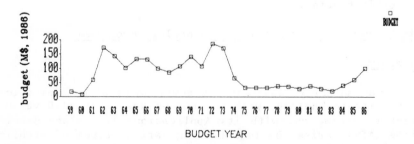

Fig. 6 NASA budget for communications.

bands for satellite communications. These led, in 1981, to NASA requesting funding from the U.S. Congress for a new programme, called the "30/20 Program", to develop and fly two experimental satellites incorporating technology for communications at 30/20 GHz, with laser inter-satellite links connecting the two satellites, and with repeaters using on-board processing at base-band to regenerate and route signals to the desired destinations.

The estimated costs for the 30/20 Program were felt to be to high by the Congress, and the NASA request was not approved. The following year, NASA requested funds for a scaled down version of their original proposal, with a single satellite and without the laser inter-satellite experiment, and Congress approved this version. The programme was re-named "ACTS" in 1983, and in early 1985, a laser communications experiment was reinstated, this time between ACTS and both a terminal mounted on a shuttle and an uplink laser terminal on the ground.

The Present and Future

NASA's present emphasis in communications is centered on activities in the advanced fixed and land-mobile services areas. The major effort is devoted to the development of the ACTS satellite, which will demonstrate the technology for a future "switchboard-in-the-sky". The technology develop in support of this is in the areas of:

. RF devices (tubes, solid state power amplifiers, and monolithic feeds)

. Interconnectivity technologies (related to on-board switching and base-band processing, as well as to optical intersatellite links)

. Large antenna developments.

The ACTS will demonstrate much of the technology developed in these areas when it is launched in, according to present planning, 1990.

In the land-mobile area, NASA is actively involved in cooperating with U.S. industry in the development of a first generation system, and in the development of the technology for future more advanced land-mobile systems. Studies and/or developments are underway on very large antennas (up to 55 meters in diameter), car-top antennas, and narrow-band digital voice technology.

Of a more general and perhaps more far-reaching nature, NASA
has a continuing study and technology development programme
for the assembly, test and servicing of large platforms in
geostationary orbit. Such platforms could, for example, be
used to provide a Direct Satellite Sound Broadcasting service, which would require much higher transmitter power
levels, and much larger antennas, than the satellites used
today for fixed, mobile, and TV broadcasting services.

NASA is also an active participant in the international
SARSAT/COSPAR emergency rescue programme, with its major
partners being Canada, France and the USSR. This programme
uses satellites to detect and determine the location of
emergency beacon transmissions from aircraft, boats and even
land vehicles. NASA has provided satellites, search and
rescue antennas, ground terminals and mission control
centres for the joint programme, and is continuing technology developments to improve the system in the future.

JAPANESE Communication Satellite Technology Programmes

History

Japan was among the first countries to become involved in
space activities, beginning in 1955 with a Research and
Development programme on sounding rockets carried out by
Tokyo University to prepare for Japan's participation in the
1957/1958 International Geophysical Year. In 1960, a permanent advisory body was set up reporting to the Prime
Minister to make recommendations on national policy on the
development and use of space science and technology. In
1969, this advisory body became the Space Activities
Commission, responsible for establishing national policy in
the space field, and co-ordinating all national space
activities. In the same year, the National Space Development
Agency, NASDA, was set up to develop space technology.

NASDA's approach to developing space technology and gaining
experience in building and operating satellites and launch
vehicles was a conservative one, which relied upon initially
buying the majority of the technology and expertise that was
needed from the United States, and only gradually developing
Japan's own technology. That this has proved to be a cost-
effective means of developing an autonomous capability is
demonstrated by Figure 7, which shows the growth trend over
the years of the Japanese space budget. It shows that
Japanese capability has been achieved with what was, at the
beginning, only a relatively modest expenditure.

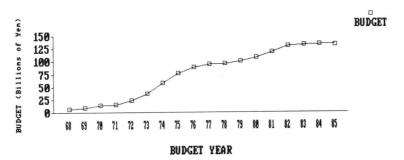

Fig. 7   Japanese space budget (155 yen is approximately US $ 1).

NASDA's charter includes the development of the technology for application satellites and launchers (another organisation, the Institute of Space and Astronautical Science, ISAS, is responsible for scientific satellites and their launchers) and for demonstrating these in orbit. NASDA is also responsible for developing operational application satellites.

In the communication satellite field, NASDA has developed and, since 1975, launched a number of satellites. These include four in the "Engineering Test Satellite (ETS) series", two Experimental Communications Satellites (ECS and ECS-b, both unfortunately lost due to launcher problems), a pre-operational and two operational Communications Satellites (CS, and CS-2a & b), and an experimental and two operational direct broadcast satellites (BSE, and BS-2a and 2b).

The Present and the Future

JAPAN's present communications technology programme includes developments leading toward two more ETS series satellites, ETS-V to be launched in 1987, and ETS-VI to be launched in 1992. In addition, second generation operational communication satellite (CS-3a and 3b) launches are planned in 1987/1988, and second generation operational direct broadcasting satellite (BS-3a & 3b) launches in 1990/1991. In addition, there are plans for the development of a much larger and more advanced Experimental Broadcasting Satellite, EBS, and an operational Japanese Data Relay and Tracking Satellite System (DRTS), both to be put into orbit in the mid-1990s.

    ETS-VI will carry an experimental aeronautical and maritime communications payload, as well as demonstrate

Japanese capability in the development of three-axis stabilised satellite bus technology. ETS-VI will be a bus design
in the two ton class (in GEO), using highly advanced technology in a number of sub-system areas (electro-thermal
hydrazine and zenon gas ion thrusters, advanced heat pipe
designs, ultra-thin solar cells, etc.). It will carry a
number of different communications payloads, including
multi-beam antennas for both fixed and mobile communications, and both S- and Ka-Band payloads to experiment with
the satellite inter-orbit communications in preparation for
the later DRTS system.

SUMMARY - DIRECTIONS AND GOALS
FOR TECHNOLOGY DEVELOPMENTS

The three programmes described above, those of ESA, NASA and
NASDA, are not, of course, the end of the story. Many other
countries and organisations around the world are also developing technology for communication satellites.In particular,
one could mention the programmes of the U.S. Military, the
budget for which exceeds that of NASA in this area by a considerable margin, those of other nations such as France
(CNES), and India (ISRO), where major development programmes
are underway, and developments in the USSR. These have not
been treated herein only due to space and time limitations,
or, particulaly with regard to the programmes of the USSR,
because the author has had only limited knowledge of the
programmes. Nevertheless, from the ongoing developments of
which one is aware, it appears that the present and future
emphasis on communication satellite development in different
organisations shows a number of similarities.

ESA Programmes

ESA has a broad based Programme covering all of the
conventional fixed, mobile, broadcasting, and data relay
services. Its emphasis, in terms of funding to be devoted to
future developments is, however, to a certain extent concentrated in two areas. The first area is associated with
inter-orbit (i.e., data relay) and intersatellite (i.e.,
satellite inter-connectivity) links and systems. The second
area is in the mobile communications area, where ESA is
building on its experience gained in the maritime field with
its MARECS satellites to develop the technology for aeronautical and land mobile services, while at the same time
developing technology for future generation maritime satellite applications.

NASA Programmes

NASA, with its ACTS programme, is concentrating on future fixed-service applications, with on-board processing and optical inter-satellite links being the means they expect will improve the performance and competitivity of satellites in this area. It is also developing technology for future land mobile applications, which could have an increasing emphasis in the future. Although not mentioned in the foregoing explanation of NASA programmes, it is also known that technology developments are underway in the data relay area, for future TDRS application, but little has been published by NASA of work in this area, presumably because TDRS is considered by the U.S. as a "National Resource".

NASDA Programmes

NASDA initially concentrated on fixed-service and broadcast satellite technology development, but, while continuing developments in these areas, is now expanding its development programme in a major way to include both mobile and data relay technology. Its industry has taken an increasingly larger and larger role in developing technology in all of these areas, and is coming close to the time when it will become a viable competitor in the world market for satellite systems.

Channel Computation for Multibeam Frequency Re-Use Maritime Mobile Satellites

N. Sultan* and P. J. Wood†

Canadian Astronautics Ltd., Ottawa, Canada

Abstract

A simple method to convert the number of users per unit coverage area or cell to a "true" number of channels per cell, has been presented for a multibeam frequency re-use, mobile satellite system. Using an antenna beam configuration and interbeam isolation characteristics, all the cell channels, initially assumed, within any one beam were "pooled" together. From this, a reduction factor based on Erlang pooling was derived to obtain the true number of channels. Both traffic distribution and beam shapes were assumed to be non-uniform.

This was applied to a three satellite global maritime and aeronautical mobile system, with an overlapping coverage area between two oceans. The traffic projection within the overlap area was varied such that all users could use one or the other satellite.

The results demonstrate that such overlap traffic sharing between the busiest two oceans plays a major role in determining the bandwidth required or the ultimate number of users that a mobile satellite system can serve with an available bandwidth.

Nomenclature

BW = Total RF Bandwidth Required per Ocean
Erl = Erlangs
FR = Frequency re-use factor
N   = True number of channels per ocean
N_{ac} = Number of allocated channels per beam

Copyright © 1987 by N. Sultan. Published by the American Institute of Aeronautics and Astronautics, Inc. with permission.
*Senior Staff Scientist.
†Senior Antenna Engineer.

n_{ac} = Number of allocated channels per 15
N_{bw} = Number of bandwidth channels per ocean
N_{tc} = Number of true channels per beam
n_{tc} = Number of true channels per 15° x 15° cell
N_u = Number of users per 15° x 15° cell
n = Percentage of overlap traffic sharing
r = Beam pooling factor

Introduction

A simple method to convert the number of users per unit coverage area or cell, to the "true" number of channels required per cell is presented, for a multibeam frequency re-use mobile satellite system. First a number of channels per cell is assumed as if those cells were separate beams, then the channels and Erlangs of all constituent cells are pooled into one beam. A reduction factor is derived to obtain the true number of channels per cell.

This is applied to a three satellite global maritime and aeronautical mobile system, with an overlapping coverage area between two oceans. The overlap traffic sharing between them is varied from 0 to 100%. The true number of channels are derived, together with the antenna Frequency re-use factor. From these, the RF bandwidth variation with overlap sharing is investigated to determine the optimum value of L-Band bandwidth and percent overlap traffic sharing.

The results demonstrate that such overlap traffic sharing between the busiest two oceans plays a major role in determining the bandwidth required or the ultimate number of users that a mobile satellite system can serve with an available bandwidth.

Consider the three geostationary satellites in Fig 1 with an L-band maritime and aeronautical mobile global coverage system. Multiple beam and frequency re-use are assumed in order to:

• increase the antenna gain and hence reduce the payload power requirements

• reduce the bandwidth needed at L-band, a rare and valuable resource to be shared between nations and allocated by the Geneva International Union WARC Conference.

Presented here are selected design considerations for a multibeam mobile satellite system. These include the Frequency Re-use capability of the antenna, the available bandwidth, and the total true number of channels required per ocean, N.

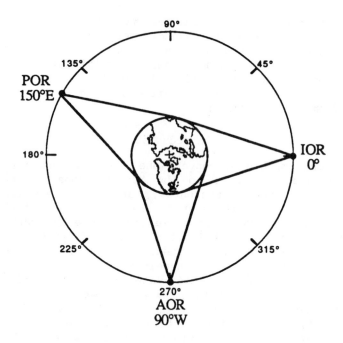

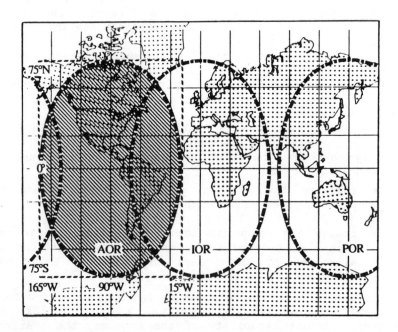

Fig. 1   Example of a global coverage by three satellites for a maritime and aeronautical mobile satellite system.

At the planning stage of such a system the only practical way of projecting future traffic requirements is to "allocate" a certain number of Erlangs or alternatively vessels and aeroplanes, here referred to as "users", n_u, at specific locations on the globe. These user locations can be defined, for example, every 15° Latitude x 15° Longitude cell, as shown in Fig 1, where the Atlantic Ocean Region (AOR) coverage is idealized to a 10 x 10 cell matrix.

The translation of n_u users or Erlangs into a true number of channels, n_{tc}, is not a straightforward process in a multiple beam system where neither the beam shapes not the traffic distribution are uniform. Normally, the Erlangs per cell can be translated into channels per cell using the Erlang Tables[1]. Any one beam covers a number of cells. It might appear that the total number of channels required per beam is simply the sum of the channels per cell within that beam, N_{ac}. However, pooling within one beam will provide more channels than are statistically required. Hence, some method needs to be developed to calculate the "true" number of channels per beam, N_{tc} (which is always less than N_{ac}).

What follows is a simple and practical methodology to evaluate the "true" number of channels needed from a given set of user and traffic requirements. It then proceeds to apply this methodology to a multiple beam reflector antenna with frequency re-use capabilities, based on an assumed isolation, to minimize the required RF bandwidth. Finally, the whole process is applied for a parametric study on the effect of traffic sharing in an overlap area between two ocean satellites.

Traffic and Beam Configurations

The conversion of a given number of users, n_u, to traffic intensity in Erlangs, Erl, is simply related to the statistical characteristics of a user's expected habit of engaging a channel for a number of minutes per day, at the peak hour. In this study, .0175 Erl/user is assumed. Fig 2 shows an example traffic intensity per 15° x 15° cell in the Atlantic Ocean Region (AOR) coverage, based of a total of 267 Erl or 15,200 users, projected for the End of Life of an AOR satellite, and an overall total of 600 Erlangs for the whole globe.

If each 15° x 15° cell was covered by a single beam then the translation of Erlangs per cell into a nominal or "allocated" number of channels, n_{ac}, would be straightforward. This is achieved using Erlang B statistical tables, for a specific quality of traffic or grade of service e.g. 2% "blocking rate" (2% of all calls cannot be set up at the peak hour). Such a translation of Fig 2 is shown in Fig 3. For example, in cell row 2, column 3, 27.8 Erlangs corresponds to 36 channels required in an individual cell. The values in Fig 3 represent the initially allocated channels per cell. They are not the true number of channels required since it is initially assumed that there is no sharing or pooling of traffic between cells before the beam configuration is considered.

The traffic requirement can only be projected within a unit area, in this case 15° x 15° cells. The grouping or pooling together of the channels in these cells is directly related to the beam configuration which is a function of the specific antenna design considered. It is this pooling that will lead to the correction of n_{ac} to n_{tc} i.e. of allocated to true number of channels.

In Fig 4, a 2.0 meter offset reflector antenna at L-band, (1.55 GHz) is designed with 6 beams, 4 transponders and 7 feeds. The beam configuration shown on the globe highlights a mix of non-uniform spot and shaped beams, intentionally designed to correspond to the heavy and light traffic regions respectively. This tends to average and

0.0	0.1	4.8	3.5	0.1	0.0	0.5	0.1	0.1	0.0
0.1	0.3	27.8	24.6	1.5	1.4	24.2	0.5	0.2	0.0
2.0	3.7	6.0	5.7	3.6	2.5	3.0	18.2	9.9	0.0
4.1	0.0	0.0	0.0	3.7	0.8	1.6	10.0	17.8	30.6
0.4	0.0	3.0	2.2	3.1	1.0	1.2	8.2	8.5	2.6
0.3	0.0	4.2	0.4	0.6	4.4	0.1	0.0	1.1	1.0
0.3	0.0	0.1	0.1	0.4	3.0	1.0	0.1	0.1	0.2
0.2	0.4	0.2	0.1	0.2	0.2	1.1	0.4	0.0	0.0
0.0	0.0	0.0	0.0	0.1	0.1	0.5	0.2	0.1	0.0
0.3	0.0	0.0	0.0	0.4	0.0	0.4	1.2	0.0	0.0

75°N (top-left) ... 0° ... 75°S (bottom-left); 165°w ... 90°W ... 15°w

Fig. 2 Example of Erlang distribution matrix, over Atlantic ocean, defined by a 10 x 10 matrix of 15° Latitude x 15° Longitude cells, highlighted in Fig. 1.

0	2	9	8	2	1	3	2	2	0
1	2	36	33	5	5	33	3	2	1
6	8	12	11	8	6	7	26	16	1
8	1	1	1	8	4	5	16	25	40
3	1	7	6	7	4	5	14	15	7
3	1	8	3	3	8	1	1	4	4
2	1	2	2	3	7	4	2	1	2
2	3	2	1	2	2	4	3	1	1
1	1	1	1	1	2	3	2	1	1
0	1	1	1	3	1	3	5	1	0

Fig. 3   Initially allocated channel distribution matrix.

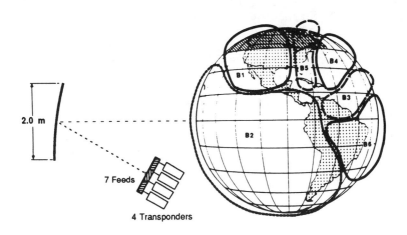

Fig. 4   Two meter offset reflector antenna design requiring 4 transponders and 7 feeds to produce 6 shaped and spot beams with frequency re-use. High gain spot beams for dense traffic, medium gain shaped beams for low density traffic. Globe shows AOR coverage at 90° W.

power per beam and hence potentially the power per
transponder amplifier, making more efficient use of the High
Power amplifiers.

Several other antenna designs and beam configurations
may be considered in order to optimize the mobile satellite
system channel capacity.[2]

Methodology for Computation of True Channels

The beam configuration of Fig 4 is shown on a 10 x 10
cell matrix in Fig 5 where the initially allocated channels
per cell, shown in Fig 3, are added up within each beam.

The following four steps outline the methodology
proposed for computing the true number of channels in a
non-uniform traffic and multibeam satellite system.

A – Deduce the initially allocated channel distribution
matrix of Fig 3, from the Erlang distribution matrix of Fig
2, using Erlang tables[1].

B – Once the beam configuration is determined, Fig 4, sum
up the initially allocated channels per cell in Fig 3 within
each beam as shown in Fig 5. For example, 153 user channels
are allocated to beam B1.

However, this cannot be a "true number of channels"
since all 153 channels, when pooled together within a single
beam, would correspond to about 143 Erlangs (2% Blocking
rate). The true traffic intensity in beam B1 adds up,
according to Fig 2 to 87.6 and not 143 Erlangs. This
discrepancy can be resolved in the next step.

C – For each beam, reduce the initially allocated number
of channels by a "ratio" or "beam pooling factor" r, defined
here as:

$$r = \frac{\text{Sum of true Erlangs per beam}}{\text{Erlangs due to sum of allocated channels per beam}}$$

$$r = \frac{87.6}{153} = 0.60, \text{ for Beam B1}$$

$$N_{tc} = 0.6 \times 153 = 93 \text{ channels}$$

This means that the true number of channels N_{tc} required
within B1 has to be reduced by the "ratio" r from 153 to 93
as shown in Fig 6, because less channels are required when 6
cells are pooled together.

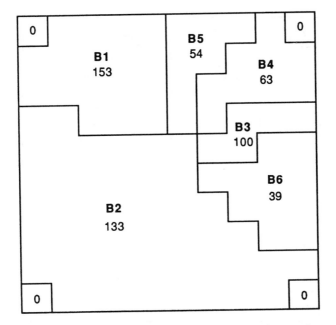

Fig. 5   Initially allocated number of channels per beam.

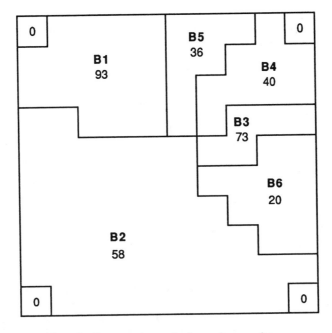

Fig. 6   True number of channels per beam.

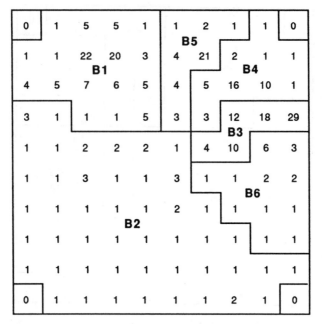

Fig. 7   True number of channels per cell.

D – Finally, the true number of channels per cell in Fig 7 is computed by multiplying the allocated channels of Fig 3 by the "beam pooling factor" r.

Table 1 lists the initially allocated channels, the true number of channels, and the beam pooling factor (referred to as "ratio") for each beam. Software was created for all the above computations.

Fig 6 and Fig 7 provide the distribution of true number of channels per beam and per cell respectively, which will be used as the basis for the computation and optimization of such parameters as:

• Frequency Re-use factor
• RF Bandwidth
• Transponder power
• Traffic Intensity
• Ocean overlap traffic

Application for Parametric Investigation

So far, a methodology was outlined for deriving n_{tc} and N_{tc}, the true number of channels per cell and per beam, from

Table 1 Initially located channels, true number of channels, and beam pooling factor for each beam.

Beam Number	Erlang Sum	Initially allocated Channels (1)	True Number of Channels (2)	Ratio
1	87.55	153	93	0.61
2	34.65	133	58	0.44
3	67.74	100	73	0.73
4	33.54	63	40	0.63
5	29.43	54	36	0.67
6	13.59	39	20	0.51
Total:	266.5	542	N = 320	0.59

Total Global Erlang = 600
Total Atlantic Erlang = 266.5

(1) Number of Channels computed on the basis of independent 15° X 15° cells or beamlets (96 cells or beamlets)
(2) True Number of Channels based on beam configuration used in antenna modelling (6 beams).

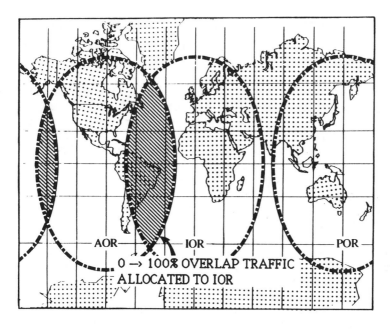

Fig. 8 Geographical coverage by mobile satellite system showing two overlap regions where users have the choice of two satellites.

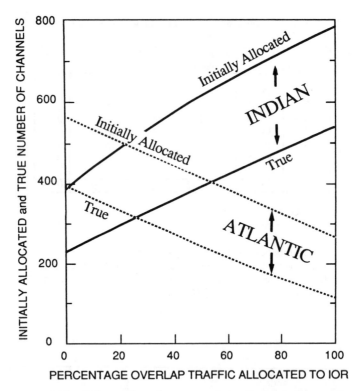

Fig. 9 Initially allocated and true number of channels vs. Overlap
Indian Sharing.

a given projected number and distribution of mobile users as
in Fig 2. Such a traffic projection is a function of the
way vessels and aeroplanes choose to operate with one or the
other satellite in the overlap regions. The ratio η of the
IOR/AOR overlap traffic was varied between 0 to 100%. Fig 9
shows the variation of the allocated and true number of
channels with percentage overlap η. The percentage of
IOR/AOR overlap is fixed at 33%/67% in Table 1 and Figs 2,
3, 4, 5, 6, and 7. Throughout this study, the percentage of
POR/AOR overlap has been fixed at 100%/0%.
     If a frequency re-use multiple beam system is
considered, as is the case here, then the frequency re-use
factor FR needs to be derived before the true RF bandwidth
BW can be ultimately deduced.
     Software was developed to compute the frequency re-use
factor for several antenna designs, based on a minimum of 20
dB isolation between any two user locations on any two
beams, whether "adjacent" or distant. Thus, "partial" or
"full beam" frequency re-use were investigated. In this

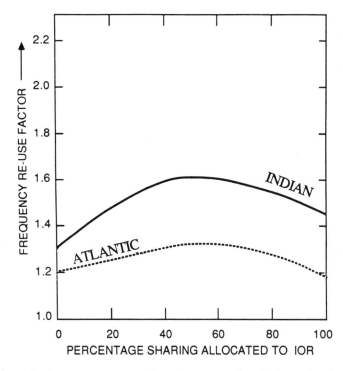

Fig. 10   Frequency re-use factor vs. overlap Indian sharing.

paper, the antenna considered has the six beams shown in Fig 4 and uses only partial beam frequency re-use. The antenna gain, patterns and sidelobes were computed in a program that identified the isolation between two points on the coverage. For a minimum of 20 dB isolation at the centre of any two cells, the program algorithms allocates the same channel frequency to both cells.

The results of Fig 10 indicate that the frequency re-use factor for both IOR and AOR peaks at an overlap traffic sharing of 50%.

At first a maximum FR of 1.6 may seem to be too low a value. This is due to the beams being adjacent and also the uneven traffic distribution across the coverage area. For the same number of beams, computations have shown that an FR of 1.6 may be equivalent to an FR of about 3.0 for a uniformly distributed traffic. Therefore, it is the best attainable value in a highly uneven traffic distribution.

In general, it is desirable to increase the frequency re-use factor, but it is more important to optimize the ultimate requirement of a Frequency Re-use system: namely the total RF bandwidth required.

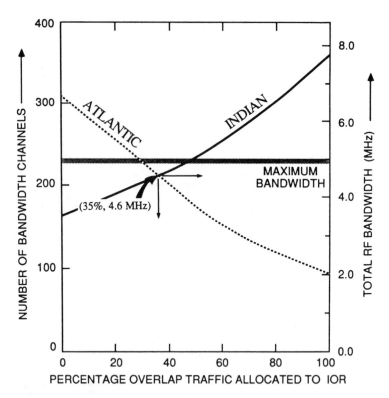

Fig. 11   RF bandwidth and bandwidth channels vs. overlap Indian sharing.

Based on an assumed 20 KHz per channel or 21.7 kHz with an 8% filter guard band, each beam occupied several sub-bands that can be shared in parts by another beam or beams, thus accounting for the system frequency re-use capabilities. The total number of "bandwidth channels" per ocean, N_{bw}, is given by:

$$N_{bw} = \frac{N}{FR}$$

The total RF bandwidths needed for any ocean coverage is then:

$$BW = N_{bw} \times 21.7 \ (KHz) = \frac{N}{FR} \times 21.7 \ (KHz)$$

The number of bandwidth channels and total RF bandwidth required are plotted, in Fig 11, against the overlap traffic sharing.   For example, for η = 0%, the IOR and AOR

satellites require bandwidths of 4.0 and 6.5 MHz,
respectively. The most efficient use of the spectrum would
ideally require that the bandwidth of the busiest two
satellites, i.e. the AOR and IOR, to be equal. Such an
optimum value, namely 4.6 MHz was found to correspond to an
overlap traffic sharing of 35%.
All conditions so far are for a fixed total global
traffic of 600 Erlangs at 0.175 Erlang per user i.e. 34,000
users. The optimum bandwidth required was found to be 4.6
MHz. On the other hand, it is more realistic to assume from
the start some value for the available RF bandwidth, say 5
MHz. In this case, to a first approximation, it is safe to
assume the optimum overlap traffic sharing will still be 35%
and the global traffic possible would be proportional to the
bandwidth i.e. an optimum of 37,000 users is possible with a
5 MHz bandwidth.

                        Conclusion

    A simple method to convert the number of users per unit
coverage area or cell to a "true" number of channels per
cell, has been presented for a multibeam frequency re-use,
mobile satellite system. Using the results of an adopted
antenna design, particularly the beam configuration and the
interbeam isolation characteristics, all the cell channels,
initially assumed, within any one beam were "pooled"
together. From this, a reduction factor based on Erlang
pooling was derived to obtain the true number of channels.
Both traffic distribution and beam shapes were assumed to be
non-uniform.
    This was applied to a three satellite global maritime
and aeronautical mobile system, with an overlapping coverage
area between two oceans, typically the Indian and Pacific.
The traffic projection within the overlap area was varied
between two extremes, whereby all users could use one or the
other satellite. The true number of channels and the
frequency re-use factor derived from the antenna design were
combined to obtain the RF bandwidth required by one
satellite at L-Band.
    The ultimate design consideration of a multiple beam
frequency re-use mobile communication satellite system is to
reduce the RF bandwidth required for a projected traffic
intensity and distribution. In practice, a projection may
be made for the relative traffic distribution but not for
the exact traffic intensity per unit area or cell.
Therefore, a mobile satellite system is better optimized on
the basis of a known value for the RF bandwidth available.
It was shown that a 2 metre offset reflector antenna, with

six spot and shaped beams, using 4 transponders and seven
feeds, can serve 37,000 users over the globe with only 5 MHz
of available bandwidth.

Acknowledgment

This paper is based upon work performed under the
sponsorship of the International Maritime Satellite
Organization (INMARSAT) Any views expressed here are not
necessarily those of INMARSAT.
The authors are grateful to Ms. Janice Neufeld for her
careful typing and David Viljoen for the graphics and
preparation of this manuscript.

References

[1] Dill G.D. and Gordon G.D., "Efficient Computation of Erlang Loss
Functions", Comsat Technical Review, Vol. 8, No. 2, pp 353-370,
Fall 1978.

[2] Sultan, N. and Ng, P., "Capacity Optimization for a Spot Beam
Advanced Mobile Satellite Systems, b"37th Congress of the
International Astronautical Federation IAF '86 Conference, paper
IAF-86-339, Innsbruck, Austria, October 4-11, 1986.

Bibliography

Ghais, A.F. "Future Development of the INMARSAT System", AIAA 10th
Communication Satellite Systems Conference, paper 0750, pp.
440-449, Orlando, FA, March 1984.

Rogard, R. "A Land-Mobile Satellite System for Digital Communication
in Europe", Land-Mobile Services by Satellite: ESA Workshop
Proceedings, ESA-ESTEC, The Netherlands, pp. 125-130, June 1986.

Zuliani, M.J., Weese, D.E. & Sward, D.J. "M-Sat - A New Dimension in
Satellite Communications", IAF 37th International Astronautical
Conference, Innsbruck, Austria, paper 339, October 1986.

Dachert, F., de Montivault, J.L., & Coirault R., "Optimization of a
High Capacity Communication Satllite for Europe", AIAA 10th
Communication Satellite Systems Conference, paper 0698, pp.
664-672, Orlando, FA, March 1984.

Sultan, N., Payne, W.F., & Carter, D.R., "Novel Approach to
Optimization of Communication Payload for High Capacity Mobile
Satellites", IEE 3rd International Conference on Satellite Systems
for Mobile Communications & Navigation, IEE Publication No. 222, pp.
60-64, London, U.K., June 1983.

McNally, J.L. & Breithaupt, R.W., "Mobile Satellite Systems: A Review", IAF 37th International Astronautical Conference, Inssbruck, Austria, paper 339, October 1986.

Sultan N., Ng P., "Multibeam Frequency Re-Use Mobile Satellite System Trade-offs", Journal Earth Oriented Applications of Space Technologies (EAST), 1987.

Sultan N., "Planning of Advanced Maritime and Aeronautical Mobile Satellite Systems with Multiple Frequency Re-Use", IAF 38th International Aeronautical Conference, Brighton, England, Oct. 1987.

Sultan N. and Wood P.J., "Adaptive Sub-bands Channelization Concept: Solution to Reconfigurability of Multibeam Frequency Re-Use Maritime Mobile Satellites", to be published in AIAA 124 International Communication Satellite Systems Conference", Crystal City, Virginia, March 1988.

Dutzi, E.J. & Nader, F., "Mobile Satellite Communication Technology: A Summary of NASA Activities", IAF 37th International Astronautical Conference, Innsbruck, Austria, paper 339, October 1986.

The Distress Radio Call System and First Results of the Preoperational Demonstration

Walter Goebel*
Deutsche Forschungs und Versuchsanstalt für Luft und Raumfahrt (DFVLR), Oberpfaffenhofen, Federal Republic of Germany

Abstract

Presently two different satellite systems are offered for distress service: The DISTRESS RADIO CALL SYSTEM (DRCS) which works through existing INMARSAT satellites (geostationary) and the COSPAS-SARSAT system which works through polar orbiting NOAA or COSPAS satellites.

While COSPAS-SARSAT allows an independent position determination due to the Doppler effect, the Distress Radio Call System needs an introduction and update of the position information from the navigation instruments (TRANSIT, LORAN or GPS etc.). On the other hand the Distress Radio Call System allows the transmission of a message within a few minutes whilst in the COSPAS-SARSAT system the message transfer is subjected to a delay of one to four hours or more.

As an international cooperation between Japan, Norway, United Kingdom, United States, USSR and FR Germany different techniques were tested for the geostationary system. From a point close to the North Cape the FR Germany system demonstrated a successful message transfer at a transmitted power of 50 mw only. CCIR recommended this system specification for use with geostationary stellites. Up from november 1986 eleven FR Germany EPIRBs are internationally and quasioperationally tested within the INMARSAT Preoperational Demonstration. First results show that the satellite coverage can easily be extended down to zero degrees due to the high margin of 10 dB inherent in the system.

To make the geostationary system applicable for small ships i.e. fishing vessels or yachts, a smaller version is now being developed and will participate in the Preoperational Demonstration during the second half of the year 1987.

1. Background

Mobile radio equipment for maritime distress alerts presently used on a terrestrial basis are subjected to a limited range by propagation effects. This is due to natural conditions such as earth curvature or antenna efficiency problems, ionospheric conditions in case of short wave propagation and in any case due to the low available DC power. Thus a successful distress message transfer depends more or less on the distance from a coastal station or a passing vessel.

Geostatic satellites cover more than a third of the total earth (**Fig. 1**). Three satellites are able to cover all three oceans and only small areas around the poles are excluded. The use of such satellites, which since 1982 are operated by the INMARSAT organization, could provide an immediate message transfer.

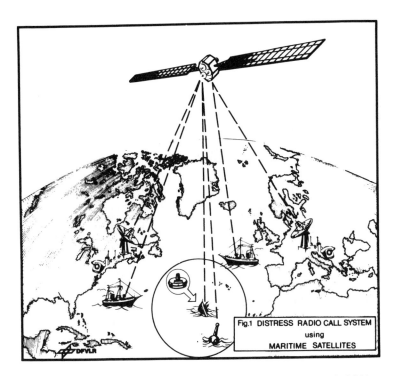

Fig. 1   DISTRESS RADIO CALL SYSTEM using maritime satellites

Presently two different satellite systems are offered for distress service: The DISTRESS RADIO CALL SYSTEM (DRCS) which works through existing INMARSAT satellites as mentioned above and the COSPAS-SARSAT system which works through polar orbiting NOAA or COSPAS satellites (Fig. 2).

The difference to the DRCS is that the moving satellites relative to the earth surface allow a position determination due to the Doppler effect. This is a great advantage but has to be compared with the missing continuous coverage over the world. If the Emergency Position Indicating Radio Beacon (EPIRB) is transmitting, it could take a delay time of hours until a satellite appears over the horizon. For higher latitudes at polar regions the so called "waiting time" may be in the order of one hour or less, at equatorial regions it may be between 2 and 4 hours. This waiting time is further increased by one to two hours if no Local User Terminal (LUT) is visible by the satellite at the time of message transfer, then the position of the EPIRB calculated by the satellite is stored and dumped to the receiving station as it comes into coverage of the satellite. The southern hemisphere of the world is not covered by LUTs until now. There are plans to reduce this second part of the delay using intersatellite links.

The DRCS, however, provides a continuous access to the geostationary satellites. Because of a missing Doppler ef-

Fig. 2 COSPAS-SARSAT system of polar orbiting satellites

fect the position information has to be introduced into the
EPIRB continuously by the navigation equipment or in intervals by hand.

The present status of the two systems is that the International Maritime Organization (IMO) gave preference to
the existing and operating COSPAS-SARSAT system, but invited
Governments to continue their participation in the presently
running Preoperational Demonstration with the view of becoming an alternative.

INMARSAT has estabished the Preopcrational Demonstration in 1986 and it will continue over the year 1987. To
support this demo the Federal Republic of Germany provided
11 EPIRBs free of charge for international testing (Fig. 3).

2. Roles of developing countries

Safety at sea is not only a matter of industrial countries. A reliable and rapid distress alert system helps saving men and is a humanitarian need for sailormen in a world
of progress in communication technology. Developing countries could participate in this demonstration with one of
the offered EPIRBs on their own vessels. They also could
develop EPIRBs themselves, the specifications under the
headline "transmission characteristics" are defined by CCIR
and shown in Table 1.

Fig. 3 New generation EPIRBs with cradle and Data Entry Device

Table I: Brief System specifications[1]

transmit frequency	1644.3 - 1644.5 MHz[+]
	1645.5 - 1646.5 MHz[++]
transmit power	1 Watt nominal
antenna gain	0 dBi
modulation	FSK
deviation	$\pm$ 120 Hz
frame length - Data	100 bits
- Sync	20 bits
- FEC	40 bits (BCH)
code	NRZ-L
modulation rate	32 baud
transmission duration	4 x 10 min

[+] INMARSAT first generation space segment
[++]INMARSAT second generation space segment

The demonstration is running within the Atlantic Ocean Area with the receiving processor (Fig. 4) positioned at the Coast Earth Station Goonhilly/England.

More developed countries especially those having an INMARSAT Coast Earth Station (CES) could develop a receiver processor which can easily be connected to an earth station at the intermediate frequency level 70 MHz. Having additional receiver processors the area of demonstration could be extended to the Indian and the Pacific Ocean. Both equipments are free of any patents claim for development, but can also be purchased from the manufactorer Dornier-System.

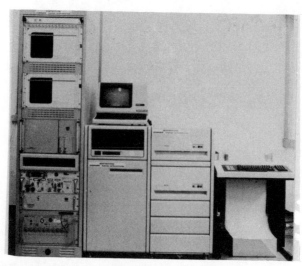

Fig. 4 Present generation of receiver processor

3. The Distress Radio Call System

In the following the system and the used technique will be described in more details. It is obvious that a slant range of about 40.000 km, with a transmitting antenna (0 dB gain) of an ommidirectional characteristic is a problem area with regard to the link budget. Furthermore deep multipath fades of more than 10 dB occur due to reflections from the sea surface or total drop-outs due to signal blockage caused by high waves at low elevation angles to the satellite. Fig. 5 shows a test near the North Cape.

To cope with the unsatisfactory link the solution was to use more time for transmission. The data message therefore is transmitted not only once but several times, up to hundred times, one data frame after the other, continuously (time diversity). For rapid acquisition a non-coherent FSK modulation at a bitrate of 32 B/s is used.

After being relayed by the satellite the appropriate frequency band is received by the coast earth station, downcoverted to audio frequencies (Fig. 6) and then sampled and digitized. For acquisation and demodulation a filterbank of a bandwidth each filter close to the bitrate is used.

After identification of a signal the filter output is put in a memory. Each message is then superimposed upon the previous. The law of physics says that the noise is added with the square root of the number of frames while the signal is linearly added. Thus for every factor 2 in the number of frames an improvement of 3 dB in signal-to-noise-density

Fig. 5 EPIRB testing near the North Cape on research vessel GAUSS

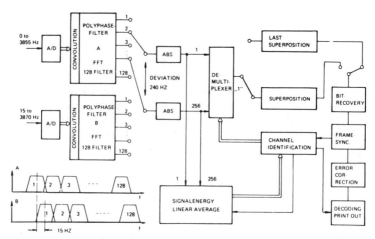

Fig. 6   Receiver processor block diagram

ratio is obtainable, in practice an improvement of totally 14 dB is obtained. Fig. 7 shows the procedure for a 10-bit-frame. The signal improvement process is made in the memory by the superposition technique. The integration can start on any sample of the frame not knowing yet the bit change or the starting point of the frame, if the framelength is a priori known very exactly. For that on both sides of the link a 10^{-6} crystal oscillator is adequate.

Thus time diversity technique has two benefits: One is a way of using battery power more efficiently, because of a lower DC-current load. The other is the extreme favourable characteristic that the fluctuations of a faded signal are totally levelled. Even complete dropouts are compensated by integrating the received fractions of a message, which are assembled like a puzzle.

4. The Coordinated Trials Program

Under the auspices of the International Radio Consultative Committee (CCIR) six countries decided to run common tests with EPIRBs in 1982 and 1983 to compare their technology: Japan, Norway, United Kingdom, United States, Union of Soviet Socialist Republics and the Federal Republic of Germany. This was an outstanding example of international cooperation.

In a first step the six different systems were subjected to a simulator at controlled fading conditions. For those simulation tests the transmitting and receiving equipments were installed at the ESA Tracking Station Villa Franca near Madrid (Fig. 8). In a second step concurrent sea trials were

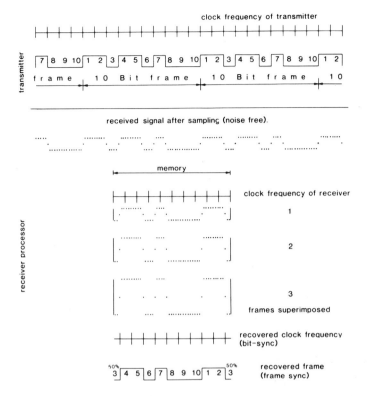

Fig. 7 Superposition technique, demonstrated for a 10-bit-frame

conducted using the German Research Vessel GAUSS which cruised from Edingburgh/UK to the North Cape along the Norwegian Coast. The receiving equipment was again at Villa Franca.

At the most northern point the EPIRB of Federal Republic of Germany transmitted a message of 50 milliwatt omnidirectionally which was sucessfully received at Villa Franca. The elevation angle at that site was 1.6° and the wave height 3 m. The transmitted message was 100 bit. Results have been accepted only if a 99 % reliability of error free message transfer could be demonstrated according to the decision of IMO.

After evaluation of the results it was concluded that a modified version of the FR Germany narrowband FSK system would provide an adequate basis for the CCIR Recommendation for a geostationary satellite EPIRB system operating at 1.6 GHz.

In laboratory tests the sensitivity of the system was measured to be 13 dB-Hz in the GAUSS-channel and 16 dB-Hz

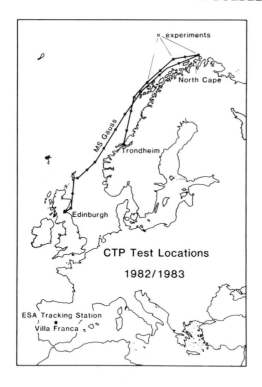

Fig. 8 Test site of the
Coordinated Trials Program
(CTP)

in a severe fading-channel (Fig. 9 a+b). Transmitting under
operational conditions a margin of 10 dB is the safety factor of the system.

5. First results of the Preoperational Demonstration

On the German research vessel GAUSS (Fig. 10) an EPIRP
of the new generation was installed on the 11th November
1986. Until the 27th November 40 tests were reported from
the ship, 39 of them were received with high signal level
and within a Message Transfer Time of maximum 1.9 min, the
mean was 1.27. One message was not even received spuriously,
therefore an operational failure is assumed.

By the end of January 1987 there were six ships
equipped with 1.6 GHz satellite EPIRBs:

RV GAUSS   F.R. of Germany
RV METEOR   F.R. of Germany
MV HUMBOLDT EXPRESS   F.R. of Germany
MV UBENA   F.R. of Germany
MV NORGE   Norway
TV EUGENIOS EUGENIDES   Greece

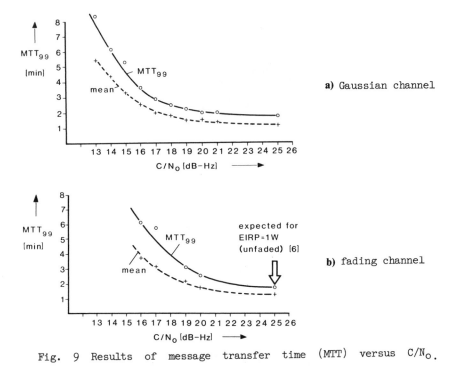

a) Gaussian channel

b) fading channel

Fig. 9 Results of message transfer time (MTT) versus C/N_0.

Fig. 10 Research vessel Gauss

The operational area of the Norwegian ship is Spitsbergen. There were messages received from positions with an elevation angle zero. Thus the operational area is much wider than the specified INMARSAT coverage area limited by an elevation angle of 5°. The extension down to near zero covers large areas even beyond shipping routes at the North Pole and South Pole.

Fig. 11 Low Power Distress Transmitter (LPDT)

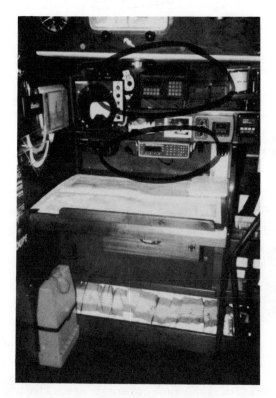

Fig. 12 Application on a yacht in combination of a satellite navigator

Fig. 13 For energizing the LPDT the cable plug of the satellite navigator is replaced by a code plug

6. Low Power Distress Transmitter (LPDT) for small ships

To make the geostationary system applicable for small ships i.e. fishing vessels or yachts, a smaller version is now being developed and will participate in the Preoperational Demonstration in summer 1987 (Fig. 11).

The position information is continuously fed into the LPDT from a satellite navigator by a cable (Fig. 12). In

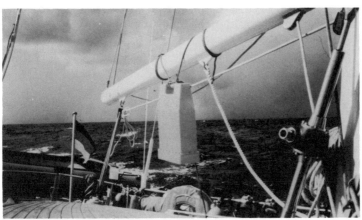

Fig. 14 Example of an operation on a yacht

case of distress the equipment is disconnected from the
cable and switched on by a special coded plug which is
fastened close to the flange socket (Fig. 13). In case of
test another coded plug is used. In the later operational
use it will be sufficient to disconnect the cable for energizing the LPDT. In one of the two "towers" the L-band antenna is positioned, in the other a Search and Rescue Transponder (SART). The equipment is floatable but should normally be used not floating, but on the deck of a ship or in
a life raft etc. (Fig. 14). The SART responds to a radar
pulse transmitting on the same frequency around 9 GHz and
is an important help for homing.

7. Conclusion

The CCIR recommended EPIRB system for the frequency
band 1.6 GHz demonstrated during the Coordinated Trials Program high performance and high reliability. For a transmitter power of only 1 Watt the system margin for extreme conditions would be about 10 dB. It transfers the distress message without delay which is important for the probability
of survival of men at the high sea. It shows a high resis-
tivity against signal fluctuations and wave blockage. In
the form of a Low Power Distress Transmitter it is a hand-
held device for small ship application.
    A very important argument is the fact that no addition-
al satellites or satellite packages and no additional earth
stations are needed. All countries, the developing ones included are invited to participate in the Preoperational Demonstration within the Atlantic Ocean Area. For those countries having a Coast Earth Station, the installation of a
receiver processor which could be developed within about
two years, would extend the test area to the Indian and the
Pacific Ocean, but would need new arrangements with INMARSAT
for that test period.

Reference

[1]CCIR, "Transmission Characteristics of a Satellite Emergency Position Indicating Radio Beacon (Satellite EPIRB) System Operating
through Geostationary Satellites in the 1.6 GHz Band", XVIth
Plenary Assembly Dubrovnik, January 1986.

Intergrated Mobile Services in Selected Areas

C. Carnebianca,* G. Solari,† and A. Tuozzi‡
Italspazio, Rome, Italy

ABSTRACT

Areas with lack of terrestrial infrastructures, would greatly benefit of a satellite-based system providing almost instantaneously a wide variety of mobile applications such as 2-D/3-D navigation, data and voice communications.

The present paper describes satellite-based continuous navigation and communications services on a selected coverage area. These services seem very attractive from the cost/performance view-point and with the potential to be gradually implemented to a global coverage.

The system is based on 4 space elements: 2 navigation packages placed in geostationary orbit on host vehicles and 2 satellites embarking navigation and communications packages in eccentric inclined orbits (EIO).

The expected navigation performances are 28 m (SEP)/ 16.5 m (CEP) for 3-D/2-D position fixing respectively. This feature combined with a low data rate service (paging) can represent a profitable business opportunity to capture a substantial portion of the potential traffic demand.

1. INTRODUCTION

In the frame of the activities performed to assess the economical feasibility for a satellite-based world wide navigation service, ITALSPAZIO has investigated different orbit alternatives.

Considering that the launch cost for such systems turned out to be the predominant factor among the cost

*Manager, Engineering.
†Navigation Engineer, Mission Analysis.
‡Communication Engineer, System Engineering.

elements, the reduction of the number of dedicated satellites was considered a primary goal.

The lesson learned indicated that a satellite constellation based on a combination of geostationary and high eccentric inclined orbits (HEIO) provides benefits which significantly enhance the system economy while keeping high level performances [1]. The major benefits include:
- Multimission suitability due to the nature of the orbits.
- Gradual implementation starting from a reduced constellation (i.e. 5 elements) to start continuous navigation service on a selected area (Fig. 1).
- Reduced number of dedicated satellites (e.g. for a global coverage like GPS, 12 HEIO satellites and 6 navigation packages on-board as many as geo-host satellites).

In this perspective, a highly precise navigation system together with a communications service by satellite would provide an important augmentation of the current service capabilities considering the rapid growth of the mobile user demand.

With regard to the communications aspects in particular, while for industrialized countries the role of a satellite would be complementary to the terrestrial system, for areas with modest infrastructures it could represent the sole mean to provide a mobile communications service capacity.

Therefore a set-up of a reduced satellite-based mobile integrated system is expected to become a profitable investment.

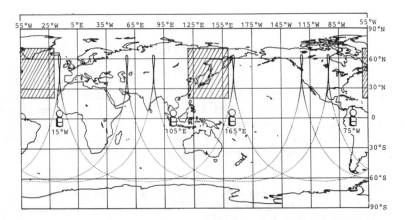

Fig. 1   Coverage zones with a constellation based on 3 HEO + 4 GEO packages

In this scenario a system using 24 hours EIO and GEO orbits seems very effective in providing a general integrated service with the potential to be further expanded.

2. SYSTEM STRUCTURE

The proposed system structure encompasses three major elements (Fig. 2):
- Space segment: it consists of two navigation packages in geostationary orbit and two satellites placed on 24 hours period eccentric inclined orbits.

Each satellite in EIO orbit will carry a navigation and a communications payload, while the geo-navigation packages can be hosted by already planned geostationary communications satellites.

The eccentric orbits, with the nodes lines separated by 180° degrees, feature the same ground-track that, due to the orbital parameters selected (25231/46340 Km of perigee/apogee altitude with 63.45° of inclination), is fixed with the Earth (Fig. 3).
- Ground Segment: it consists of a suitably located master station (MS), augmented by monitoring stations

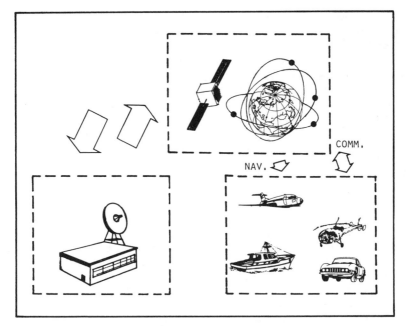

Fig. 2 System structure

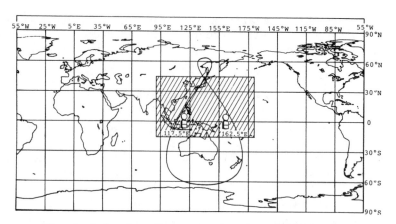

Fig. 3　Reduced configuration based on GEO+EIO orbits

which continuously track all satellites. The MS provides for the overall system control, for navigation and
communications services.
User Segment: it consists of different classes of receivers to support navigation, paging and voice communications services.

3.　NAVIGATION PERFORMANCES

As reference to GPS and other planned global navigation systems (e.g. NAVSAT, GRANAS), the user position
determination is based on passive triangulation technique.
It derives that at least 3 satellites have to be continuously in visibility.

Considering the time synchronization aspects a
fourth satellite has to be added to solve the four unknown: (x,y,z, time).

In this context the navigation receiver measures
the pseudoranges to four satellites, evaluates satellites
position by processing the ephemeris data and estimates
the user position and time.

However, due to cost reasons, the planned systems
tend to keep at the minimum level the number of the constellation satellites, also tolerating limited outages in
coverage areas and in time occurrences.

It is worth recalling that for a good position
fixing the satellites have to be in good geometric relationship with the user. A good geometry is usually expressed by the Position Dilution of Precision (PDOP),
which can be considered proportional to the inverse of
the tetrahedron volume formed by the four satellites and
the user.

Good performances of a navigation system imply low PDOPs.

Outages occur when the PDOPs exceed predetermined thresholds.

To overcome outage occurrences, various aiding techniques (e.g. altimeter, clock) have been envisaged to enhance the navigation performances when less than four satellites are available.

As concerns the number of satellites in visibility in the selected service area using the proposed constellation, Fig. 4 provides a quick look of the general evolution and Fig. 5 the relevant statistics. It can be seen that 4 satellites are at the most in visibility, reduced to 3 only for limited periods.

As anticipated before, 3 or 4 satellites in view, depending on the specific application, is not a sufficient condition to assure good navigation performances, if not substantiated by PDOP statistics for 3-D naviga-

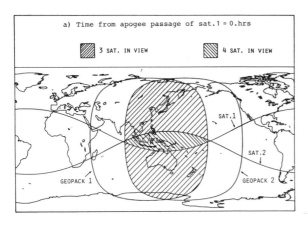

Fig. 4 Satellites visibility pattern

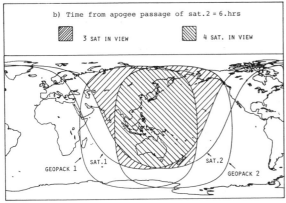

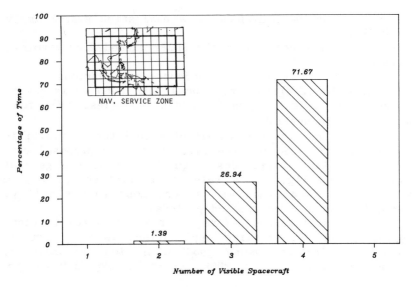

Fig. 5 Satellites visibility statistics

tion and horizontal DOP (HDOP) statistics for 2-D navigation.
        These statistics using altimeter or clock aiding
techniques are presented in Fig. 6. It can be seen that for
the proposed system the following values at 50% of probability are expected:

PDOP = 5.6 (SEP),
HDOP = 3.3 (CEP).

        Examples of time evolution of PDOP parameters are
presented in Fig. 7 for selected locations inside the
service area.
        As concerns the expected position accuracy, it
can be estimated for 3-D and 2-D applications simply multiplying the PDOP and HDOP by the user receiver pseudorange error (1-sigma).
        Therefore, the proposed system, with a typical
pseudorange error of 5 m (1-sigma), would provide position accuracies such as:

3-D position error: about 28 m (SEP)
2-D position error: about 16.5 m (CEP)

        For comparison purposes, GPS/Navstar, with the
intentionally degraded C/A code to be for civil applica-

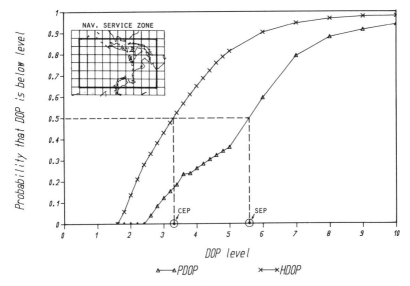

Fig. 6 Dilution of precision statistics

tions, will provide a 3-D position accuracy of the order of 100 m.

        In conclusion the proposed constellation offers, with a very limited infrastructure, a large area of service with performances comparable to a global satellite based system, such as the GPS/Navstar with a precise position determination.

Navigation Payload Aspects

        The payload mass and power budgets will mostly depend on the signal structure and the navigation signal generation.

        The present Navsat approach, (Fig. 8) with the satellite receiving the navigation signals generated on ground in C-band and retransmitting to the mobile users in L-band, allows the use of a simple transparent configuration for the on board transponder.

        A TDMA mode of the navigation signal structure only requires power during the navigation burst transmission. Therefore, the average RF power demand is expected to be modest.

        This approach will ease a multimission configuration feasibility (i.e., navigation-communications mission, sharing the cost for the satellite bus, launch and operations).

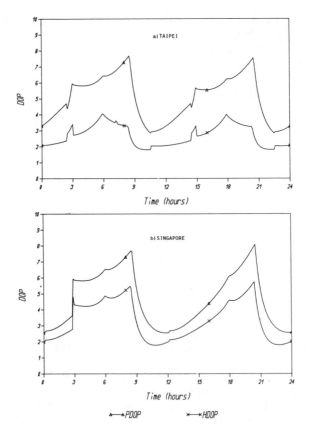

Fig. 7   DOP evolution from selected locations

4.   COMMUNICATIONS SERVICE

The major indication from a broad spectrum of investigations is that a satellite-based system, at the present state of the art, is not competitive with the cellular terrestrial system, which can rely on a far higher frequency reuse.

The satellite in this case could play an important role in covering gaps or holes left by the ground system.

In areas with poor terrestrial infrastructures a satellite-based system would have a primary role both for fixed and mobile communications.

At the current costs of space hardware, the primary limitations comprises the on board power availability and the antenna dimensions. Considering a Radiotelephone service (RT), the satellite-based system could aspire to capture only the "high value" segment of the mobile service inhabit.

In this scenario, a satellite-based system mainly providing paging services or other short two-way messages, seems an effective solution in order to make a profitable capital investment.

In fact, on the basis of the market forecast analysis, it is expected that the major user demand segment is relevant to paging service.

A limited number of voice channels could be retained to cope with specific high value services demand such as emergency and disaster alleviation, which at a certain extent might transcend pure socio-economic values.

Hence the proposed approach is deemed suitable to set up with a limited capital investment mobile communications service that has the potential to capture a substantial portion of the growing user demand.

Communications System Architecture

As described before, the system structure envisages a communications payload on board each EIO satellites, sharing the bus with the navigation payload.

With these two payloads, due to the selected orbits, it is possible to cover continuously a large area in the northern hemisphere with high elevation angles

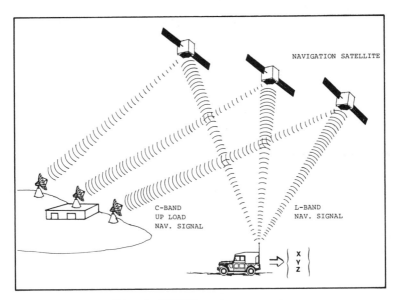

Fig. 8 NAVSAT system concept

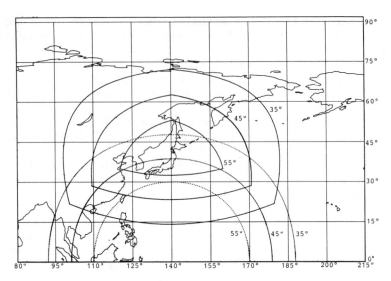

Fig. 9 EIO's coverage boundaries with a guaranteed minimum elevation angle, compared to GEO

(Fig. 9). In addition the system is also capable of covering higher latitudes of the southern hemisphere, although
with a reduction of the system availability (Fig. 10).
     The major benefits deriving from the proposed
configuration include the extension of maritime and other
mobile services to the polar areas (Fig. 11) and improvement of all the mobile communications for higher northern
and, to a certain extent, southern latitudes, where the
users would experience greater elevation angles compared
to a geostationary solution (Fig. 12). These features,
combined with a directive antenna at the users terminal,
are also capable of providing for a multipath reduction,
thus enhancing the system economy.

Operational aspects

     Considering that the described system could be
implemented on an international cooperation basis, a certain number of gateways stations shall be foreseen. At
each gateway a slice of the satellite capacity would be
assigned. The band availability will be equally shared
between the envisaged communications payloads.
     In this scenario (Fig. 13), each gateway station
will be in charge of the traffic routing of the relevant
subscribers. The channel allocation may be operated on a
demand assignment basis.

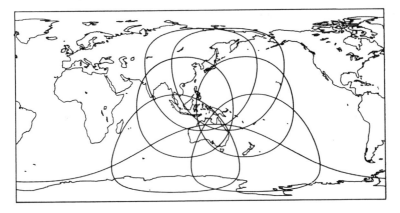

Fig. 10 Coverage evolution on southern hemisphere with a 35°
mask angle (2 hrs step)

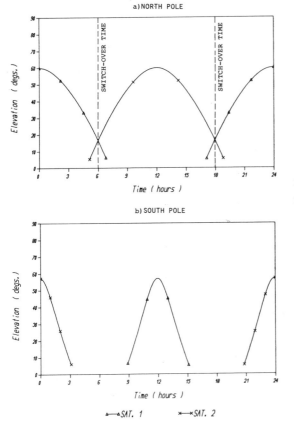

Fig. 11 Elevation angle
plots from earth poles

302   C. CARNEBIANCA ET AL.

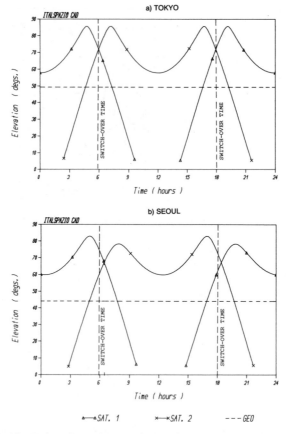

Fig. 12 EIO's elevation angle evolution compared to GEO

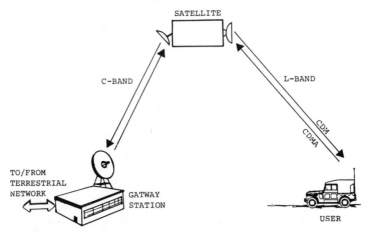

Fig. 13 Paging service scheme

Table 1 Communications system general characteristics

. Frequency	L/C-band
. L-band bandwidth	6 MHz
. Satellite ant. gain	25 dB
. Access scheme	CDM/CDMA
. On-board power req.	1000 W/dc
. Service capacity (mono/bidirect.)	1 million
. Waiting time	about 2 sec.

5. COMMUNICATIONS PAYLOAD ASPECTS

With the proposed system the communications payload can be based on a transparent configuration and can rely on conventional technology.

The service capacity trade-off analyses versus on board power requirements, carried out for similar concepts 2 , can be easily extrapolated to the present conditions. The proposed system capacity for 2-way paging service (Table 1) seems attractive to assess the mission economic feasibility.

6. CONCLUSIONS

The demand of ever more integrated navigation and communications services is expected to grow substantially in the near future on the part of a wide spectrum of user classes.

In the scenario of the potential alternatives to cope with the market forecast, a configuration has been identified, which offers the opportunity to set up an integrated system on a regional basis with the potential to be further augmented to a global coverage.

This system could be implemented on the basis of an international cooperation for civil applications.

The salient features of the proposed system are:
- 2-D/3-D navigation services in a large coverage area with a minimum satellite-based infrastructure.
- The constellation consists of two dedicated navigation satellites in eccentric inclined orbit (EIO) and two navigation packages in host geo-satellites.
- The EIO satellites will also embark a communications payload capable of extending the maritime aeronautical and land mobile services to the polar areas and of providing enhanced services at higher northern latitudes,

exploiting the best link geometry conditions.
- The system has the capacity to be implemented gradually to a full coverage.

The service nature, i.e., 2-way short alphanumeric messages and navigation signal bursts in a TDMA format, with a power requirement typical of a medium class satellite, could support an unlimited number of navigation users and about one million of paging users.

These elements appear promising enough to encourage further investigation to assess the technical and economical feasibility of the proposed system.

In conclusion, with a minimum satellite-based infrastructure it seems possible to set up a continuous service in a large coverage area to provide passive navigation augmented with low data rate service.

In this context it is deemed that the proposed configuration can represent a contribution to enhance the economy of a world-wide navigation and communications system for civil applications.

References

[1] C. Rosetti, and C. Carnebianca, "NAVSAT: A Global Satellite Based Navigation System," CH2365-5, IEEE Plans, 1986.

[2] C. Carnebianca, B. Pavesi and A. Tuozzi, "Land Mobile Communications Satellite," Proceedings of ESA Workshop on Land Mobile Services by Satellite, ESA SP-259, Sept. 1986.

Galactic Communication with Small Duty Cycles

Jörg Pfleiderer
University of Florida, Gainesville, Florida

Abstract

It is shown that SETI strategies based on omnidirectional and continuous emission of the sender may fail even if ETIs do exist because they could prefer to do communication such that the combined effort of sender and receiver is more or less minimized. An extremely large effort is necessary for omnidirectional continuous emission. The problems include energy consumption, thermal pollution, and radiation pollution. On the other hand, discontinuous illumination of targets with a narrow beam and a small duty cycle, one target after the other, is technically rather simple. It can be done with power self supply from the illumination by the central star and poses no pollution problems. The corresponding SETI strategy should then be near-omnidirectional and near-continuous. However, very simple emission characteristics are necessary in this case which must be guessed by the searching community. That such guessible characteristics are indeed possible is illustrated by a consistent example. The technical requirements on the receiver's side are high but within our present technology. It would need many receiver dishes at different places and a combined effort of many countries including the developing ones.

1. Introduction

The problem of CETI (communication with ETIs, extraterrestrial intelligences) is twofold: The sending side must provide enough energy for a sufficiently far-reaching signal while the receiving side must guess what to look for. There are no simple and straightforward answers on how to do CETI.

*Visiting Professor of Astronomy, on leave from the University of Innsbruck. Present address: Institut f. Astronomie d. Universität, Technikerstr.25, A-6020 Innsbruck, Austria.

Consequently, the problem is a highly speculative one, and
so are many arguments, assumptions, estimates, even order-
of-magnitude considerations, which are therefore open to
doubt. Different opinions on the relevance of facts or the
validity of arguments are possible and even necessary.

At present, we are mainly concerned with SETI (search
for ETIs), i.e., with the receiving-side problems. Most scientists agree that the message carrier would be electromagnetic waves, and that the medium-wavelength radio region
would be the best choice because it is relatively noise-
free. Good arguments were put forth by many authors, as by
Cocconi and Morrison[1] as early as 1959, or more recently by
Oliver[2], for selecting the "waterhole" radio-wavelength
range (λ 18 - 21 cm). This view shall be adopted here, but
many arguments of the subsequent sections will apply to
other kinds of carriers as well.

No definite agreement has been reached on the probable
wavelength of the message within the waterhole. This is,
therefore, one of the questions that should be discussed.Si-
milarly, present SETI strategies try to avoid a decision on
the specific form of the carrier modulation which would be
the signal proper. The question of which target to look on
is at present mainly determined by the fact that most SETI
programs were limited in time. Under the assumption that the
sender has no infinite energy supply, this meaning that any
incoming signal would probably be extremely weak, some proposals for correspondingly large receiver efforts have been
put forth. One of the most spectacular and most expensive
ones was the CYCLOPS project[2,3].

A common feature of most SETI strategies hitherto proposed is the assumption that the sender uses an omnidirectional beacon (or perhaps a narrow one directly aiming at our
Earth) and continuous emission. This implies that the time
at which one looks, within a SETI project, on any specific
target is not relevant. The preponderance of such views is
well illustrated, for instance, by Brin[4] who, in a very ela-
borate discussion on various SETI aspects, says: "Although
SETI efforts have been extremely modest so far, the project
has at least eliminated the nearest dozen or so candidate
stars...". His statement implies that a target not detected
at one time can be expected also not to be detected at any
later time unless the sensitivity of the receiver is increased.

However, those views place the burden of technological
effort more or less wholly on the side of the sender, asking
for extremely advanced ETIs. Far-reaching omnidirectional
beacons are excessively energy consuming so that only Kardashev II civilizations (KIIC) could afford it[5]. No good argu-

ments have ever been put forward as to why a KIIC should not
make use of the enormous economic advantage of using directional and intermittent emission beamed successively towards
each target star. Moreover, even the possibility of the existence of KIIC has not been proven beyond doubt while the
possibility for a Kardashev "zero" civilization (KOC) such
as ours is proven by our very existence. For the purpose of
the present paper, a KOC is defined as being in an early
technological stage and having not yet gained access to all
advantages of the fully developed Kardashev I (KIC) stage.

It will be shown that, for CETI within the waterhole,
several considerations of quite different kinds lead con-
sistently to the use of conventional antennas of $10^2 - 10^3$ m
diameter (Effelsberg or Arecibo type) both for the sender
and the receiver, making both sides of CETI accessible to
KOCs. Then intermittent emission is unavoidable, thus transmitting, towards each target star of interest in succession,
with a small duty cycle. More or less continuous reception
would consequently be necessary for any SETI which is not
intended to minimize the chances of success. While such
strategies would put some additional technological burden on
the receiver's side, the combined effort of sender and receiver would approach a minimum.

Whenever a signal does not arrive continuously, it becomes mandatory to have a good guess on the signal characteristics. Trial and error will not do. Considering the fact
established by ethology that higher intelligence can be defined as having a learning ability going beyond the trial
and error stage, one might wonder whether any ETI would not
expect us to be able to arrive at a good guess.

While most scientists agree that ETIs probably would
inhabit a planet, little is known about how an inhabitable
planet would look like. For the sake of the argument, the
ETI planet shall be assumed to be similar to Earth so that
solar-system quantities may be used for order-of-magnitude
estimates.

2. Power Requirements

The equivalent isotropic radiation power EIRP necessary for emitting a detectable signal is[6]

$$EIRP = 4\pi R b d^2 C kT / q Q A \sqrt{B} t \qquad (1)$$

where R is the signal-to-noise ratio at the receiver, b
and B the effective bandwidth of emission and reception,
respectively, d the distance travelled by the signal, C
the receiver constant of the order of unity (C>1), k Boltz-

mann's constant, T the total system noise temperature of the receiving system, q and Q the aperture efficiencies of the emitting and receiving telescopes, respectively, A the area of the receiving telescope, and t the integration time on the receiver's side.

The bit rate r of the message cannot exceed b/2, B cannot exceed b (if the actual receiving bandwidth would, then one would partly observe a void band of signal-free frequencies and so effectively reduce the bandwidth and also the sensitivity), and t cannot exceed the duration 1/r of one bit of message. The aperture efficiencies are <1 for conventional telescopes but an omnidirectional emitter would have q = 1.

The minimum power follows from optimum conditions: R = 3 (limit of significance), T = 3 K (microwave background), b = 2r, B = b, t = 1/r, q = Q = C = 1. It is

$$EIRP_{min}(W) = 2.5 \times 10^{12} \, r(Hz) \, d^2(kpc) \, A^{-1}(km^2) \ . \qquad (2)$$

For r = 10 Hz (which is about the bit rate used for the 1974 Arecibo message[7,8]), d = 10 kpc (average galactic distance), and Q A = 0.04 km^2 (Arecibo-type mirror), a minimum EIRP of 6×10^{16} W results which is about as much as the total solar power $P_\odot$ reaching the Earth's surface ($P_\odot = 2 \times 10^{17}$ W outside the atmosphere and, with an albedo of about 1/3, $= 6 \times 10^{16}$ W on the ground).

One might try to reduce the power requirements by using a small bandwidth and, correspondingly, a low bit rate. However, excessively small bandwidths are difficult to handle and pose serious technical problems. The receiver might, e.g., have to cope with frequency shifts during the duration of one bit. It is also difficult to detect an excessively slow modulation in a very weak signal. One cannot gain many orders of magnitudes in power from the bit rate or bandwidth without increasing the technical burden far beyond our present capabilities.

Another possibility would be to increase the size of the receiving telescope. As von Hoerner[9,10] demonstrated many years ago, the size of a fully steerable telescope on Earth is limited to less than 1 km^2 with presently available construction materials. Much larger telescopes would, furthermore, pose pointing problems because the effective solid angle of reception, W, is reciprocally proportional to the telescope size - approximately $W = \lambda^2 / A$ (λ = wave length). Large telescopes would tend to have deteriorated surfaces as compared to smaller ones. This does not ease the pointing problems but may severely diminish the relevant effective telescope size (i.e., decrease Q). A sizable power decrease by use of a very large receiving telescope would, therefore, again put a heavy technical burden on the receiving side.

Finally, as the only possibility left, the actual radiation power P_{rad} needed can also be reduced by emitting only into a narrow cone of effective solid angle w, giving

$$P_{rad}(W) / w(sr) = EIRP(W) / 4\pi . \qquad (3)$$

The solid angle is related to the area a of the emitting telescope of conventional construction by (approximately) $w = \lambda^2 / a$. An emitting area of 1 km^2 ($w = 4 \times 10^{-8}$ at 21 cm) which on Earth is technically feasible even today, would reduce the required power by a factor of 3×10^8 to a still high but reasonable KOC level.

If one therefore considers to avoid excessive effort on either side of the communication, one is left with a single choice, viz., a small emitting solid angle. It implies, of course, that different parts of the sky can only be illuminated one after the other.

A special case is the possibility of "eavesdropping": The inadvertent near-omnidirectional leakage into space of internally used (commercial, military, etc.) radiation. This is not a communication in the usual sense. It is rather a kind of "empty message" with $r \to 0$ at a finite bandwidth. It can, nevertheless, transmit, by its very existence, at least one bit of information. In terms of eq.(1), the case is characterized as radiation with a bandwidth that is much larger than the reciprocal integration time, or the time in which essentially one bit is observed. The necessary power is larger than the minimum, and it decreases, contrary to eq.(2), only with the square root of the effective bit rate $1/t$. Correspondingly, eavesdropping on a KOC is impossible beyond a maximum distance of a few tens of parsecs at best.

3. Thermal Pollution

Not only may the heavy energy consumption of omnidirectional emission be a problem by itself but it also gives rise to pollution problems which shall now be discussed.

The total power requirement P_{tot} on the sender's side is higher than given in eq. (1) by a factor of $1/e$ where e is the efficiency of transforming energy into beamed emission. When the total effort is taken into account, e can be assumed to be very small. In estimating $e = 1\%$, one is probably quite on the optimistic side. However, accessible values of e depend heavily on the technological status. The excess power $P_{tot} - P_{rad} = P_{rad} (1/e - 1)$ will consist mainly

of thermal energy which necessarily must be radiated away
for a steady state condition. Thermal radiation is generally
omnidirectional.

Whenever the ETI message is emitted omnidirectionally,
the waste energy might easily surpass the energy input from
the central star and will, therefore, pose a severe pollution problem. It can be avoided only if the sender is put into space. If one wants to have the thermal input on the home
planet (radius r_p) not to exceed the fraction f of the central star's input P_s, the distance D to the sender would
have to be

$$D > \frac{1}{2} r_p \, (P_{rad} / e \, f \, P_s)^{1/2} \, . \tag{4}$$

For omnidirectional and continuous emission, one would have
$D > r_p$, implying a sender in space, while for narrow-beam
emission, $D < r_p$ would be possible, that is, a sender located
on the planet.

The above example is again useful for finding an order-
of-magnitude estimate. Let the Earth's thermal pollution be
restricted to less than 1% of $P_\odot$. Then one would find, with
$w = 4\pi$, $P_{rad} = P_\odot$, and $e = 1\%$, that P_{tot} is about 10^4 times
larger than what would be tolerable. The planet as seen from
the sender should then cover not more than 10^{-4} of the
sphere. The distance from the planet to the sender should be
at least 50 times the planet's radius, or roughly the Moon's
distance in the case of the Earth. That is, the sender could
still be a satellite of the home planet. With more stringent
requirements on the maximal thermal pollution, or on P,
this would no longer hold. Troitsky[5] suggested a necessary
distance of not less than several AU.

If the energy input were to come from the radiation of
the central star, the solar-cells area would have to surpass
the size of the planet by a large factor, depending on the
efficiency of the solar cells. While such requirements are
not entirely impossible to be met, they certainly require a
KIIC.

For narrow-beamed emission, thermal pollution would be
small or even non-existent. The sender can be placed on the
home planet. Assuming that it receives its power directly
from the central star, and that the area of solar cells is
about as large as the emitting area, we find that the
obtainable EIRP is proportional to the fourth power of the
telescope's diameter. With the above given example, an
emitting area of 1 km^2 needs a beamed-radiation power of
150 MW. The corresponding pollution of 15 GW equals the
average solar input on 30 km^2 and is, thus, well within
the present technology realm.

4. Radiation Pollution

Another kind of pollution, not yet discussed in the
literature, occurs when the SETI receiver picks up emission
from its own sender. Considering the fact that the emission
power may exceed the receiver's sensitivity by perhaps up to
30 powers of ten, even a very low efficiency in picking up
unwanted signals may completely bury another ETIs signal.

The simplest possibility to avoid spill-over radiation
is, of course, to switch off the emission during SETI times,
i.e., to emit and receive alternatingly. It implies, however, that strictly continuous emission would not be expected even in the case of omnidirectional emission, but
rather a duty cycle of about 1/2 or less. One might
also think of shielding off the emission. A very effective
shielding method is to put the receiver on the other side of
the planet, implying that any SETI telescope can be used
only part-time and cannot be put into space.

Another possibility is to reduce spill-over by the use
of narrow-beamed emission which however implies necessarily
a very small duty cycle for the gradual illumination of the
whole sky, or as the other alternative, the illumination of
only a small number of targets: ETIs are, however, barely as
numerous as to allow an interested civilization to be so
selective.

A rough estimate of the resulting spill-over will show
the point. With typical aperture efficiencies of 1/2, one
can - very approximately - describe the beam pattern of both
the emitting and the receiving telescopes as half of the
emission going into (or: reception coming from) the main
beam, and the other half being emitted (received) more or
less omnidirectionally. Then the SETI telescope (beam W, on
the planet or in space) measures the fraction $W/4\pi$ of the
radiation falling in from the emitter (assumed to be in
space) which itself is only the fraction $w/4\pi$ of the radiation in the main beam w of the emitter. The radiation is,
however, stronger by a factor $(d/D)^2$ in the distance D
(distance between the two telescopes) than in the galactic
distance d where it is supposed to be just detectable. That
is, simultaneous emission and SETI reception without severe
radiation pollution is possible if

$$W w / 16\pi^2 < (D/d)^2 / s \qquad (5)$$

where s<1 is a factor describing frequency spill-over from
the emitter frequency to the receiver frequency - one would
of course avoid observation exactly at the emitter's frequency.

Inserting $D = 1$ AU (characteristic planetary distance), $d = 10$ kpc, and $s = 10^{-3}$ (30 db), we find $W < 3 \times 10^{-15}$, corresponding to a telescope diameter of >4000 km, for omnidirectional emission $w = 4\pi$. That is, the SETI receiver would have to be excessively large. Whether or not it would be feasible even for a KIIC to overcome the difficulties encountered in correctly pointing an extremely narrow-beamed telescope of this size to any given target, shall be left open.

On the other hand, the assumption $W = w$ (equally sized telescopes for emission and reception) leads, with the same numbers, to conventionally sized telescopes of >0.3 km diameter. Such telescopes need not be put into space, especially since a planet effectively shields off radiation so that D may be much decreased. It would be essentially sufficient to place the two telescopes out of mutual sight. Also, frequency spill-over would not be much of a problem.

That is to say, the simplest remedy against radiation pollution seems to be the use of Kardashev zero technology.

5. Omnidirectional Receivers

We have found that energy consumption and pollution pose problems with omnidirectional emission which can, if at all, only be solved by KIICs, i.e., by extremely far-developed civilizations. The problems are reduced to conventional levels for narrow-beamed emission which is, in any given direction, intermittent with a small duty cycle if more than a few targets are to be illuminated. To detect such emission, it is mandatory to observe essentially continuously, since there seems to be no way of finding out when the illumination will take place. Also, as long as one does not know where to look, omnidirectional observation of the sky is the only remedy. There is no such thing as a true omnidirectional high-sensitivity receiver, but a large number of narrow-beamed receivers will do as well. The sensitivity of a receiver increases inversely with the beam size. It is therefore not possible to have a separate telescope for each direction of observation. Instead, each telescope must be equipped with a large number of receivers.

Such devices, barely realizable at present for technological reasons, seem feasible on Earth within the next decades, thus still belonging to a KOC technology. If the signal reduction is done with microcomputers at the amplifier exit, receivers can be made quite small, and each receiver dish can be equipped with many of them, thus covering a good part of the sky. For instance, the presently

used field of view of the Arecibo dish is about 2×10^3
square degrees or about 10^6 times the beam size. A million
21-cm horns of 3×10^{-2} m^2 each would only cover approximately 5 % of the dish area.

Telescopes would have to be placed, for full-sky coverage, on many different locations, thus demanding cooperation
of many countries.

6. Emission Characteristics

While continuous emission facilitates detection tremen-
dously because all characteristical parameters can be found
by trial and error, a small duty cycle necessitates correct
guesses on the SETI side at least on the frequency or its
possible range.

Starting from the hydrogen line, it would at least save
energy if the frequencies of interstellar hydrogen emission
were avoided. One of the emission characteristics then is
the frequency shift off the hydrogen line. This does not,
unfortunately, dispense of broad band SETI observations
because a KOC would not particularly aim at the Earth or
another more interesting target, and would be unable to
correct for the Doppler effect.

It would certainly facilitate detection if the signal
contained different levels of information, increasing levels
being detectable with increasing sensitivity of the receiver. The lowest level 0 should contain just one bit: "We
Are Here". One of the simplest ways for encoding a single
bit is a periodic frequency jump in the signal. Even an integration over the whole message would reveal two spikes in
the spectrum. The jump period is another parameter.

For the message proper, one would reasonably expect at
least a low bit rate (level 1, for very weak signals) and a
high one (level 2, for stronger signals from nearby ETIs),
notwithstanding the possibility of a hierarchical structure
of the message with information levels between 1 and 2. The
highest bit rate determines the necessary band width.

One simple way of having a hierarchical structure which
will also facilitate detection and decoding by providing for
different levels of sensitivity on the receiver's side would
be to emit at two frequencies simultaneously with different
bandwidths and bit rates.

Finally, a duty cycle <1 involves as further parameters
the duration of the message, its cycle time, and a shorter
time span for a single repetition of the message (first in
far-reaching mode with small bandwidth and level 1 only,
then with larger bandwidth and - at the same EIRP - lower
range for level 2).

All these parameters must be chosen from sufficiently simple and general premises. Otherwise, wrong guesses on the SETI side which were of course fatal to CETI, would be too probable. All parameters should be interrelated, so that the guess of one implies others. This is also the only way to communicate them within a few bits of the message in order to clear up possible ambiguities, which would help to fully recognize at least the repeated message.

7. A Proposal For The Characteristic Parameters

One of the possible ways of guessing is to leave physics at the main frequency and only use the very simplest mathematics. It was shown by Gruber and Pfleiderer[11] that this is indeed possible even if the result is just one guess within many possible ones. The authors used a number series defined only by the very simplest number pertinent to information transmission, viz., the number "2" (binary unit, or unit of a bit with the two possibilities 0/1, or on/off, or +/-), together with one of the simplest operations with this number, viz., the power operation which is just the squaring operation. Each member n_i is the square of the former, i.e.

$$\log_2 n_i = 2^i \; ; \; i = 2,3,4,5. \tag{6}$$

In the slightly extended version presented here, use is made of exactly $4 = 2^2$ members and also of the combinations of each 2 members.

The result of dividing the members and their combinations into the hydrogen laboratory frequency f_0 is summarized in Table 1. Note that all members and combinations can be used and, given the requirements of the last section, must be used. Note also that higher members would be irrelevant because $n_6 = n_5 n_5$ and because $n_7 = 2^{128}$ would lead to a time which is larger than the age of the universe.

With these numbers, the message proper is characterized by 256 bits on the lowest level and about 10^7 bits on the highest level as well as a few million different targets which can be illuminated within the cycle time.

The interpretation given is, however, not unique and is only intended to show that, interestingly enough, none of the resulting frequencies or periods is completely outside a reasonable range.

8. Technical Requirements

Following Table 1, one needs, for smallband communication at the lowest level (r = 0.3 Hz, b = 0.7 Hz) over 10 kpc with Arecibo dishes on both sides, a radiation power of about 100 MW. At this power, the highest-level communication (b = 22 kHz) still reaches as far as 60 pc. Arecibo-type dishes are about the smallest for which CETI over galactic distances is feasible within technologies comparable to our present one. With $4\pi/w = 2 \times 10^7$ and a ratio of cycle time to message duration of 10^7, two or three such senders would suffice for whole-sky coverage within the cycle time. Two senders are, of course, in any case needed on a round planet.

As pointed out above, the considerable ease in the requirements for the sender implies some - relatively moderate - increase in the requirements for the receiver. The number of receivers necessary in omnidirectional SETI is $4\pi/W$ ($=2 \times 10^7$). According to section 5, about 20 receiver dishes would be the minimum, each equipped with about 10^6 receivers. Use of smaller (and thus less sensitive) dishes, such as Effelsberg-type ones, would reduce the number of receivers by an order of magnitude, but not the number of dishes, because the usable field of view cannot be much increased. A good distribution of the dishes in planetographical longitude is, of course, essential while the latitude coverage is of less importance. These requirements are certainly high. Nevertheless, while being different they are certainly not higher in total than those of the CYCLOPS project[2] which was quite seriously suggested in 1973.

Table 1 A consistent set of emission parameters: frequencies (f) and periods (p)

n	value	f	p	meaning
–	–	1420 MHz		laboratory frequency f_0 of hydrogen line
n_2	2^4	89 MHz		frequency shift off f_0 ($\rightarrow$1509 MHz)
n_3	2^8	5.5 MHz		level 0 frequency jump ($\rightarrow$1509/1514 MHz)
$n_3 n_2$	2^{12}	0.7 MHz		freq. difference of simultaneous emission
n_4	2^{16}	22 kHz		highest bandwidth (highest bit rate 11 kHz)
$n_4 n_2$	2^{20}	1.3 kHz		high bandwidth at other frequency
$n_4 n_3$	2^{24}	84 Hz		low bit rate at first frequency
n_5	2^{32}	0.33 Hz	3.0 s	lowest bit rate (smallest bandwidth 0.7 Hz)
$n_5 n_2$	2^{36}		48 s	frequency jump period
$n_5 n_3$	2^{40}		13 min	duration of message
$n_5 n_4$	2^{48}		2.3 d	single repetition span (low and high level)
$n_5 n_5$	2^{64}		412 a	cycle time

In order to cover all possible Doppler shifts, say up to 200 km/s, the total receiver bandwidth should be at least 1 MHz. The spectral resolution must be high. In order to accomodate the smallest bandwidth suggested by Table 1, it should then be 3×10^{-7}. Following Table 1 in more detail, one would need a bandwidth of about 7 MHz with two separated high-resolution bands.

As far as the numbers are concerned, the SETI receiver requirements can be considerably eased by looking only at small parts of the sky. In that case, considerations as those of Balázs[12] on the best choice of directions of investigation become highly relevant.

9. Concluding Remarks

In most present SETI strategies, it was tacidly assumed - or sometimes even explicitly stated - that "we" are not interested in detecting ETIs which are of lower than KIIC type. The question whether "they" would at all be willing to place gigantic and - as this paper shows - completely unnecessary efforts into a possible communication instead of trying to somehow minimize the combined efforts of both parties, thus reducing either effort effectively to KOC standards, has barely been touched.

It is not the proper place here to dwell on the question of whether or not SETI is useful or promising by any means. I personally doubt it very much for many reasons of which the probability of ETI existence is but one. If, however, SETI efforts are to be continued, one should at least spend some thought on all possible lines of strategies. The lines indicated in this paper would necessitate a close longtime cooperation of many countries including developing ones - which can be considered as being valuable by itself.

Acknowledgment

Georg M. Gruber, Rhodes University, contributed heavily to the ideas presented in this paper.

References

[1] Cocconi, G., Morrison, P., "Searching for Interstellar Communications", Nature, Vol. 184, 1959, pp. 844-846.

[2] Oliver, B.M., "Project Cyclops Study: Conclusions and Recommendations", Icarus, Vol. 19, 1973, pp. 425-428.

[3] Oliver, B.M., "Project Cyclops", NASA Publ. CR 114445, 1973.

[4] Brin, G. D., "The 'Great Silence': The Controversy Concerning Extraterrestrial Intelligent Life", Quarterly Journal of the Royal Astronomical Society, Vol. 24, No. 3, 1983, pp. 283–309.

[5] Troitsky, W. S., "Ausserirdische Zivilisationen und das Problem der Ausstrahlung von Rufzeichen: Warum werden keine Signale ausserirdischer Zivilisationen beobachtet?", Die Sterne, Vol. 56, No. 1, 1980, pp. 21–26.

[6] Gruber, G. M., Pfleiderer, J., "ETIs in an Early Technological Stage", South African Journal of Physics, Vol. 8, No. 1/2, 1985, pp. 43–44.

[7] Sagan, C., Drake, F., "The Search for Extraterrestrial Intelligence", Scientific American, Vol. 232, No. 5, May 1975, pp. 80–89.

[8] The Staff of the National Astronomy and Ionosphere Center, Arecibo Puerto Rico, "The Arecibo Message of November 1974", Icarus, Vol. 26, 1975, pp.462–466.

[9] von Hoerner, S., "Design of Large Steerable Antennas", Astronomical Journal, Vol. 72, 1967, pp.35–47.

[10] von Hoerner, S., "Homologous Deformation of Tiltable Telescopes", Journal Structural Division Proceedings American Society Civil Engineers, ST5, 1967, pp.461–485.

[11] Gruber, G. M., Pfleiderer, J., "Eine Suchstrategie für ETI-Signale mit kleinem Tastverhältnis", Sitzungsberichte der Österreichischen Akademie der Wissenschaften, Mathematisch-Naturwissenschaftliche Klasse, Abteilung II, Vol. 191, No. 10, 1982, pp.495–504.

[12] Balázs, B. A., "Galactic Position of Our Sun and the Optimization of Search Strategies for Detecting Extraterrestrial Civilizations", Acta Astronautica, Vol. 13, No. 3, 1986, pp.123–126.

Advanced Technological Aspects of a Wideband Communications Satellite for Europe

F. Arzpayma,* X. Henrion,† and J. P. Marre‡

ALCATEL Espace, Toulouse, France

Abstract

Marketing studies have shown that, within the next decade, there will be a need for a high capacity satellite for Europe. In order to face such a new system, it has been necessary to investigate quite early the organization of a WIDEBAND communications satellite network and the associated payload diagram to identify the necessary new concepts and technological predevelopments.

Within a study contract sponsored by the European Space Agency (ESA), a WIDEBAND satellite system study has been performed under ALCATEL ESPACE (ATES) having a critical technologies identification purpose.

This paper presents two candidate architectures and payload designs for a high capacity WIDEBAND Communication Satellite planned for the end of 1990's in the 20/30 GHz frequency band, covering following concepts.

- TDMA multibeam hopped antenna system or scanning spot beam phased array antenna as an alternative
- On board multicarrier demodulation with SCPC or MF/TDMA access techniques
- On board high speed baseband processing.

The system analysis has led to the identification of the following critical technologies and equipment, needing industrial predevelopment

*System Engineer, Digital Satellite Communications.
†Department Head, Transmissions Department.
‡Wideband Study Manager.

- 30/20 GHz scanning active phased array antenna
- direct microwave frequency modulator and demodulator
- baseband processors
- special efforts to be done for hybrid or MMIC, custom VLSI and optical technologies for both electronic and microwave modules.

A mass and power consumption budget has been evaluated for the satellite with above technological assumptions. This budget has been investigated with reference to OLYMPUS European satellite platform and ARIANE IV launcher capabilities.

Finally, ALCATEL ESPACE has been able to demonstrate that with a minimum of technological predevelopments on some particularly critical items, it will be quite possible to master the WIDEBAND communication satellite field between the years 1990 and 2000.

Introduction

This paper reports briefly the future needs of the high speed digital communications for Europe and describes their evolution towards a wideband integrated services digital network (ISDN).

The main objectives of the study sponsored by the European Space Agency (ESA) on WIDEBAND space communication systems are then described.

The first conclusions of the study are presented in this paper for two advanced conceptual design hypothesis for the WIDEBAND space segment in the 30/20 GHz frequency bands:
- multibeam antennas with complex waveguide beam forming networks (BFN) working in a user   oriented SCPC access technique
- electronically scanning active phased array antennas working in a TDMA system.

As a conclusion of this paper, ALCATEL ESPACE (ATES) outlines the technological predevelopments needed to master the WIDEBAND communication satellites field of the year 2000.

Future digital networks

In Europe, high bit rate digital networks are being rapidly developed carrying both 2 Mbits/s, 34 Mbits/s,

and 140 Mbits/s data. These data will be transmitted
through a new type of digital network : the WIDEBAND ISDN
(wideband integrated services digital network). They are
mainly made of :
- trunk telephony
- videoconference or visiophone data
- high quality TV
- high speed electronic mail
- high definition screens.

   Present commercial satellite technologies only allow
traffic throughputs of 1.5 Gbits/s for EUTELSAT II and 4
Gbits/s for INTELSAT VI.

   These satellites will be completely obsolete in a few
years when the above future digital needs will have to be
handled. For example, NASA is preparing an advanced
satellite project (ACTS project) with 8 Gbits/s
throughput. The European needs can even go further, as
marketing studies have shown.
In order to face such capacities, it has been necessary to
investigate quite early the organization of a WIDEBAND
communications satellite network. It is in this technical
and economical context that the European Space Agency
(ESA) has awarded ALCATEL ESPACE with a study contract of
a WIDEBAND satellite system having a critical technologies
identification purpose.

            System aspects of the WIDEBAND ISDN

   The ISDN network is first of all a high rate network.
It has been designed to face the future needs of digital
networks in terms of capacity: 2 Mbits/s for
videoconference and medium quality picture transmission as
well as trunk telephony;   34 Mbits/s for high quality TV;
140 Mbits/s for high trunk telephony.

   The ISDN is also an homogeneous network : it handles
a wide variety of services using a unique support of
transmission (for the earth segment) and a unique
signalling protocol. A wide range of different terminals
are connected to the ISDN, allowing restoration of the
basic information in the users premises. So, all the users,
wherever they are in Europe, whatever they wish to
transmit, can access and communicate through a unique
integrated network. Beyond a certain distance (called
critical range) between two users of the network, the

satellite contributes to cheaper communications.
This range can be computed as about 500 km. Such a
critical range suggests that introducing a satellite in
the European wideband ISDN would be quite a good
economical operation. Moreover, a satellite suppresses the
cost of the major part of the earth cables; it improves
the flexibility of the network thanks to its on board
switching capacity; it covers all the territory whereas a
terrestrial ISDN still keeps shadowed areas with no
connection and very expensive equipment for hilly regions;
it allows both point to multipoint communications; it
can handle very easily mobile terminals such as TV
registering studios going from football field to football
field; it can also manage a sudden increase of traffic in
a given country for a limited period (because of world
football championships for example) without needing
particular infrastructures (see ref. 1).

     After having investigated the economical interest of
satellites in the wideband ISDN, ALCATEL ESPACE and the
European Space Agency have made an assessment on what
should be the main technical characteristics of such a
WIDEBAND ISDN SPACE SEGMENT.
     As a trade off between economical requirements and
technical performances, ALCATEL ESPACE has been sponsored
by ESA to study a WIDEBAND communications satellite:
- having the maximum digital throughput
- compatible with all the characteristics of the
terrestrial WIDEBAND ISDN; intergrating the maximum
number of earth stations in Europe, these stations being
as simple as possible for economic reasons
- having a maximum flexibility in the switching of
traffics independently of their type and rate
- giving a guarantee of minimum bit error rate to the user.

     Those requirements have led to the following
technical conclusions :
- use of 30/20 GHz frequency band in order to dispose of
sufficient bandwidths to handle 4.5 Gbits/s; moreover,
this frequency band needs minimum frequency coordination
with fixed terrestrial networks
- use of multibeam reconfigurable antennas allowing
frequency reuse in the allocated frequency bandwidth
- on board demodulation and switching capable to perform
on board cross traffic handling; it will simplify 2
Mbits/s stations
- TDMA access technique for a large number of users; this
will decrease the number of on board receivers and
regeneration channels

- coding of information in order to minimize bit error
rates; - direct access to the network for low rate users
at 2 Mbits/s.

WIDEBAND mission objectives

Following important features have been given to ATES
by the European Space Agency, which have a great impact on
the dimensions of the system:

The relevant traffic is a mix of video conferencing
and trunk telephony which are distributed within a
terrestrial network composed of :

-     5 earth stations of 140 Mbps
-   50 earth stations of 34 Mbps
- 500 earth stations of   2 Mbps

Given the high number of user terminals, it is
fundamental to keep them as simple as possible with a low
cost. For this reason an architecture must be retained to
permit a "user oriented" approach for customers generating
2 Mbps and 34 Mbps data rates with an eventual complex
satellite system.
Service zone of the satellite is shown in Fig. 1.

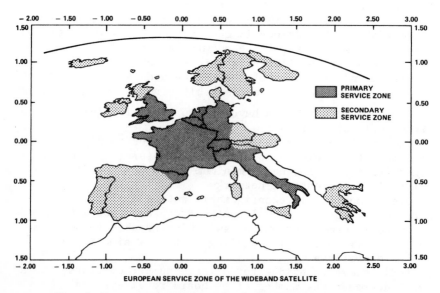

Fig. 1 European service zone of the WIDEBAND satellite.

Satellite throughput in term of traffic (erlang) are :
- for central Europe 100 erlangs at   2 Mbps
                      35 erlangs at   34 Mbps
                       5 erlangs at 140 Mbps

- for greater Europe 282 erlangs at   2 Mbps
                      15 erlangs at   34 Mbps

which results in a total throughput of the satellite of
4.4 Gbps.

The allocated bandwidths are :
- for central Europe up-link   : 29   - 30   GHz
                     down-link : 18.2 - 19.2 GHz
- for greater Europe up-link   : 29.5 - 30   GHz
                     down-link : 18.7 - 19.2 GHz

The above allocation has been deliberately chosen to
keep the coordination with terrestrial digital radio relay
links.

Modulation is QPSK with an uncoded BER objective of
10^{-3} corresponding to theoretical Es/No of 9.8 dB in case
of a transparent repeater (in case of a regenerative
repeater BER objective is 5.10^{-4} corresponding to a
minimum theoretical Es/No of 10.4 dB).

The objective of link availability must be close to
99.9 %.

The worst case propagation effects deduced from CCIR
world wide map (L zone) during 99.9 % of the time are :
- up-link 10.9 dB
- down-link 5.7 dB

Blocking probability is 10^{-2} with the possibility of
on board routing at 2 Mbps channel (30 voice channels
multiplexed or one videoconference channel).

With these requirements it seemed more wise as a first
step to distinguish within Europe two types of users
and to optimize the system architecture design for each
type of them. As a second step, ALCATEL ESPACE proposed a
satellite structure allowing a global TDMA technique for
all customers of the network.

Satellite system architecture using
TDMA and SCPC techniques

A double coverage has been defined for Europe : one
covering the central Europe with high concentrated

traffics (primary zone) and the second one covering the
great Europe with low concentrated traffics (secondary
zone). Fig. 1. shows those two zones.

Within the primary zone, users through the
terrestrial links are connected to high bit rate gateways
(500 Mbps) to access the satellite with TDMA technique
using large antenna diameters of high cost. However in
secondary zone, 2 Mbps and 34 Mpbs users are directly
accessed to the satellite with SCPC technique using small
antenna diameters of low cost.
The satellite has also the duty to guarantee the
crosstraffic handling between primary and secondary zones.

Payload configuration using TDMA and SCPC techniques

Fig. 2. shows a simplified block diagram of the
communication payload. It is basically constituted of two

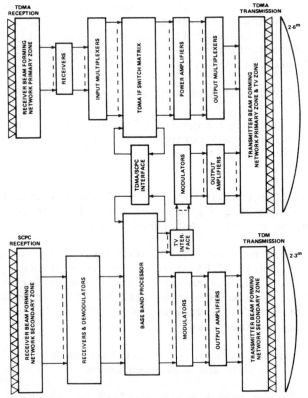

Fig. 2 Simplified payload block digram handling TDMA and SCPC
traffic.

subsystems, each processing separately the traffics of primary and secondary zones.

Since it is meant that the 25 % of high traffic (primary zone) and 75 % of low traffic (secondary zone) should be routed by the satellite from one zone to another there is an interface between the two communication subsystems to guarantee the cross-traffic handling.

The traffic within the primary zone is received and transmitted through a transparent repeater which incorporates extensive point to point interconnectivity using satellite switched TDMA concept as principle. The zone is covered by 14 fixed spot beams (see Fig. 3.). The traffic, however, is routed through the ensemble of 3 spot beams which are "hopped" simultaneously and cycled sequentially over a frame period to repeat the configuration. The "hopping" operation is accomplished within the receive/transmit beam forming networks using latching circulators as principle elements. The receive/transmit antenna subsystem consists of multifeed horns using 14 beam forming networks based on a "1/3 overlap septet" concept to minimize side lobe effects (27 dB), thus improving beam to beam isolation (20 dB).

The receivers for primary zone repeater set the noise figure for the whole system and directly receive at 30 GHz the sub bands affected to each spot beam.

The channelization of the received spot beams signals are performed in the 3 accesses input multiplexers at 30 GHz, each carrying 3 synchronous TDMA frames at 500 Mbps burst rate. To allow the interconnection of traffic bursts from any up-link beam to any down-link beam in an appropriate assigned time slot, the repeater uses a 12 X 12 TDMA microwave switch matrix operating at 12 GHz. The transmit section consists of 9 up converters 12/20 GHz, 9 TWTA's with 3 output multiplexers each combining 3 frequency sub bands at 20 GHz. Since the key designs of this communication subsystem are "beam hopping" and SS TDMA microwave switch matrix, a complete synchronization is primordial using onboard central processors and digital control units.

The traffic within the secondary zone (34 Mbps quality TV and 2 Mbps video conference) is received by a regenerative SCPC satellite communication subsystem suitable for maximum use of the output power of earth station HPA's as well as on-board TWTA's. The regenerated signals are switched in base band, multiplexed by time

For each spot, ∅3dB = 0.47°

Beam hopping implementation

3 beams are formed simultaneously
at each interval of time.

1st interval of time : beams nr 1, 4, 10
2nd interval of time : beams nr 2, 8, 13
3rd interval of time : beams nr 3, 9, 11
4th interval of time : beams nr 5, 7, 12
5th interval of time : beams nr 6, 14.

There are 3 TDMA frames in each beam.

Fig. 3 Fourteen spot beams coverage for central Europe and the beam hopping principle.

division multiplexing (TDM) and then transmitted to earth stations (see Fig. 2.).

The secondary zone is covered by 14 fixed spot beams which reuse a set of 3 frequency sub-bands. Adjacent beams will be on two different frequencies but non adjacent beams can use the same frequency all over again (see Fig. 4.) with appropriate spacing. The antenna subsystem consists in receive/transmit multi feed horns using 14 beam forming networks based on "1/3 overlap septet" concept (to improve side lobe performances). Since the receive beam forming network is frequency selective, the traffic in each spot beam is selected around its own frequency. Receivers down convert the SCPC signals at 30 GHz to a unique intermediate frequency at 4 GHz. The demodulation is accomplished at 4 GHz, which is quite an efficient frequency to demodulate coherently 36 SCPC carriers per spot beam.

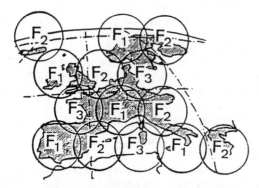

For each beam, ∅3dB = 1°

Each beam is implemented by a septet.

The 14 beams are shaped simultaneously.

One frequency is permanently assigned
to each beam.

There are 34 users at 2 Mbps and 2 users
at 34 Mbps in each beam.

Fig. 4 Fourteen spot beams coverage for greater Europe and frequency allocations for each beam.

The demodulators can be of multi-carrier type. The satellite comprises 504 demodulators on board. The base band processor is a type of time-space-time switch which synchronizes all SCPC regenerated data streams with an on board reference clock, sending them at 2 Mbps channel level.

Among secondary zone traffic, the high quality TV (34 Mbps), after routing through the base band switch, is separated from communication subsystem and connected to the transmission section of primary zone through common horns having double accesses.

Satellite system architecture with mixed TDMA and hybrid FDMA/TDMA techniques

An alternative to "user oriented" approach is currently being studied by ALCATEL ESPACE to interconnect the individual customers at 2 Mbps and 34 Mbps data rates of the secondary zone.

Although the previous SCPC/TDM solution satisfied the direct access of the users to the satellite, the spacecraft became more complex mainly due to the on board SCPC demodulators. The system presented in this section takes advantage of a dual bit rate mixed TDMA and hybrid FDMA/TDMA transmission technique to integrate the high traffic (primary zone) and low traffic (secondary zone) of the WIDEBAND system with full interconnectivity. Both traffics are transmitted on a single repetitive frame. The first portion of the frame is devoted entirely to the primary zone. The users of this zone through terrestrial links are connected to high bit rate gate ways (480 Mbps) to access the satellite with TDMA technique.

During the second part of the frame a hybrid FDMA/TDMA technique is used in which multiple carriers are time-shared between the users of secondary zone transmitting 120 Mbps bursts. The system uses the "single user per burst" concept in a way that each 2 Mbps or 34 Mbps customer premises terminal is equipped with a TDMA equipment transmitting bursts of data at 120 Mbps at a multiplicity of time and frequency slots in a time-frequency map. If the network operates on N frequencies, up to N users can burst simultaneously.

The technique mentioned above significantly increases the satellite efficiency during the low burst rate period

and provides the full connectivity between low traffic
burst and high traffic burst stations. It is also believed
that the space segment complexity will be reduced in case
of on board regeneration.

Payload configuration using mixed TDMA and hybrid
FDMA/TDMA technique

ALCATEL ESPACE is currently studying an optimum space
segment configuration for WIDEBAND satellite based upon
the use of a scanning active phased array antenna which is
more compatible with TDMA technique. This phased array
antenna has also an interesting trend towards new
technologies (MMIC and optical) when it is compared with
multi beam hopped antenna systems using complex waveguide
BFN's.

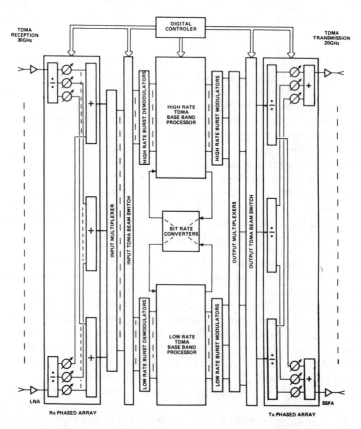

Fig. 5 Simplified payload block diagram with phased array antennas.

Fig. 5. shows a simplified block diagram of the communication payload. The antenna receive/transmit system consists of scanning beams covering six distinct geographical zones.

Conceptual burst scheduling and transponder loading plan is shown in Fig. 6. For the first time interval of the frame duration, the 1 GHz transponder band is accessed in TDMA mode by 3 carriers at 480 Mbps. For the remaining time interval, 6 carriers access the 500 MHz of the transponder bandwidth in hybrid FDMA/TDMA mode with 120 Mbps bursts directly originated from low traffic users equipped with TDMA equipment.

The satellite is regenerative in a way that the digital bursts are demodulated for each zone. Up to 27 burst demodulators are needed on board (without redundancy). High bit rate bursts are routed within high rate base band processors and low bit rate bursts within low rate base band processors.

The speed-conversion process requires on board memories to interconnect high bit rate bursts of primary zone to low bit rate bursts of secondary zone and vice versa. In the

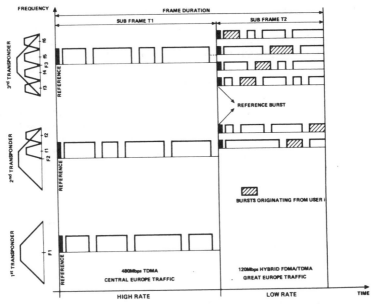

Fig. 6 Burst scheduling and transponder loading plan.

transmit section the digital bursts in their appropriate assigned time slot are remodulated on 9 carriers and are sent down to the ground stations via the scanned spot beams following a time sequence schedule. The scanning spot beams generated by the phased array antennas and base band switch matrix are completely synchronized between themselves by an on board digital controller receiving information from a main control earth station.

Overall assessments on investigated configurations

Table 1 depicts the mass and power consumption of all subsystems (including redundancy) within the first payload configuration using TDMA and SCPC techniques. Following technological predevelopments forecast in 1990 have been estimated for the equipment of the payload as the basis of the evaluation:

- RBFN/TBFN with waveguide latching circulators and classical waveguide couplers at 30/20 GHz- LNA with hybrid FET chip technology at 30 GHz
- Up and down converters with hybrid technology
- IMUX with classical invar cavity technology
- OMUX with classical invar and manifold technology
- Microwave switch matrix with 0.5 μ MMIC FET for switching elements and strip line MIC for input/output coupler
- DCU with CMOS and TTL using hybridation technology
- TWTA with magnesium body for TWT and HEXFET unitrod for switching elements- Local oscillators with synthetizers using PLL and dielectric resonators for VCO's and hybridation technological efforts for loop elements
- Demodulators with direct demodulation handling up to 36 carriers at 4 GHz within one heavy structure integrated equipment using high dielectric resonators (VCO's with ε~80), MIC for front ends and custom HCMOS IC's for base band processors

- Modulators with direct modulation at 20 GHz using hybrid couplers on silicon substrates and driver circuits with hybridation technological efforts
- Redundancy switches on waveguides
- Baseband processor with high speed ECL custom VLSI for baseband switch matrix and CMOS VLSI for digital controller
- Bit rate converters with VLSI ECL/CMOS and CMOS/ECL technologies.

Table 1 Mass and power of all subsystems within the first payload configuration using TDMA and SCPC techniques.

	MASS (kg)	POWER (W)
Primary sub system	100	815.15
Secondary zone subsystem	311	2355.8
TV zone subsystem	37	517.2
Cross traffic subsystem	6	29
Power supply subsystem	30	30 (η=85%)
Communication and Data Handling subsystem (TTC)	35	50
Antenna & Feed horns subsystem	60	−
Payload harness	65	−
TOTAL	644	3900

Although the above mass and power evaluation has been compatible with European Olympus platform capabilities, the technological constraints have led to a very complex payload mainly due to the large number on board SCPC demodulators and very complex waveguide RBFN/TBNF structures.

For these reasons the second configuration of the payload is being currently investigated by ALCATEL ESPACE. This configuration takes the advantages of regenerative satellite using TDMA technique to reduce the number of on board modems as well as scanning phased array antenna with technological assumption of implementing one MMIC SSPA on each individual feed to increase the flexibility of the system.

Conclusion

As a further step, ALCATEL ESPACE is presently envisaging consideration of optical technologies for future developments of receive/transmit phased array elements. These elements are RF combiners/dividers which can be replaced by optical waveguides ; RF phase shifters can be eliminated by ROTMAN's lenses (see ref. 2). The RF/optical interfaces would be realised by optical modulators and photodetectors.

As a conclusion, it is believed that the investigated range of solutions will considerably reduce the complexity of the WIDEBAND payload resulting then to a lower mass and power budget and therefore to an economical interest of WIDEBAND high capacity satellites.

References

[1] Wideband communications via satellite : prospects in Europe for the 90's; by R. COIRAULT (ESTEC Noordwijk), F. DACHERT (THOMSON-CSF), and B. CAMOIN (SAGATEL) in space communication and broadcasting (North Holland) 1984.

[2] Final report of phase 1 of the WIDEBAND study ESA contract 4931/81 THOMSON-CSF and SAGATEL, August 1983.

[3] Wideband communications on an ISDN integrated satellite by G. COFFINET (Alcatel Thomson Espace); 7th IDATE international workshop on "European Communications"; 21 Nov. 1985 Montpellier.

[4] Optimization of a high capacity communications satellite for Europe; by F. DACHERT, JL. de MONTLIVAULT (THOMSON-CSF), and R. COIRAULT (ESTEC Noordwijk), AIAA 1984.

[5] Wideband communications : trends in technology; by F. GERIN and C. VEYRES (DGT France); Space communication and broadcasting 1984 (North Holland).

[6] A 30/20 GHz multibeam array antenna suitable for future high rate TDMA European satellite systems; by R. LENORMAND, JP. MARRE, D. RENE (Alcatel Espace), H. EBBESEN (ESTEC Noordwijk); IEEE symposium on Antennas and Propagation; June 1987.

[7] A dual frequency band, dual bit rate payload ocncept for communications satellites; by G.K. SMITH (Inmarsat) and A. VERNUCCI (Telespazio).

[8] A demand-assigned mixed TDMA and FDMA/TDMA system; R.J. FANG COMSAT technical review Vol. 12; Spring 1982.

[10] A 20/30 GHz multibeam antenna for European coverage ; by G. DORO, A. CUCCI, M. DI FAUSTO (SELENIA), and A.G. ROEDERER (ESTEC Noordwijk); IEEE Antennas and Propagation International symposium; 1982; Albuquerque.

[11] An experimental scanning spot-beam satellite system implementing 600 Mbps TDMA; by A.J. RUSTAKO, G. VANUCCI, and C.B. WWODWORTH (Bell Laboratories).

[12] On board antennas with multi-feed reflectors; by A.G. ROEDERER (ESTEC Noordwijk); Annales des Télécommunications; vol 39; number 1-2; 1984.

[13] Microstrip antennas for millimiter waves; by M.A. WEISS; IEEE Transactions on Antennas and Propagation vol. AP-29; number 1; January 1981.

[14] Compact optical beam forming system for large phase array antenna "MILCOM 84" C.A. KOEPF and JL. MOLZ.

Communication Satellite and
Related Advanced Technologies in Japan

Kiyoshi Takahara*
Mitsubishi Corporation, Tokyo, Japan

Abstract

This paper discusses the possible applications of
future developments in satellite communications to a
variety of new industries such as new communications media,
biotechnology, space technology, new materials, and
electronics.

1. Introduction

After the launch of the geostationary satellite,
"Early Bird", in 1965, satellite communications have become
very important world-wide both for domestic and international telecommunication uses.

As will be discussed later, the characteristics of
satellite communications can be summarized as being of low
cost, having a wide band as well as covering a wide area.

Internationally, the INTELSAT and the INMARSAT are
the typical satellites. Domestically, more than one hundred
satellite systems are now in orbit or are going to be in
orbit belonging to more than ten countries, including
Japan.

In Japan, an experimental communication satellite
called the "CS" was launched in 1977, ten years ago.
Then in February and August of 1983, commercial communication satellites called the "CS-2" were launched from
Tanegashima Island off the main Island of Kyushu.

In these satellites, digital communication systems,
using quasi-millimeter waves (K band, 20/30 GHz), were
used for the first time in the world.

---

*Managing Director, Department of Technology Affairs.

Although C band (6/4 GHz) and Ku band (14/11 GHz) had been used internationally up till now, plans to utilize Ka band are going to be discussed at NASA in the United States and at ESA in Europe as countermeasures for frequency congestion in C band and Ku band.

Another characteristic of satellite communications to be stressed is the reliability of communication systems against disasters on land.

In particular, after the cable fire accident which occured in the Setagaya area in Tokyo November 1984, the CS-2 satellite and transportable earth stations were fully activated for communication between people in the Setagaya area and people in the other parts of Japan.

The aims of satellite communications in the future are to develop multibeam satellite communication systems and increase the mass of satellites. As will be explained in the figures later, these two technologies contribute to the reduction in the communication cost and in the effective utilization of geostationary orbit.

The application of satellite communications to mobile communications is another subject for future satellite communications. Although the INMARSAT is already in commercial use, the application of satellite communications to automobile communication systems and even to portable telephone systems can be expected in the future by the use of multibeam satellite communications.

2. Principle of Satellite communications

Figure 1 shows the principle of satellite communications. Repeaters (transponders) and antennas for radio wave communications are mounted on the geostationary satellite, which moves synchronously with the earth at a height of about 36,000km above the equator. Using these repeaters and antennas, radio waves transmitted from, for example, earth stations on the west coast of the United States are received, amplified and transmitted to earth stations in Japan. The earth stations in both Japan and on the west coast of the United States are connected directly by radio waves. In the same way, earth stations in Japan and those in Australia are also connected directly by radio waves. This is the principle of satellite communications.

Figure 2 explains the geostationary satellite. When a satellite is launched from the launch pad, either by expendable rocket or a space shuttle, and the satellite is controlled so that the centrifugal force and centripetal force of the satellite in relation to the earth will become

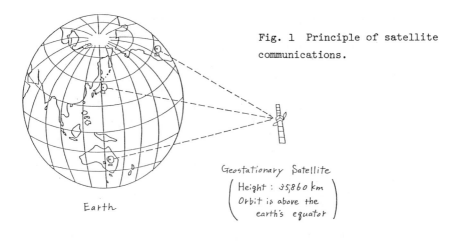

Fig. 1 Principle of satellite communications.

Earth

Geostationary Satellite
(Height : 35,860 km
  Orbit is above the
      earth's equator)

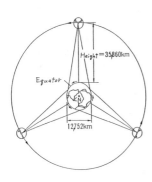

Fig. 2 Geostationary satellite. Velocity = 11,070km; rotation time = 23 hours 56 minutes.

Table 1 Characteristics of satellite communication.

Fundamentals	Utilization
Wide Area Capability: Any earth station within antenna pattern area can communicate with each other	(1) Transmission cost and quality are independent of distance between two earth stations.
	(2) Capability of one point to-multipoint transmission.
Wide Bandwidth Capability: Only one radio repeater in space	(3) A multiple-access facility is capable by carrying all types of signals on a demand basis.
Long distance causes large delay time (300 ms) and large propagation loss.	(4) Capability of speedy link set up and simple maintainance.
	(5) Reliable against disaster on land.

equal to each other, then the angular velocity of the
satellite orbiting around the earth will become the same
as that of the natural rotation of the earth. This will
result in the orbit of the satellite being synchronized
with the rotation of the earth. When the orbit of the
satellite is synchronized with the rotation of the earth,
the height of the satellite from the equator, velocity of
the satellite and the orbit time of the satellite are
estimated as 35,860km, 11,070km per hour and 23 hours
56 minutes, respectively.

Figure 3 explains the fundamental configuration of
satellite communications. Input signals to the transmit
earth station are modulated, frequency converted, amplified
by a power amplifier, filtered by the output multiplexer
and then radiated to the satellite through an antenna
In the satellite, radio waves from the transmit earth
station are amplified by a power amplifier and then
radiated to the receive earth station through the transmit
antenna. In the receive earth station, radio waves from
the satellite are received by the antenna, filtered,
amplified by a low noise amplifier, frequency converted
and then demodulated to the same original input signals as
to the transmit earth station.

Table 1 illustrates the characteristics of satellite
communications and the ways to utilize these characteristic
for telecommunications.
The fundamental characteristics are:
Firstly, a wide area capability, which means that any
earth station within the antenna coverage area can
communicate with each other.
Secondly, a wide band capability, which means that the
transmission medium consisting of radio propagation and
only one repeater is the best medium for obtaining a wide
band.
Finally, the long distance causes a large time delay,
300ms, and a large propagation loss.
The above characteristics are utilized for telecommunication systems in the following ways:
Firstly, transmission cost and quality of transmission are
independent of the distance between two earth stations.
Secondly, single point to multipoint communication is
possible.
Thirdly, capability of multiple-access brings about the
capability of carrying all types of signals on a demand
basis.
Fourthly, capability of a speedy set-up of communication
links and the capability of easy and simple installation
and the maintenance of communication systems.

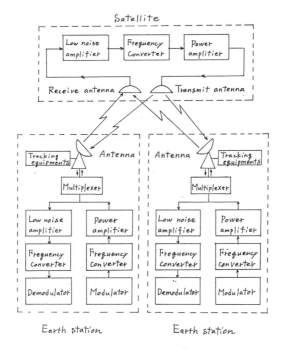

Fig. 3 Fundamental configuration of satellite communications.

And finally, reliable communication system to prevent
potential disasters on land.

3. Development of Satellite Communications in Japan

Now, the author wishes to talk about the history of
development of satellite communications in Japan.
An experimental communication satellite, "CS" was launched
in December, 1977, about ten years ago, from Cape Canaveral
by NTT. Utilizing the experiments, commercial communication
satellites CS-2a and CS-2b were launched in February and
August, 1983, respectively, from Tanegashima Island by NTT.
In these satellite communication systems, C band and
Ka band (20/30 GHz) were utilized.
    Succeeding the CS-2s, the CS-3s, which hold larger
channel capacities than CS-2s, are going to be launched
from Tanegashima Island in 1988 by NTT.
    On the other hand, "Space Communication Corporation
(abbreviated by SCC)" - a new common carrier, established
about two years ago as the subsidiary of Mitsubishi

Table 2 Fundamental parameters of SUPER BIRD satellite system.

Type of Stabilization	Three Axis Stabilization
Mass (On Geostationary Satellite Orbit, Initial)	1,400kg
Electrical Power (At The Life End)	4kw
Geostational Orbit	East Longitude 158° (SUPER BIRD a) 162° (SUPER BIRD b)
Station Keeping	South/North ±0.05° East/West ±0.05°
Life	10Years
Launch Vehicle	Ariane 4

Corporation – is going to launch a new communication satellite, called "SUPER BIRD" in Nov. 1988, by the launch vehicle of Ariane Space.

4. SUPER BIRD (Communication Satellite of SCC)

Table 2 illustrates the fundamental parameters of the satellite system of SUPER BIRD. Type of stabilization is three axis, initial mass on geostationary satellite orbit is 1,400Kg, electrical power at the life end is 4KW, geostational orbits are 158 and 162 degrees east longitudes, respectively for "a" and "b", station keeping accuracy is within ± 0.05 degrees, life expectation is about 10 years and the launch vehicle is expected to be Ariane 4.

Figure 4 shows the appearence of the SUPER BIRD, Table illustrates the fundamental parameters of the communication system of SUPER BIRD. Frequency band for the up links are 14.0 GHz – 14.4 GHz for Ku band and 27.5 GHz – 29.55 GHz for Ka band, frequency band for the down links are 12.35 GHz – 12.75 GHz for Ku band and 17.7 GHz – 19.45 GHz for Ka band. The number of transponders are 19 for Ku band and 10 for Ka band. Operating channels during eclipse are the same as the normal conditions discussed later.
Band widths are 36 MHz for Ku band and 100 MHz for Ka band. Antenna coverages are all Japan both for Ku band and Ka band, having one spot beam to Tokyo area for Ka band as shown in Fig. 5..Saturation output power of transponders are 35W for Ku band and 29W for Ka band. EIRPs are 53.7dB for Ku band and 50.7dB for Ka band with 57.0dB for spot beam.

As discussed in Table 2, the fundamental features of the
SUPER BIRD communication satellite are summarized as
follows:
(1) High power of SUPER BIRD, which allows minimize earth
terminal size of antenna (1.2 - 1.8 meters for Ku band,
0.6 - 0.8 meters for the Ka band), is especially
suitable for private data networks with very small
aperture terminals (VSAT).
(2) Use of interference free from terrestrial microwave
bands (Ku band and Ka band) operation allows no on premise
terminal siting restraints.
(3) High power allows mobile up-link for events or news
gathering with small mobile units as shown in Fig. 6.

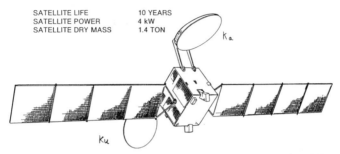

SATELLITE LIFE   10 YEARS
SATELLITE POWER   4 kW
SATELLITE DRY MASS   1.4 TON

Fig. 4   SCS with KA-KU band transponders. SUPER BIRD.

Fig. 5   Antenna pattern of SUPER BIRD.

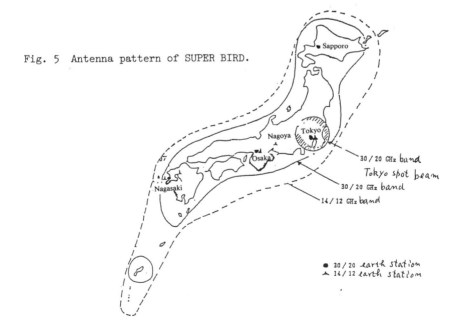

K. TAKAHARA

(4) High power and interference free Ku band and Ka band
transponders distribute video programs directly to offices
and homes as shown in Fig. 6.
(5) SUPER BIRD provides full eclipse operation with 70%
depth of discharge.
(6) Very large usable band width (total 3,368 MHz with
a and b) of SUPER BIRD provides economical transmission
systems compared with other satellite communication
systems.
(7) The redundancy configuration for the Ku band power
amplifier is that any amplifier may fail without the loss
of operations in any channel.
　　On the other hand, in the satellite communication
systems, there are some unique characteristics to be
solved in designing them. The following are the
problems and their corresponding technical countermeasures
in the design of SUPER BIRD satellite communication
systems.

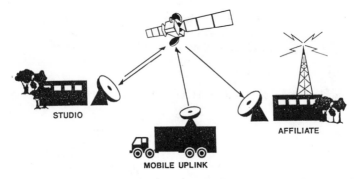

Fig. 6　TV distribution.

Table 3　Fundamental parameters of SUPER BIRD communication system.

Frequency	Ku	Ka
Up Link	14. 0 GHz～14. 4 GHz	27. 5GHz ～29. 25GHz
Down Link	12. 35GHz～12. 75GHz	17. 7GHz ～19. 45GHz
Channels	19	10
Operating Channels During Eclips	19	10
Band Width	36MHz	100MHz
Antenna Coverage	All Japan	All Japan ,
		One Spot Beam To Tokyo Area
Saturation Output	35W	29W
Power of Transponders		
EIRP	53. 7	50. 7, 57. 0 With Spot Beam

(1) As for the first problem of large propagation loss, technical countermeasures are to use shaped beam satellite antennas, low noise receivers and high power transponders.
(2) Second problem is the propagation delay time, and technical countermeasures to it are the use of an echo canceller against voice echo and a protocol capablity against data errors.
(3) Third problem is the satellite eclipses during the autumnal and vernal equinoxes, and technical counter-measures to them are the use of high efficiency and light satellite batteries as mentioned before.
(4) The fourth problem is the sun transit interference, and technical countermeasures to it are to switch over from a normal satellite to a redundant satellite and to increase transmitting power temporarily according to the prediction, if necessary.
(5) The fifth problem is the rain attenuations, and technical countermeasures to them are to use shaped beam satellite antennas as shown in Fig. 5 and to increase the transmitting power temporarily according to the prediction, if necessary.
(6) The last problem is the security, and technical countermeasures to it are to use cryptograms such as DES (Data Encryption Standard) etc.

5. Service Planning with SUPER BIRD

SUPER BIRD a and b are going to be launched in November 1988 and May 1989, respectively.

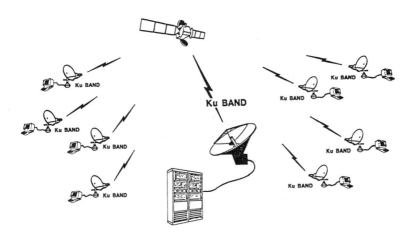

Fig. 7 VSAT application.

With the SUPER BIRD, the following services are
scheduled as the satellite communication systems.
(1) Television distribution services:
Distribution of broadcast programings to the television
broadcast stations, distribution of TV programming to the
stations of cable TV, up links and down links services on
demand, distribution of educational channels for remote
areas, special schools and that of special interest
channels such as horse racing, motor boat racing and
bicycle racing are scheduled as the examples of TV
distribution services.
(2) Private communication networks:
As the services for data, automated teller machine systems,
reservation systems, point of sales systems, data
collection platforms, data base services, electronic mail
services, remote sensing and monitoring systems are
scheduled as the examples of private data communication
network services. As for the data base services, stock
prices, medical informations, information of electronic
components and questions in entrance examinations are the
examples of the services. Fig. 7. illustrates the concept
of VSAT application as an example of private data
communication networks.

Examples of the private video communication network
are services such as video training, new product
introductions of electronic equipment, motor cars, etc,
and sales or engineering meetings. Figure 8 illustrates
the concept of private video communication networks.

As the services by intercompany communication networks,
flight reservations, insurance, just-in-time manufacturing
for the manufacturing of motor cars, electronic equipment,
etc, and the electronic document interchanges are scheduled
as examples of intercompany network services.

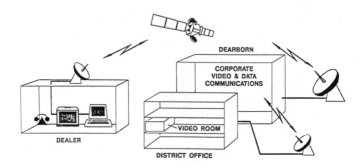

Fig. 8 District video conferencing and corporate data communications.

(3) Common Carrier Services:
As the services by common carriers, thin rout voice and
data services serving remote areas, back up for high
density and medium density routes, emergency restoration
of terrestrial backbone facilities and communications
between very distant (e.g., longer than 2000 km) areas are
scheduled as the examples of common carrier services.

6. Multibeam Satellite Communication Systems as a Future
Project by NTT (Nippon Telephone and Telegraph Corporation).

Now let us go on to the subject of a multibeam
satellite communication system as a future project by NTT.
Figure 9 explains the principles of multibeam
satellite communication systems. It is well known that,
in the case of the single beam – on the left side of the
Fig. 9,-when the diameter of the transmit antenna is
increased ten times, the diameter of the antenna coverage
decreases one tenth at the receiving points in relation to
the former diameter. As a result, the electric wave
strength is one hundred times greater and the transmission
capacity is also a hundred times greater at the receiving
points. That means that the diameter of the receiving
antenna can be reduced one tenth of that of the former,
if a constant transmission capacity is required.

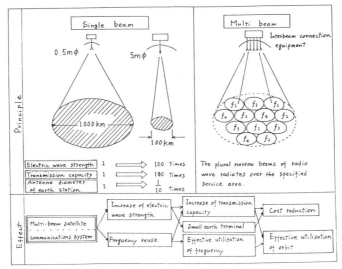

Fig. 9 Principles and results of multibeam satellite communications
system.

Accordingly, if we use radio wave radiators of plural
narrow beams and radiate these plural narrow beams over
the specified service area, and if we connect the beams
on each satellite, allocating a seperate frequency to each
on satellite, allocating a separate frequency to each
beam and reusing the frequencies as shown on the right
side of the Figure, we can increase the transmission
capacity of the satellite communication systems.
In addition, we can use smaller diameter antennas on earth
stations as well as using the frequencies effectively.

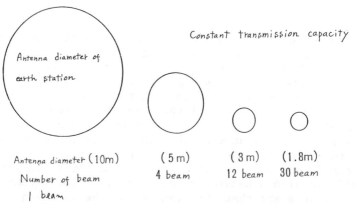

Fig. 10 Antenna miniaturization of earth station.

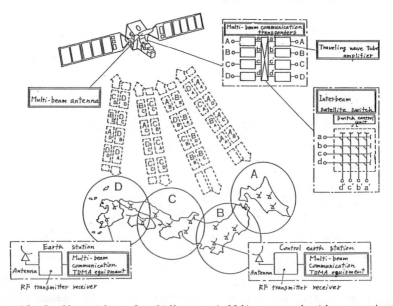

Fig. 11 Configuration of multibeam satellite communications system.

This results in a reduction of cost of satellite communication systems and the effective utilization of orbits.
     Figure 10 illustrates the miniaturization of
antennas on earth stations and Fig. 10 shows the relation
between the diameter of the antenna and the number of beams
necessary, in order to keep a constant transmission
capacity in the service area.
     Figure 11 illustrates a configuration of multi-
beam satellite communication systems. In this figure,
mainland Japan is covered by four multibeams. The multibeam antenna, multibeam communication transponders and
the interbeam satellite switch are mounted on the
satellite. The destinations of radio wave signals from
each Time Division Multiple Access earth station of any
area are assigned by the TDMA control earth station. As a
result, signals from areas A, B, C and D are repeated and
transmitted to the destination area, and it is able to
communicate among each earth station as shown by the dotted
lines in this Figure.
     Figure 12 explains the effectiveness of the increase
in the mass of satellite on the utilization of
geostationary satellite orbits. In this Figure, the
transmission capacity of the satellite communication system
is assumed to be 100,000 telephone channels. If we use a
satellite of 550kg class, we need a geostationary satellite
orbit range of 20.4 degrees, since we need seventeen

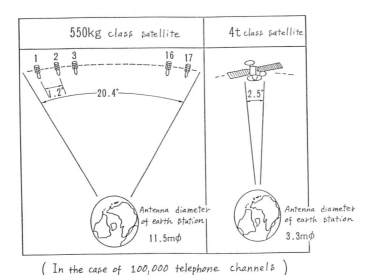

(In the case of 100,000 telephone channels)

Fig. 12 Effective utilization of geostationary satellite orbit.

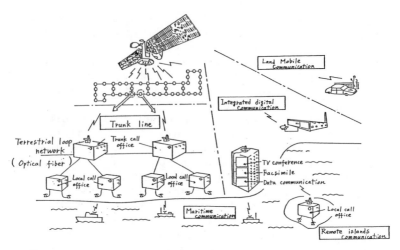

Fig. 13   Role of satellite communications in ISDN in Japan.

satellites with the orbit spacing of 1.2 degrees, in order
to carry 100,000 telephone channels. We also need an
antenna diameter of 11.5m for the earth stations.

On the other hand, if we use a satellite of the four-
ton class, we need a geostationary satellite orbit range
of only 2.5 degrees, since four-ton class satellite is
able to carry 100,000 telephone channels with one satellite.
We also require antenna diameter of only 3.3m for the
earth stations.   This is one of the reasons why we are
going to use a large-mass satellite.

7. Role of satellite communications in ISDN

Figure. 13. illustrates the role of satellite communications in ISDN in Japan.

According to this concept, in the middle of the 1990's
communication satellites weighing several tons will be
launched. In every prefecture in Japan, one or two trunk
call offices, of which earth stations for satellite
communications are to be installed will be arranged, and
between trunk call offices, communication circuits will
be connected with each other both through satellite
communications and by optical fiber communications as
shown in the Figure.

Trunk call offices and local call offices, and local
call offices and subscribers will be connected by optical
fiber circuits or copper wire circuits.

The larger offices or factories and local call offices
on remote island and trunk call offices on mainland Japan

will be directly conneted by satellite. Furthermore, ships in local Japanese waters and trunk call offices will be connected by satellite as domestic communications.

In addition, cars and the terrestrial network will be connected by satellite as land mobile service. The service items in this communication network -ISDN- will be TV transmissions, TV conferences, data communications, facsimile communications, electronic mail, videotex services, as well as telephone and telegraph services.

8. Conclusion

In conclusion, the author has explained the principles and configurations of satellite communications, development of satellite communications in Japan, fundamental features and service schedules of SUPER BIRD - communication satellite of SCC - multibeam satellite communication systems as a future project by NTT, and role of satellite communications in ISDN.

I believe that at present and in the future, advances of satellite communications will contribute very much to the advance of communication systems both in domestic and international uses.

Chapter IV.   Remote Sensing

Transportable Tracking and Receiving Station for Polar Orbiting Remote Sensing Satellites

Peter K. Pleitner*

GeoSpectra Corporation, Ann Arbor, Michigan

ABSTRACT

The very high cost and long lead-time required to install a conventional tracking and data receiving station for remote sensing satellites has denied or delayed the practical availability of this valuable technology for many developing countries, and has significantly limited the timely availability of vital information about resources and environmental conditions in many parts of the world. GeoSpectra and Space Data Corporation, with a complementary and combined experience in this technology of over three decades, have formed a joint venture with Jaeger Aerospace and others specifically to provide lower cost satellite tracking and data receiving stations with efficient image processing hardware and extensive software, including an optional transportable design version. Near real-time image data display and automated image quality analysis capabilities are integrated with a real-time geographic reference capability, which allows for intelligent allocation of image data processing and imaging resources to receive and/or process only quality image data of high priority geographic areas.

This paper presents an overview of our design philosophy, the system's configuration, and a description of it's mode of operation. The system's benefits include lower cost, effecient operation, and rapid deployment with lease options for the transportable version.

351

INTRODUCTION

On-board tape recorders and data relay satellites have
not delivered their promised share of the data obtainable
from remote sensing satellites. Therefore, image data from
many parts of the world without ground receiving stations
are not being acquired on a timely basis. For example, one
of two tape recorders failed on SPOT-1[1]; and TDRSS-WEST,
the second of NASA's data relay satellites, was destroyed
in the tragic launch failure of the Space Shuttle
CHALLENGER in 1986[2].

The upgrade of existing LANDSAT MSS receiving stations
in most of the world is progressing much too slowly to take
advantage of X-band data available from the new generation
high spectral and high spatial resolution scanners on
LANDSAT 5 SPOT-1[3]. The reasons are many, ranging from
satellite program and launch uncertainties to high cost and
long delivery/commissioning lead-time for ground stations.
Too many years of potential coverage from the current
generation of civilian remote sensing satellites have
already been lost, especially in those parts of the world
where the data is vitally important for resource and
disaster monitoring, planning and economic development. The
net result is that current and prospective users (as well
as the value added industry) who are planning programs
dependent upon the availability of high resolution
satellite data, are repeatedly frustrated by the lack of
timely and cloud free data. Ten years ago LANDSAT MSS
technology application was oversold by NASA; this led to
serious user frustration due to the inadequacy of the data.
Consequently, the French SPOT system, LANDSAT Thematic
Mapper and TDRSS were created. Now we have an equally
damaging situation - the promise of world wide, on demand,
high resolution data that solves many of the technical
shortcomings of MSS data, yet a severely limited means to
capture adequate data coverage for many parts of the world
as well as for many prospective users. The full benefit of
this highly advertised new generation remote sensing
technology is not equally distributed around the world
because satellite data can not routinely be acquired from
any place in the world. A rapidly deployable tracking and
image data receiving station is the most economical
solution to the problem of access to timely data.

Since 1974, GeoSpectra Corporation's business has
depended upon the availability of timely, not resampled,
and technically sound remote sensing data from around the
world. From that time, to the present, the lack of suitable
data has been a perennial problem. A transportable
derivative of a previously developed, relatively low cost,

S/X-band data receiving station has recently been designed in the U.S. by Space Data Corporation and GeoSpectra Corporation for Jaeger Aerospace Engineering, Incorporated. Rapid deployment and commissioning, relatively low yearly lease rates and trained operating personnel will enable a host country or organization to benefit from the technological benefits of space remote sensing in a matter of a few months anywhere in the world, thus able to be completely self supporting in terms of data access and data exploitation.

Discussed below is the functional design philosophy, technical summary, and data processing capacity of this transportable S & X band data receiving station for polar orbiting remote sensing satellites.

FUNCTIONAL DESIGN PHILOSOPHY

The satellite tracking and data receiving station was designed from the ground up based upon a compact and highly efficient data capture and data processing architecture. Consequently, the entire computer facility, including air conditioner, crew working quarters, data storage, and electronic equipment racks are deployable in a self contained standard size overseas shipping container that is

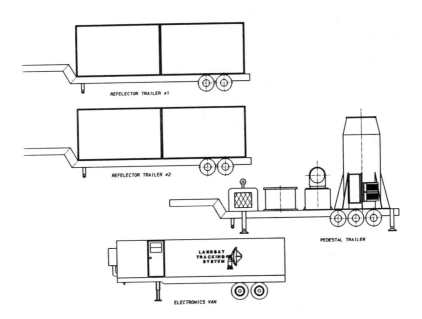

Fig. 1 Deployable shipping containers and pedestal hardware.

modified to function as a computer room. Two identical containers are used to transport sections of the high performance 7 meter diameter antenna dish; a fourth container is used to transport the antenna pedestal, electric motors, gimbals assembly, counter weights, and a diesel powered electric generator. The complete system is transportable by ship and flat bed trailer, or optionally by military transport aircraft.

Site preparation consists of a thirty square meter concrete pad with imbedded tie-down hardware for attaching the antenna pedestal or pedestal trailer hard-points, and a graded surface for parking the computer facility and storage of two reflector containers. Temporary use of a crane for lifting the 7 meter dish assembly, gimbals and counter weights to the top of the pedestal is also required.

Figures 1, 2, and 3 illustrate the various components of the transportable GeoSpectra/Space Data/Jaeger Aerospace 7 meter diameter satellite tracking and data receiving antenna, and image data processing system.

TRACKING AND RECEIVING SYSTEM

A high-performance dual S/X-band telemetry tracking and receiving system was developed to provide a low-cost earth station for receiving high-resolution data from current and future LANDSAT/SPOT types of polar orbiting satellites. Performance evaluation of a fielded 10 meter system, and the addition of electrically cooled extremely low-noise preamps, has allowed the dish size to be reduced from 10 to 7 meters diameter. The antenna system consists of a dual S/X-band telemetry tracking feed in a Cassegrain configuration with a 7 meter parabolic reflector designed for over 100 mph wind loading, reliable tracking in over 40 mph winds, and over 10 deg/sec/sec accelerations. The monopulsed antenna system is mounted to a newly developed elevation-over-azimuth tracking pedestal, which incorporates the latest technology in a dual brushless d.c. servo motor torque-biased drive train for each axis. This drive train provides an exceptionally wide dynamic range in tracking velocities for very slow horizon tracking as well as very fast velocities for near-overhead passes. A microprocessor-based servo control system, using the latest state variables feedback and adaptive control techniques, is used to provide accurate tracking for both slow and fast rates. Various control modes are selectable including autotrack, manual position command, manual rate command, programmed acquisition mode, and programmed overhead drive

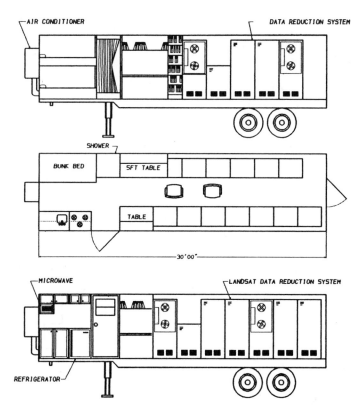

Fig. 2 Computer module details.

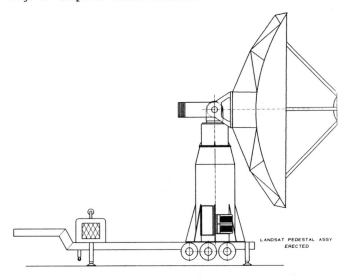

Fig. 3 Assembled antenna (optional trailer pad shown).

mode. Automated built-in test equipment is provided, along
with an S-band and X-band test target, for systems testing
and boresighting. Simultaneous reception of both S- and
X-band downlinks is provided with this system. The system
can be left unattended for automatic acquisition, track,
and data reception. Automatic boresighting and systems
testing can also be performed without operator
intervention.

ARCHITECTURE OF DATA PROCESSING HARDWARE

Figure 4 illustrates the data processing system's
hardware architecture. The system is based upon computing
hardware available in the United States in 1987, configured
into an architecture which concentrates the data transfer
rates and computational power precisely where it is most
needed.

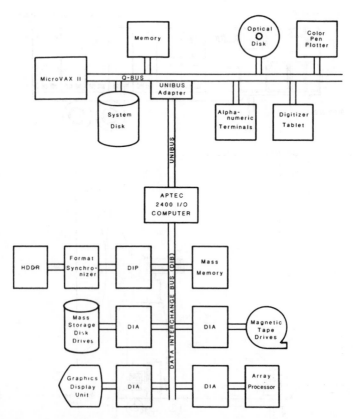

Fig. 4 Overall computer system architecture for satellite image data
processing facility.

In particular, the CPU has been downsized to a MicroVAX II, but the input/output (I/O), memory, computation, and archival capabilities are maximized. This is accomplished by using a very high speed data bus for the foundation of the architecture, specifically the Aptec I/O computer. Exceptionally high computation performance is provided by a FPS 5105 array processor. This specific model is selected based upon its compatibility with the Aptec I/O computer, and the radiometric and geometric image processing requirements of the receiving station.

A high resolution (1024x1280) moving window display is provided for quick-look monitoring of a decimated video channel when a satellite is within range of the station. When this image display unit is not in the B & W quick-look mode, it is usable as an operator work station for interactive multispectral (color) image processing.

All of the image data displayed by the moving window device can be recorded by a hard-copy camera system or on a 12 inch WORM (write once, read many) optical disk. WORM optical disks can also be used to build an archive of processed data, or traditional computer compatible tapes (CCT) can be written. WORM data storage technology provides the most compact, rapid random access, and permanent storage of image data. It is ideally suited for an image browse library of quick-look image data and for bulk storage of processed data.

Specifically selected and speed compatible magnetic disks are used for temporary storage of all telemetry data or formatted, converted and processed image data. Any of the currently available hard-copy image plotters can be linked to this architecture either through the Q-bus or the Unibus. However, we recommended that the hardcopy image generation device be configured as an off-line stand alone subsystem with its own tape drive so it can be operated independently of the receiving and image processing schedules simply by mounting processed image tapes.

Figure 4 shows the Q-bus and the DIB (high speed bus) as open to signify that this architecture can be readily expanded at these locations with additional peripherals to increase data processing throughput, and/or system management support devices. The throughput of the base processing system comfortably exceeds four full TM scenes and five panchromatic or multispectral SPOT scenes per 16 hours of operation. The base processing system can be easily expanded to linearly increase the throughput by adding a second or more array processors and associated devices to the DIB.

FUNCTIONALITY OF IMAGE PROCESSING SOFTWARE

Three of the unique attributes of this LANDSAT
receiving station's capability are: installation of
GEOSPECTRA's proprietary and very efficient batch image
processing software; its ability to prioritize geographic
regions for conversion and processing directly from the
high density digital tape (HDDT); and the ability to make
real-time data quality decisions from LANDSAT thematic
mapper telemetry data.

The system is furnished with geo-reference software
that will enable the operator to prioritize geographic
areas (as small as 10 x 10 km cells) to be acquired and
processed within the receiving station's data reception
area. This allows the operator to concentrate his system's
hardware and image data conversion resources on the highest
priority areas. In addition real-time TM data quality
assessment eliminates (at the operator's discretion) all of
the unnecessary and time consuming geometric and
radiometric processing of cloudy and/or over-water image
data.

Cells are automatically defined by the geo-reference
system along all ground tracks of each satellite. When
image data of a priority cell is acquired and the data
quality meets or exceeds the operator's quality parameter
vis-a-vis percent cloud or snow cover (for TM data
acquisition cycles), the scan lines that contain the target
image data cell(s) are automatically passed through the
system for format conversion and image data processing. The
output is a set of CCTs for distribution that conform to
the LTWG standard adopted by the Landsat Ground Station
Operators Working Group (LGSOWG), that can be optionally
backed up on WORM for the data archive. During a data
acquisition cycle the operator has the option to view in
near real-time a decimated single channel image of data
recorded on HDDT using the high resolution Image/Graphics
display monitor. He also has the option to print out a
graphic data coverage map of political and physical
boundaries annotated with high priority 10 x 10 km cells,
data quality symbols i.e. cloud/snow cover symbols, scan
line numbers, sensor, date and HDDT catalog number. He can
then write some or all quick-look and graphic overlay data
on a WORM for the "browse library" of the data archive.

Almost all of the image processing software was
developed and optimized during the past thirteen years of
GEOSPECTRA's commercial LANDSAT and SPOT data processing
service activity in the UNITED STATES. Consequently, the
operator can produce many types of enhanced, thematic or
otherwise value added image products using one of over

thirty programs; or he can customize an image product using the many image processing software "tools" that are delivered as part of this receiving station. Additional image processing tools are delivered to perform various format conversions, various types of spatial enhancement, radiometric corrections for cosmetic repair of noisy data, and for resampling the image data to any of over twenty map projections with the aid of ground control acquired from maps in any of over twenty map projections. All image tapes and hardcopy products can be automatically annotated with latitude/longitude identified along the margins (their intersections identified by small markings within the image), an alpha-numeric grid along the margin, and mile/kilometer scale bars with scene identification, date, and image processing data information along the bottom margin.

Most of this receiving station's hardware can be used 24 hours of the day. A great fraction of this time the system can be assigned to sophisticated, yet unattended batch radiometric, geometric, and multispectral image processing when a satellite is not within tracking range. When a satellite comes within tracking range, the system automatically reverts to its default stand-by tracking and recording mode, runs through its internal system status checks, and prepares for recording and processing of new data.

OPERATIONAL SCENARIO

Check out tracking, recording and quick-look systems; verify operational readiness, and activate system for operation.

Review ground track of approaching satellite and list of relevant priority areas.

The equipment aims the tracking dish, captures its signal and tracks the satellite, records the telemetry on HDDT and produces quick-look softcopy and optionally produces hardcopy imagery of decimated systematically corrected data.

System manager compares quick-look images with geographic referenced cloud cover strip chart. He makes preliminary determination of data quality in priority areas and marks the strip chart with first, second and third priority HDDT processing tasks.

System manager compares the system's CPU available access
time before the next satellite tracking and data
acquisition event, and compiles the batch data conversion
and image processing job priorities.

CCT production and image processing commences and continues
until the time of the next scheduled system check out and
data acquisition event occurs. HDDT, quick-look images on
WORM, and strip charts from the graphic plotter are
archived.

References

[1]"Satellite Update,"SPOTLIGHT, The Quarterly Newsletter from SPOT
Image Corporation, Volume 1, Number 3, October 1986, page 6.

[2]"Shuttle Destroyed, Killing Crew; Manned Space Flights Halted,"
Aviation Week & Space Technology, February 3, 1986, page 18, and
Dornheim, M. A., "TDRS Program Delayed Further Due to Loss in
Challenger Accident," Aviation Week & Space Technology, February
10, 1986, page 58.

[3]"Landsat Receiving Station Capabilities," EOSAT; LANDSAT Data
User Notes; Volume 2, Number 1, page 4.

Application of Landsat MSS and TM Data to Geological Resource Exploration

Ken M. Morgan* and David G. Koger†
Texas Christian University, Fort Worth, Texas
and
Donald R. Morris-Jones‡
KRS Remote Sensing, Kodak Inc., Landover, Maryland

Abstract

Remote sensing is becoming widely accepted by energy companies as an exploration tool. Computer enhanced Landsat MSS and TM data have proven useful in delineating structural and geochemical information often associated with petroleum exploration.

Integrating MSS and TM data with geophysical information production maps, and geochemical studies can lead to a better understanding of subsurface relationships important to exploration studies. This can be easily accomplished by use of computer systems equipped for multi-level data integration. Examples of computer oriented remote sensing geological investigation are presented in this study. This includes the utilization of Landsat data for detecting diapirs, growth faults, fracture patterns, and surface alteration related to hydrocarbon and mineral deposits throughout the world.

Over the past few years, we have found Landsat data extremely helpful in delineating target areas for further investigation. Proper computer enhancement of satellite data often provides the geologist with important information about surface and subsurface conditions in an exploration area.

---

*Director, Center for Remote Sensing and Energy Research.
†Research Associate, Center for Remote Sensing and Energy Research.
‡Vice President, Applications Division.

361

Introduction

Over the past few years, satellite remote sensing data
have become increasingly important in the search for energy
resources (Bailey et al.[1]). Many exploration companies now
now recognize the usefulness of satellite data for pin-
pointing that warrant further geologic investigation.
Because of their accurate and broad coverage, satellite
images can facilitate the worldwide search for mineral and
petroleum deposits (Kinnucan[2]).

There are many types of sensor platforms currently
being used in the industry. One of the most popular
sensors is the multispectral scanner (MSS) on board the
early Landsat satellites. Landsat MSS satellites have
circled the earth since 1971 at an altitude of 920 km (570
mi), recording reflected radiation in two visible (band 4,
0.5-0.6 m, and band 5, 0.6-0.7 m) as well as two near
infrared (band 6, 0.7-0.8 m, and band 7, 0.8-1.1 m)
wavelength bands (Goetz and Rowan[3]).

Each band of MSS data is stored on computer-compatible
tapes (CCT's) as digital data and can be used to produce
Landsat imagery (79 m resolution). Computer processing of
Landsat data can modify the image so that geometric, tonal,
and spectral characteristics are enhanced. The synoptic
view of large areas, spatially registered multispectral
information, and repetitive imaging make Landsat MSS data
especially useful as an exploration tool (Morgan et al.,[4]).

Recognizing the popularity and potential of sensors
like Landsat MSS, NASA recently launched (1982 and 1984)
Landsats 4 and 5, which differ from their predecessors in
orbit, design, and remote sensing capabilities. These
advanced satellites carry a payload that includes higher
resolution multispectral scanner (MSS) and a thematic
mapper (TM). Landsat TM is relatively new and, with the
advanced sensor technology, may prove to be exceedingly
valuable as a remote sensing tool for exploration over the
next few years.

Comparison of MSS and TM Data

According to Sabins[5], major improvements of TM over
conventional MSS data are: (1) increased spatial resolution, (2) improved spectral separation, (3) expanded
spectral range, and (4) collection of both reflected and
emitted (thermal) information. Table 1 shows a detailed

Table 1   Comparison of Landsat MSS and TM Data

	Landsat MSS	Landsat TM
Altitude	920 km	705 km
MSS on board	Yes	Yes
TM on board	No	Yes
Spectral bands	4	7
Resolution (visible & IR)	79 m	30 m
Resolution (thermal)	--	120 m
Image coverage	34,225 km^2	34,225 km^2
Pixels per scene	29x10^6	266x10^6

Table 2   Spectral Bands for Landsat TM

Bands	Spectral Range (m)	Characteristics	MSS Equivalent
1	0.45 to 0.52	Some water penetration	None
2	0.45 to 0.60	Green wavelength for vegetation mapping	Band 4
3	0.63 to 0.69	Red wavelength for soils, geology, vegetation, and urban mapping	Band 5 Portions of
4	0.76 to 0.90	Reflected IR for mapping drainage, vegeta-	Bands 6 & 7
5	1.55 to 1.75	Reflected IR, soil moisture, geobotanical, and vegetation mapping; penetrates thin clouds	None
6	2.08 to 2.35	Reflected IR, used with band 5 for hydrothermal alteration mapping; also, soils geology, and tonal pattern identification	None
7	10.4 to 12.5	Emitted Thermal IR	

comparison of these two sensor platforms. Note the improved resolution of TM over MSS (30 m vs. 79 m) and the increase in the number of spectral bands (four to seven). Table 2 gives the spectral location and general characteristics for each of the TM bands. Spectral TM bands 2,3, and 4 are similar to MSS bands 4, 5, 6 and 7, respectively. TM bands 1,5,6, and 7 were not available of previous Landsat systems.

Figure 1 is a comparison of two image types. This is a split screen image over a portion of San Antonio, Texas, showing the higher resolution capability of Landsat TM. The two half-scenes are at approximately the same scale (1:60,000) and have the San Antonio International Airport in the center for comparison. An urban scene was chosen to illustrate the greater mapping detail of the TM data.

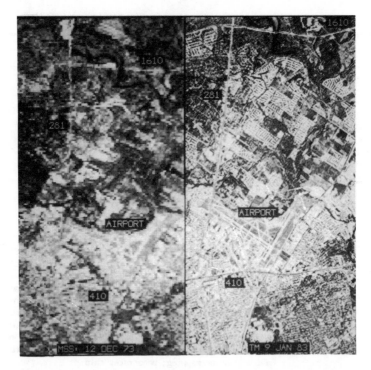

Fig. 1   Split screen comparison of MSS (left) and TM (right)
imagery at a scale of 1:60,000 over a portion of San Antonio, Texas.

Both scenes are multispectral composites of equivalent
bands (i.e., TM 2, 3, and 4 and MSS 4, 5, and 7).

      A potential drawback to the widespread use of TM data
is that it costs up to six times more than MSS information.
MSS data continue to have a very useful role as a reconnaissance exploration tool to be followed by more detailed
TM analysis.

              Image Processing and Interpretation

Image Processing

      One of the most important aspects of a remote sensing
study is processing the digital data to produce images with
improved characteristics for interpretation and analysis.
Techniques include geometric, radiometric, and enhancement
procedures to insure an accurate and high quality image.
Computer enhancement is extensively used to facilitate
geologic mapping, permitting easier interpretation of: 1)

lithology, 2) lineaments, 3) structures, and 4) tonal
anomalies. In the hands of a skilled geologic remote
sensing specialist, a digital image processing system
becomes a very powerful tool.

It is important to recognize that detailed remote
sensing analysis is often based on the type of imagery used
in a study. Proper selection of enhancement techniques can
sometimes be critical to the success of a project based on
image interpretation. When presented with a full compli-
ment of computer enhancements, the interpreter is more
fully equipped for image analysis.

Interpretation Principles

Geologic interpretation of Landsat data is often based
on lineament analysis, tonal patterns, and stream deflection. The information gained from each interpreted feature
is given below:

1) Lineaments--Potential faults and fractures
2) Tonal patterns--Stress, alteration and leakage
3) Stream deflection--Surface and potential
   subsurface structure

Lineaments are often due to fractures, faults and/or
joints that occur within the geologic material present in
an area. Sometimes these can be indicative of buried
structures or faulting at depth. It has also been shown
that some lineaments may line up with subsurface trends.

Tonal patterns can sometimes be related to hydrocarbon
seepage and subsequently alter surface materials or induce
vegetation stress. Mathews et al.,[6] report that tonal and
textural changes have been observed and mapped over known
oil and gas fields from Landsat data. Tonal anomalies are
not often seen clearly in the visible wavelengths, but some
times can be detected in the infrared region of the electromagnetic spectrum.

Both structural and stratigraphic traps may be detected by anomalous drainage or lineament patterns where drap-
ing or differential compaction occurs. These subsurface
features may not be detectable by normal surface indications, but by using the multi-band Landsat data, they can
often be enhanced for recognition (Morgan[7]).

Image Analysis for Exploration

In a recent report by Morgan et al.,[8] analysis of
Landsat data proved successful to date in the detection of
diapirs, faults, anticlines, and geobotanical anomalies
important to exploration programs. Although often subtle
in appearance careful study of the surface expression of
many of these features has provided important information
for planning seismic lines or conducting other surface
investigations in an area (i.e., aeromagnetic, geochemical,
and electrical techniques).

Proper analysis of Landsat imagery often calls for the
interpreter to map the following important geologic
features:
   1) Lithology
   2) Lineaments
   3) Structures
   4) Anomalies

Lithology

The property of spectral reflectance enables one to
differentiate between different materials using Landsat
imagery. Spectral reflectance curves represent the percent
of incident light reflected by materials as a function of
wavelength. Most interpreters think the spectral reflectance of rocks and soils is determined mainly by the oxida-
tion-reduction state and coordintion of the constituent
transitional metal ions. Electronic transitions in the
metal ions cause broad conspicuous absorption bands for the
individual metal-bearing minerals.

Because Landsat studies have determined that the
spectral radiance values of most rock types are surprisingly uniform, some very practical uses for spectral
reflectance curves can be used to provide comparison standards for identifying spectral characteristics of unknown
materials, and to recongnize spectral regions in which
various materials can be differentiated.

Lineaments

A lineament is defined as a mappable simple or composite linear feature on the surface differing distinctly from
the patterns of the adjacent features and presumably
reflecting a subsurface phenomenon. Although many lineaments are controlled by structural displacement, they may

also represent geomorphic features or tonal features caused
by contrast difference. A typical geomorphic lineament
would be a straight stream valley, whereas a tonal lineament could be caused by differences in vegetation, moisture
content or soil and rock composition (Sabins[5]).

Linear features that represent a line or zone of
structural offset are called faults. Newly discovered
linear features mapped on an image must be checked in the
field to establish the presence of structural offset before
they can be designated as faults. Because of the effectiveness of Landsat imagery for regional geologic studies,
it is not uncommon to detect a large number of features,
many extending tens and hundreds of kilometers, that do not
appear on published geologic maps.

Those linear features that do not show substantial
structural displacement are described simply as fractures.

Fig. 2   Split screen comparison of MSS (left) and TM (right)
imagery in southern Oklahoma showing that lineaments are often
more easily mapped from TM data (scale = 1:60,000). On this
figure IGN = igneous rocks, AG and PO = sedimentary rocks, and MF
= Meers Fault.

The origin of most fractures can be related to zones of
crustal weakness or, quite commonly in very rigid litho-
logies, they are found associated with features of structural displacement, such as faulting or folding. By using
these features to make regional geologic interpretation,
many applications for mineral, petroleum, and groundwater
exploration have been found. As the interest has developed
in using fracture porosities to increase oil and gas production, many recent petroleum exploration projects have
centered on finding and interpreting fractures. In order
to intersect the fracture zone, it is necessary to determine the dip of the fracture and offset the well location
appropriately.

Figure 1 is a split image comparing the resolution of
MSS and TM data for fault and fracture mapping where lineaments are very subtle and difficult to map. The images
cover the same area in the Frontal Fault Zone of the Ana-
darko Basin in Oklahoma. Both scenes are approximately the
same scale (1:60,000). It is easy to see how the higher
resolution TM data is superior to MSS in depicting the
fault/fracture systems in the area. At this scale, it is
difficult to trace out the faults on the MSS scene.

In many areas, lineaments, such as those in Figure 2,
may be controlling secondary porosity for shallow reservoirs. High density fracture zones often become potential
sites for shallow drilling operations or for deeper prospects if projected to lower horizons.

Structures

The tremendous potential of many subsurface structures
makes them especially attractive prospects. Examples include anticlines, plugs, reefs, and diapirs (salt domes).
In many cases, near surface structural feaures can be
recognized on imagery by obvious or subtle surface expressions. When examining Landsat imagery, these structures
often appear as circular features caused by differential
compaction, radial drainage, deflected stream patterns, or
tonal variations, (i.e., soil and vegetation anomalies).

Doeringfield, Jr., and Ivey[9] state that for structural
mapping, geomorphic analysis is of primary importance.
This includes drainage, landform, fracture, and tonal
pattern recognition of which drainage analysis is usually
the most important. Structurally controlled drainage
patterns include rectangular, trellis, radial, arcuate and
deflected.

Fig. 3   Landsat MSS image over a buried serpentine plug (depth
= 1,480 meters) in south Texas (scale = 1:80,000). The letter
"S" marks the center of the ancient volcanic plug.

        Figure 3 is an example of a Landsat MSS image over a
"serpentine plug" in Zavala County, Texas. "Serpentine
plugs" are ancient volcanic intrusions that occurred along
a fault/fracture belt extending over 1200 kilometers along
the Gulf Coast is in the U.S. These "plugs" are, today,
subsurface highs that often provide structural and stratigraphic traps for hydrocarbons. Note the display of
radial, arcuate, and deflected drainage patterns in the
area and over the known plug. When compared to existing
topographic maps of this area, many of these structurally
controlled drainage patterns are not mapped.

Tonal Anomalies
        One of the most interesting and rapidly advancing
areas of image interpretation is the recognition of tonal
anomalies associated with oil and gas deposits. Recent
studies by Texaco and Gulf (Mathews et al.,[6]) concluded
that preferential pathways of hydrocarbon leakage can often

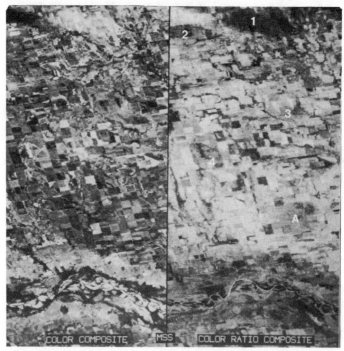

Fig. 4   Split screen comparison of a standard MSS multi-band
composite (left) and an enhanced ratio composite (right). The
numbered areas (1-4) are known producing oil fields. Location A
appears spectrally similar, but has not been drilled.

be recognized by spectral and textural analysis of Landsat
imagery.

A popular technique to highlight soil and vegetation
anomalies is to generate a color ratio image. Ratio
processing consists of dividing the spectral values of one
band by another band. Scaling the resulting bands over the
color range and assigning colors to each ratio allows a
color composite ratio image to be generated (i.e., 4/6=
red, 5/7=blue, 6/7=green).

Morgan and Morris-Jones[4] found ratio images are useful
because they exaggerate color differences over a full
scene. In effect, ratio images suppress topographic
details but emphasize color boundaries.

At present, there are no absolute or unique anomaly
patterns associated wiht hydrocarbon leakage. Soils,

vegetation, and climatic factors change drastically from
area to area. Often what is done, is to check out the
ratio response over known fields and look over the scene
for other area that show up similar tonal patterns. Using
the proper ratio of bands, this technique has proven
successful if the conditions in the area are conducive to
producing anomalies.

Figure 4 is a split scene example of MSS data showing
the advantage of a ratio composite over a standard mulit-
band composite image. This scene is located in southern
Oklahoma and illustrates how subtle tonal differences on
the standard composite (left) can be enhanced and displayed
by ratioing (right). The numbered areas (1-4) are known
producing fields. The location labeled A has not been
drilled, but a preliminary ground survey has shown anomalous geochemical readings. The important concept illustrated here is the ability of ratio processing to exaggerate spectral reflectance differences and produce
anomalous tonal patterns that otherwise might not have been
mapped. We find that varying geologic conditions make the
selection of appropriate ratio combinations difficult to
predict, but if microseepage is occurring in the area,
ratioing Landsat data appears to be an important technique
for displaying surface reflectance variations of the soil
and/or vegetation.

Summary and Conclusions

Many governments and exploration companies now utilize
Landsat imagery in conjunction with existing data, or when
other sources of information are inconclusive, when searching for hydrocarbons. Often the interpretation of Landsat
data is performed on basic single or multi-band imagery
with little attention given to resolution or processing
techniques. This paper illustrates the importance of
selecting both the appropriate resolution and enhancement
technique for mapping surface indications of subsurface
geologic features associated with hydrocarbons.

It is important to remember that Landsat analysis is
usually only one step in a successful exploration program.
When Landsat interpretation is combined with other tools,
detection and confirmation of potential sites can be more
accurately located. When used correctly, Landsat data can
be a valuable asset to global exploration of oil and gas
deposits.

References

[1]Bailey, G.B., Dwyer, J.L., Francica, J.R., and Feng, M.S., "Update on the Use of Remote Sensing in Oil and Gas Exploration," in Davidson, M.J., and Gottlieb, B.M., eds., Unconventional Methods in Exploration for Petroleum and Natural Gas III, Southern Methodist University Press, Dallas, 1984, p. 231-253.

[2]Kinnucan, P., "Earth Scanning Satellites Lead Resource Hunt," High Technology, March/April 1982, p. 53-60.

[3]Goetz, A.F., and Rowan, L.C.,"Geologic Remote Sensing," Science, v. 211, no. 2, 1981, p. 781-791.

[4]Morgan, K.M. and Morris-Jones, D.R.,"Basic Concepts in Digital Image Processing and Geologic Data Integration," Proceedings of the International Symposium on Remote Sensing of Environment, Second Thematic Conference, Fort Worth, Texas, 1982, p. 11-23.

[5]Sabins, F. F., Jr., Remote Sensing - Principles and Interpretation, Freeman, San Francisco, 1978, 428 p.

[6]Mathews, M.D., Jones, V.T., and Richers, D.M.,"Remote Sensing and Surface Leakage," Proceedings of the International Symposium on Remote Sensing of Environment, Third Thematic Conference, Remote Sensing for Exploration Geology, Colorado Springs, Colorado, 1984, p. 663-670.

[7]Morgan, K.M., "Satellite Data Aids Exploration Efforts," The American Oil and Gas Reporter, April, 1983, p. 53-61.

[8]Morgan, K.M., Koger, D.G., and Dees, D.A.,"Landsat Image Enhancement for Tonal Anomalies," Bulletin of the Association of Geochemical Explorationists, v. 2, no. 1, 1986, p. 63-70.

[9]Doeringsfeld, W.W., and Ivey, J.B.,"Use of Photogeology and Geomorphic Criteria to Locate Subsurface Structure," The Mountain Geologist, v. 1, no. 4, 1964, p. 183-195.

The Landsat Sensors:
EOSAT'S Plans for Landsats 6 and 7

Jack L. Engel*
Santa Barbara Research Center, Goleta, California

ABSTRACT

This psper describes the recent configuration of the
Enhanced Thematic Mapper (ETM) sensors that are presently
under construction for Landsats 6 and 7. The paper begins
with a brief introduction to EOSAT and a status report on
the Landsats 4 and 5 operations, followed by an overview
description of the Thematic Mappers that are currently flying on Landsats 4 and 5. The enhancements to the Landsats
6 and 7 Thematic Mappers are then described in some detail,
including the implementation of a panchromatic band of detectors providing 15-m spatial resolution for both ETM sensors. The Landsat-7 ETM may include as many as five bands
of thermal detectors with 120/160-m spatial resolution; the
implementation and the performance of this operation will
be discussed.

The paper also provides a brief description of two
new sensors that are being considered for Landsats 6 and 7:
a low-resolution (500 m) wide-field sensor (a Regional
Mapper) with the Thematic Mapper's spectral coverage; and a
high-resolution (10/20 m) narrow-field pointable Advanced
Landsat Sensor (ALS) using multispectral linear array
technology.

ABOUT EOSAT

The Earth Observation Satellite Company (EOSAT) is a
joint-venture partnership formed by the Hughes Aircraft
Company and the RCA Corporation for the express purpose of
establishing a private sector U.S. operational land obser-

*Associate Manager, Engineering Laboratories.

vation and data service program. EOSAT was formed in
accordance with the provisions of the Uniform Partnership
Law of the State of Delaware. Hughes and RCA each have an
equal interest in this joint venture.

EOSAT is operating under the authority of The Land
Remote-Sensing Commercialization Act of 1984 (Public Law
98-365, signed by the President on 17 July 1984) which
empowered the Secretary of Commerce "to solicit proposals
from United States private sector parties for a contract
for the development and operation of a remote-sensing
space system capable of providing data continuity for a
period of six years and for marketing the unenhanced
data." EOSAT was the successful competitor for the right
to operate the current U.S. Government system, and
continue development of the U.S. civil land remote
sensing, or Landsat program, and was awarded a 10-year
contract on 27 September 1985. The System Contract
Program Schedule is shown in Fig. 1.

The basic mission of EOSAT is to establish a commercially viable U.S. land remote sensing program. In order
to meet this mission objective, EOSAT has established the
following goals: (1) to provide global continuity of
Landsat data and user services, (2) to operate the
Landsat-4/5 systems and market the data products, (3) to
build, launch, and operate the Landsat 6/7 systems, (4) to
expand the line of products, (5) to provide quality
products and services in a timely manner, and (6) to
continue to develop enhanced sensors and data products.

EOSAT presently has three facilities in the United
States. The EOSAT Headquarters are located in Lanham,
Maryland, just outside Washington, DC, and two Field Sales
Offices are located in Houston, Texas and Norman,
Oklahoma.

LANDSAT-4/5 OPERATIONS

The Landsat-4 spacecraft continues to operate in a
degraded mode, and is used exclusively to provide Multispectral Scanner (MSS) imagery to international ground
stations. Two of the spacecraft's four solar panels have
failed, severely limiting the available power. Both
x-band transmitters have failed, precluding the direct
transmission of Thematic Mapper (TM) data, even though the
Thematic Mapper is completely functional. Landsat-4 was
launched on 16 July 1982. It will continue operation until a major system failure occurs which precludes further
use.

The Landsat-5 spacecraft is fully functional (except
for a minor failure in one of the s-band telemetry chan-

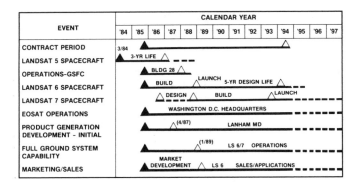

Fig. 1. System contract program schedule.

nels), and is being used to provide TM image data world-wide, and MSS data, primarily for U.S. domestic use. Approximately 100 Thematic Mapper scenes are acquired daily for U.S. processing, and another 150 scenes per day are transmitted to international ground stations. Landsat-5 was launched on 1 March 1984. It is presently expected to provide useful data into 1988, well beyond the end of its 3-year design life.

LANDSAT-6 AND BEYOND

Mission Objectives

Landsats 6 and 7 will provide image data compatible with the TM data presently available from Landsats 4 and 5 and will also provide enhancements that will increase the utility of the data. The new spacecraft will fly in orbits identical to those of Landsats 4 and 5 (circular, sun-synchronous, altitude of 705.3 km, 98.2 degree inclination, 98.88-min period, 9:45 a.m. equatorial crossing time for the descending node). The Landsat-4/5 World Reference System (WRS) will be retained with a repeat cycle of 16 days and 233 orbits per cycle.

The space segment will be designed with the capability (power subsystem, thermal control, etc.) to acquire up to 200 minutes (approximately 400 scenes) of Enhanced Thematic Mapper (ETM) imagery every day. Up to 70 scenes may be recorded for subsequent transmission to the EOSAT ground station. The ETM data format will be compatible with the TM data formats of Landsats 4 and 5. Image processing systems designed to process TM data from Landsats 4 and 5 will be able to process ETM data with only minor

TABLE 1   Thematic mapper parameter summary

Orbit	9:30 a.m. at 705 km (380 n mi)
Period	99 min
Swath Width	185 km (100 n mi)
Revisit Period	16 days
Telescope	16 in. f/6 Ritchey-Chretien

Bands (µm)	Detector Type and Amount		IGFOV (µrad)	Resolution (m)
0.45-0.52	SiPd	(16)	42.5	30
0.52-0.60	SiPd	(16)	42.5	30
0.63-0.69	SiPd	(16)	42.5	30
0.76-0.90	SiPd	(16)	42.5	30
1.55-1.75	InSb	(16)	42.5	30
2.08-2.35	InSb	(16)	42.5	30
10.4-12.5	HgCdTe	(4)	170	120

Radiative Cooler	90K, 95K, 105K
Minimum Field Size for Classification	10 acres
Scan Frequency	7.0 Hz bidirectional
Scan Efficiency	85%
Bit Rate	85 Mbps
Size	79 × 26 × 43 inches
Weight	535 lb
Power	332W
Duty cycle	30%

Fig. 2. Protoflight Thematic Mapper on shipping container base, aperture view.

changes. The Enhanced Thematic Mappers will provide data
of a quality equal to or better than that of the
Landsat-4/5 Thematic Mappers.

Landsat-6/7 Enhanced Thematic Mappers

Before describing the enhancements that are being
added to the Thematic Mapper for Landsat-6, we will review
the Thematic Mapper configuration as it existed for Landsats 4 and 5. A photograph of the aperture side of the
Landsat-4 Thematic Mapper is shown in Fig. 2. The Thematic Mapper is an object-space line scanner that collects
radiometric data in seven bands in the spectral region between 0.45 and 12.5 μm. The six spectral bands that sense
reflected solar energy (Bands 1 through 5 and Band 7)
utilize 16 detector elements each and provide 30m ground
resolution. Band 6, the thermal band, has four detector
elements and provides a ground resolution of 120m. There
are 100 signal channels in the Landsat-4/5 Thematic Mappers. The signals from the 100 channels are, after suitable amplification, sampled and converted into 8-bit digital words, which are then multiplexed into a serial
85-megabit data stream. Table 1 provides a summary set of
information about the Thematic Mapper.

Figure 3 illustrates the configuration of the TM
optical system, the location of the two focal plane assemblies, and their relationship to the ground track. An

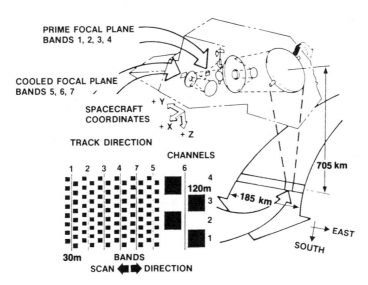

Fig 3. Thematic Mapper operation.

object-space scan mirror sweeps the TM line of sight back
and forth at a frequency of 7 Hz in a direction normal to
the orbital ground track to form a raster of 16 lines in
Bands 1 through 5 and 7, and four lines in Band 6. Data
are collected during both the forward (west to east) and
reverse (east to west) directions of scan with a scanning
efficiency of 85 percent. An image plane scanner called a
scan line corrector rotates the line of sight backward
along the ground track during the active scan time of the
scan mirror to compensate for the orbital motion of the
spacecraft. This action generates scan swaths that are
perpendicular to the ground track and parallel to each
other.

The Landsat-6 ETM

     The Enhanced Thematic Mapper for Landsat-6 will re-
tain all of the features of the Landsat-4/5 Thematic Mappers and will add the following capabilities: (1) a
panchromatic band of detectors will be added with 15m
ground resolution and spectral coverage from 0.5 to
0.9 μm, (2) a second multiplexer will be added to enhance
the reliability of the ETM and allow the recording or
direct transmission of two 85-megabit data streams, (3) a
partial aperture solar calibration capability will be
added to allow a periodic update of the calibration of the
internal calibration lamps, (4) a second power supply will
be added (each supply is fully redundant internally) to
allow the simultaneous operation of all spectral bands and
both multiplexers, (5) the dynamic range (the radiance
level at saturation) of each spectral band is being reviewed and will be changed as appropriate, (6) and two

TABLE 2   Anticipated performance of new thermal bands

Band No.	IGFOV (m)	$\Delta\lambda$ (μm)	NEΔT at 300K (K)	NEΔε	MTF
M	120	3.53-3.93	0.29	0.012	0.44
8	60	8.2 -8.75	0.48	0.009	0.32
9	60	8.75-9.3	0.47	0.008	0.31
10	60	10.2-11.0	0.54	0.008	0.29
11	60	11.0-11.8	0.57	0.008	0.28

gain states providing for a factor of two gain change will
be implemented in each spectral band (independently selectable by ground command).

The panchromatic band detectors will be located in
the prime focal plane just outside the Band-1 detectors.
An illustration of the projection of all the Landsat-6 detectors on the orbital ground track is shown in Fig. 4.
The panchromatic band of thirty-two 15m detectors will
generate data at the combined rate of four 30m bands.
Mode selection and switching will be necessary to select
which spectral bands are processed by which multiplexer.
Three data formats will be available for Landsat-6, the
basic seven-band set (as available from Landsats 4 and 5),
and two formats which include the panchromatic band.

The Landsat-7 ETM

As many as five additional bands of detectors may be
added to Landsat-7. The additional bands would be located
in the 3.5 to 4.0 μm and 8 to 12 μm spectral regions and
have 120m and 60m ground resolutions, respectively. A
joint NASA/EOSAT committee was formed to recommend which
precise spectral regions would provide the maximum data
utility. The committee's spectral and spatial recommendations are shown in Table 2 along with the anticipated
radiometric performance of the new bands. Figure 5 illustrates the ground track projection of the Landsat-7 detector assemblies. The four detectors providing 120m resolution for the 3.5 to 4.0 μm band (Band M) replace the
Band-6 detectors of Landsats 4 to 6, while the detectors
for Spectral Bands 8 to 11 (8 to 12 μm) are offset in the
track direction from the detectors for the other bands by
exactly two array lengths. This is necessary due to optical system field angle and baffling limitations. The additional spectral bands will necessitate new data
formats. Six formats will be provided for Landsat-7. The
data content of the Landsats 6 and 7 formats is shown in
Table 3. The decision to implement thermal IR enhancements for Landsat-7 will be made before the end of 1986.

The Low-Resolution, Wide-Field Sensor for Landsat-6

On 30 May 1986, EOSAT announced plans for a new generation of land and ocean sensor capabilities for
Landsat-6. The decision to pursue market studies for the
new land and ocean data, and the R&D for a sensor which
will be designed to accompany the Enhanced Thematic Mapper
(ETM) on the new series of Landsat spacecraft, was made by
the EOSAT Executive Committee on 1 May 1986.

J. L. ENGEL

TABLE 3   Landsat-6 and 7 enhanced thematic mapper formats

Format		P	1	2	3	4	5	6	7	8	9	10	11
	1		✓	✓	✓	✓	✓	*✓	✓				
L-6	2	✓				✓		*✓	✓				
	3	✓				✓	✓	*✓					
L-7	4		✓		✓	✓		*	✓	✓	✓	✓	✓
	5		✓		✓	✓	✓	*		✓	✓	✓	✓
	6	✓						*		✓	✓	✓	✓

*L-6 only (M replaces 6 for L-7)
- Two multiplexers allow any two formats simultaneously
- Landsat-6: Three formats
- Landsat-7: Six formats

Wide-field data, with spectral resolution similar to
that of the ETM, will serve users involved in regional
monitoring activities. Applications include agriculture,
hydrology, mineral exploration, environmental monitoring,
and disaster assessment.

Wide-field data, easily co-registered to the high-
resolution ETM data, greatly enhances the cost-effective-
ness of the high-resolution data. It is EOSAT's plan to
provide wide-field data to users within 48 to 72 hours of
acquisition.

The Wide-Field Sensor will provide image data in
either of two sets. One data set will be "land" image

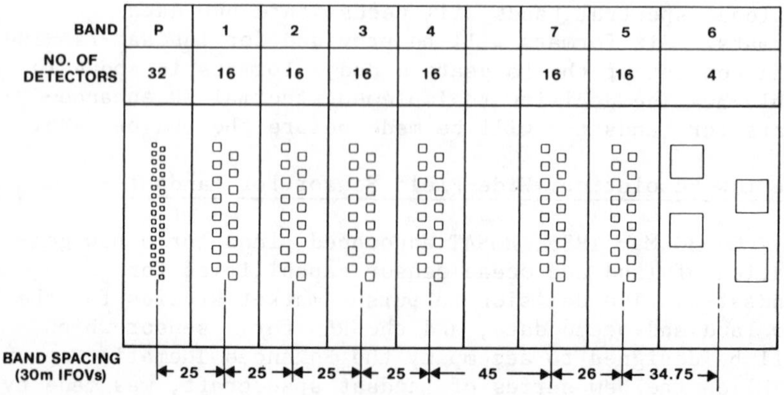

BAND	P	1	2	3	4		7	5	6
NO. OF DETECTORS	32	16	16	16	16		16	16	4

BAND SPACING (30m IFOVs)   ⊢ 25 ⊣⊢ 25 ⊣⊢ 25 ⊣⊢ 25 ⊣⊢ 45 ⊣⊢ 26 ⊣⊢ 34.75 ⊣

Fig 4.   Focal plane projection on ground track for Landsat-6.

data with a ground resolution at nadir of 500m. The
second data set will be "ocean" image data with a resolution at nadir of 1000m. Both data sets will cover a swath
width of 1500 km. The 1500-km swath will provide a 2-day
revisit period for equatorial imagery and a 1-day revisit
for scenes at latitudes greater than approximately 55 degrees.

The Land Data Set will provide image data in five of
the Thematic Mapper's spectral bands [three visible/near
infrared (VNIR), one short-wave IR (SWIR), and one thermal
IR (TIR)], while the Ocean Data Set will provide image
data in six spectral regions (four VNIR and two TIR). The
Land Data Set will be encoded with 9-bit resolution and
will generate data at 4 megabits per second. The Ocean
Data Set will be encoded with 8-bit resolution and will
generate data at 1.1 megabits per second. The two data
sets will be selectable by ground command and will not be
provided simultaneously.

Table 4 summarizes the Wide-Field Sensor's current
technical parameters. Other design variations are being
considered, including wider swath coverage for daily revisit over the entire earth and additional spectral bands
for the Ocean Data Set.

The Advanced Landsat Sensor (ALS)

The Advanced Landsat Sensor (ALS) is a third-genera-
tion multispectral remote sensing instrument presently

TABLE 4   Landsat-6 wide-field sensor (regional mapper)
parameter summary

Orbit	Landsat orbit (705.3 km), sun synchronous
Swath Width	1500 km
Ground Resolution	500m land, 1000m ocean (at nadir)
Scanning Efficiency	85%
Weight and Power	175 lb (80 kg), 220W

Land Data Set Parameters		Ocean Data Set Parameters	
Spectral Bands (µm)		VNIR Spectral	
	0.45-0.52	Bands	443, 20
	0.63-0.69	(Center λ	500, 20
	0.76-0.90	and Vλ in nm)	565, 20
	1.55-1.75		765, 20
	10.5-11.5	LWIR Spectral	10.5-11.5
		Bands (µm)	11.5-12.5
Encoding Resolution	9 bits	Encoding Resolution	8 bits
Data Rate	4.0 Mbps	Data Rate	1.1 Mbps

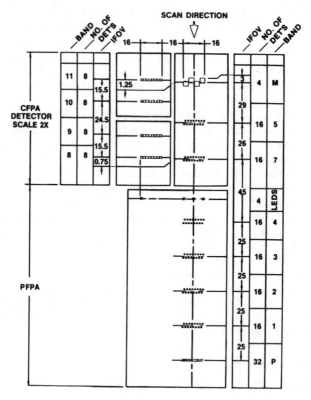

Fig. 5.   Focal plane projection on ground track.

being considered by EOSAT for the Landsat-7 spacecraft to
provide high-value data for scenes of opportunity. It
would provide a data set complementary to that of the ETM
and the Wide-Field Sensor. It is conceivable that all
three sensors could fly on Landsat-7. The ALS will utilize multispectral linear array technology to provide
high-resolution imagery in eight spectral bands (four VNIR
and four SWIR) simultaneously over a swath width of
41 km. Each of the four VNIR and four SWIR detector arrays will be illuminated through one of four selectable
bandpass filters. This provides for 32 spectral band
choices (16 VNIR and 16 SWIR), from which eight can be selected by ground command for any given scene acquisition. The ground resolution for the VNIR bands will be
10m, while the resolution for the SWIR bands will be
20m. The resulting data rate for the ALS will be 116
megabits per second.

10m VISIBLE AND NEAR INFRARED
20m SHORT-WAVE INFRARED

STEREO AND CROSS-TRACK VIEWING

41 km SWATH

32 SPECTRAL BAND SELECTIONS
 • 4 VNIR/4 SWIR SIMULTANEOUSLY
 • 4 FILTER CHOICES PER DETECTOR
   ARRAY

116 Mbps DATA RATE

24,576 DETECTORS

122 × 87 × 132 cm, 193 kg, 250W

Fig 6. Advanced Landsat Sensor (ALS).

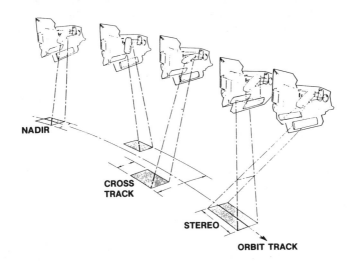

Fig 7. ALS imaging modes.

The ALS will provide a pointing capability of ±41 degrees, both along track and across track. The along-track
pointing capability will allow for the acquisition of a
stereo pair of imagery within approximately 2 minutes in
the same orbit, thereby eliminating the possibility of
temporal changes between the pair. The capability for
across-track pointing will allow the acquisition of high-
priority scenes with an average revisit time of 2 days (as
compared to 16 days without across-track pointing).

Figure 6 illustrates the configuration of the ALS and also provides a summary of the technical parameters. Figure 7 illustrates the imaging modes that can be obtained using the pointing capabilities.

SUMMARY

EOSAT is interested in providing the worldwide remote sensing community with the best image data possible within the financial constraints of a commercial entity. The existence of EOSAT and the commercial Landsat system allows the continuity of high-quality data to users through the 1990's. We sincerely believe that the Enhanced Thematic Mapper (ETM), the Wide-Field Sensor, and the Advanced Landsat Sensor (ALS) described in this paper will be capable of providing that data.

EOSAT will rigorously adhere to the nondiscriminatory policy set forth in Public Law 98-365 by making Landsat data equally available to all requestors. Specifically, any request for data will be filled at published prices.

BIBLIOGRAPHY

Engel, J.L., Thematic Mapper - An Interim Report on Anticipated Performance, AIAA, October 1980.

Engel, J.L., and O. Weinstein, The Thematic Mapper - An Overview, IEEE-IGARS, June 1982.

Engel, J.L., J.C. Lansing, D.G. Brandshaft, and B.J. Marks, Radiometric Performance of the Thematic Mapper, ERIM, May 1983.

Engel, J.L., Tne Thematic Mapper - Instrument Overview and Preliminary On-Orbit Results, SPIE, August 1983.

Engel, J.L., The Land Satellite (LANDSAT) System, Earth Observation Satellite Company (EOSAT's) Plans for LANDSAT-6 and Beyond, SPIE, April 1986.

Engel, J.L., The LANDSAT Sensors: EOSAT's Plans for LANDSATS 6 and 7, IAF, October 1986.

Alternatives for Mapping from Satellites

G. Konecny*

University of Hannover, Federal Republic of Germany

Abstract

The paper summarizes the aims and experiences of using remote sensing data for cartography. Satellite images of the more recent space missions (Space Shuttle, SPOT satellite) are suitable for cartographic missions. Also costwise they prove as attractive alternatives.

1. Need for Cartography

Maps represent a model of the earth's surface, on which it is possible to plan human activities. Since early history of man maps have been compiled for different purposes: *transportation* with maps at small scales (1:200 000) to be able to reach distant places; *defence* to protect the interest of rulers and later those of nations. Defence measures depend on ease of transportation at greater detail (1: 50 000); *planning of natural resources*: as nations moved from a pure agricultural to a production oriented industrial economy, maps formed the basis for planning measures on a regional scale (1:50 000) and to plan agricultural activity on a local scale (1:2000 to 1:10 000); *public works* required surveys and maps for large objects of construction, such as those of highways, railroads, dams (1:500 to 1:5000). *Urban environment* has a very complex setting. To plan the urban infrastructure maps are required at large scale (1:500 to 1:10 000).

As we have moved forward from the agricultural into the industrial age, and in some parts of the

---

*Professor, Institute for Photogrammetry and Engineering Surveys.

world even into the service oriented post- industrial age our mapping requirements have increased.

We need a map 1:50 000 for regional planning needs, and a large scale map (1:500 to 1:50 000) for the urban and suburban rural environment. Our mapping technologies by photogrammetry are sufficient to create the mapping requirements at small (1:200 000) and medium scales (1:50 000) for the first mapping coverage of densely populated countries, but information is lacking in the developing continents of the world. Likewise urban environments greatly lack the large scale map coverage to follow the rapid changes. Therefore two new technologies are required to deal with the mapping challenges of today.

The present status of world cartography according to United Nations studies published in 1984, valid for 1980 shows that only 42 % of the earth's surface is covered by 1:50 000 maps; it takes more than 40 years to cover the lacking areas at a progress rate of 1.2 % per year with photogrammetry, and it also takes a rate of 2.7 % per year to update all existing maps at that scale, which means also 40 years. In larger scale mapping the progress is even much slower.

Only a new maping technology can improve the progress rate. has been shown in the past: For the Federal Republic of Germany of an area of 250 000 km^2 it has taken 100 years to compile a map 1:50 000 or 1:25 000 from the year 1800 to 1900 by plane table ground surveys. Today photogrammetry is capable to newly survey or to update such a territory with existing facilities in 10 years. Only a new technology, which still needs to be developed can help with progress. For this reason **Satellite Mapping** is of interest to the Mapping Community.

2. Remote Sensing and Satellite Imagery

Photography is a classical remote sensing tool. However, remote sensing has reached its world-wide significance only after satellite systems have been able to propagate it as a discipline since Landsat in 1972. *Optical sensors* in the visible spectrum (400 to 700 nm = λ) are classically black and white, color or multispectral photography with photographic cameras. Photography has the advantage of very high resolution, but the disadvantage that the emulsion must be physically carried from the platform to the evaluation. Scanners, which direct radiation via rotating mirrors on a moving platform to a photosensitive element permit to convert radiation into electrical signals, which may be transmitted via communication channels (Landsat MSS and TM). More recently several photosensitive elements could be coupled in the form of photosensitive arrays to deduce electrical signals for a limited amount of sample elements (pixels), usually arranged along a line pependicular to the platform movement (Spot 1). *Near Infrared Sensors* are within narrow limits (700 to 800 nm) able to utilize photographic technology. Beyond this range scanners must be utilized (700 to 1000 nm and beyond (Landsat MSS &TM). Arrays for the near infrared range (1 to 2 µm) are under development (Spot 3).

The *thermal range* requires a scanner. It can be operated day and night (Landsat TM). The *microwave range* finally requires an active system, in which the energy must be generated. Radar systems are common in the weather independent X-band (3 cm = λ), the C-band (λ = 6 cm) or the L-band (λ = 30 cm) (Seasat, SIR-A,B).

The remote sensing information is usually treated in digital picture elements generated from high digital density tapes (HDDT) after reception on computer compatible tapes (CCT). Such digital information subjected to the rules of the Nyquist-sampling theorem of signal processing, must be compared in resolution to that of photographic systems, which usually express resolution in line pairs per mm (lp/mm). Photographic resolution is defined as the crossing of the total modulation transfer function

$$(M_{total} = M_{optics} \cdot M_{film} \cdot M_{image\ motion})$$

with the limiting function of the observaton system, with the modulation

$$M = \frac{\rho_1 - \rho_2}{\rho_1 + \rho_2}$$

dependent on the reflectances of two adjacent objects; the MTF being the curve of modulations for objects of different size (lp/mm). Due to sampling the value for 1 lp becomes $2 \cdot \sqrt{2}$ pixels, in order to compare photographic resolution with digital pixel size. The sensors may be used on various platforms:
Aircraft is usually limited to low altitudes limited by oxygen requirements to 3 km, by pressurization requirements to 5 km and by motor operation to 8 km.

From then on until 15 km (in exceptions higher) jets must be utilized. *Orbital laboratories* operate between 150 to 300 km, and are suitable for manned operations for a limited duration (Space Shuttle, Sojuz-Saljut, Mir). *Sun-synchroneous satellites* useful to carry optical sensors at equal illumination conditions, operate at about 800 km. *Geo-synchroneous satellites* orbit with the rotation of the earth on its equator at an altitude of 36 000 km. Satellite systems of the sun-synchroneous or the geo-synchroneous type have a usual lifetime of 3 years. Sensors and platforms lead to systems for a particular purpose, requiring either a high repetitivity or a high resolution. *Meteorology* and *oceanography* require only low resolution but high repetitivity. Most *remote sensing systems* for agriculture or mineral resources require intermediate resolution and intermediate repetitivity and *cartography* requires high resolution and low repetitivity. Therefore 3 types of remote sensing satellites are in use: *meteorological satellites* such as Meteosat, GMS and GOES operating every 30 min at 5 km spatial resolution from a geostationary orbit. These satellites operating worldwide are supplemented for the polar regions by the NOAA and Meteor-satellites operating twice daily (NOAA has 1 km resolution); *remote sensing satellites* are typically represented by the Landsat-series, with the MSS sensor operated at 80 m pixel size and the TM at 30 m in 6 visible and near IR-channels, as well as at 120 m in a thermal channel;

cartographic satellites have the characteristics of high spatial resolution, stereo-viewing capability for better interpretability and for 3D-measurement at a relatively low repetitivity. Camera systems from space (Space Shuttle, etc.) with 5 m pixel equivalent fall into this category, as well as SPOT with 10 m panchromatic pixel size which also attempts to increase repetitivity by directed viewing with a mirror systems at the cost of coverage.

The current space effort is divided between principally the USA and the USSR, but also other nations are beginning to operate space systems, such as the European Space Agency ESA (Meteosat), France (SPOT), the Federal Republic of Germany and India, which have already launched remote sensing systems from space and China, Japan, Brazil and Canada which intend to do so in the near future.

3. Alternatives for Mapping from Space

Let us examine the capabilities of mapping from space:
In order to map at a particular scale three requirements must be met:
(1) the image must have a resolution suitable for the
    interpretation of details to be shown in the map
(2) it must have a geometric accuracy within the
    range of map accuracy standards
(3) if topographic mapping of height information is
    required, the range of height accuracy require-
    ments must be met.

It is easy to see why the first Landsat satellites,
which up till now were generally the satellite information available world wide, could not meet 1:50 000
mapping standards in all 3 respects:
A topographic map shows significant detail, on Landsat
image at same scale no roads are visible. If one takes
a 1:50 000 stereo image suitable for mapping by photogrammetry at 1:50 000 and if one digitizes the imagery
at intervals of 80 m, 40 m, 20 m, 10 m, 5 m, 2.5 m etc.
one can notice loss of resolution the coarser the
digitization is done. The models from the digitized
imagery may be used to plot visible details in a stereo-
plotter. The result shows that a 2.5 m pixel size is
required in monoscopic plotting and of 5 m in stereo-
plotting to show the details of a 1:50 000 topographic
map. This is why only the most recent cartographic
satellite systems come close to meet topographic
mapping requirements, even though remote sensing
satellite systems may be useful for thematic mapping:
1) The coverage of Meteosat, every 30 m with 5 km
   pixel size permits for example only temperature
   or vegetation index mapping.

2) Landsat MSS has become the most popular satellite
   remote sensing system due to the many receiving
   stations capable to receive MSS-imagery. Already
   in 1980 an almost full global Landsat coverage
   was achieved; in most parts at several seasons.
   This is more than the mapping coverage. Where
   receiving stations did not exist, tape recorded
   images were utilized, which were read out over
   the USA. In many parts of the globe, e.g. Africa,
   generally only photographic reproductions produced on a recorder in the US exist as a standard
   product.

When digital technology is available in form of CCT's and image processing equipment, only then it becomes possible to utilize the image content to the full.

There is a different requirement for geometric restitution of remote sensing satellites, which only wish to relate image features to those of a map and which wish to use multitemporal information based on the same geometry, such as is the case with Landsat MSS. Cartographic possibilities have, however, more stringent requirements. Due to the higher resolution of cartographic satellites restitution in 3D is required, to remove planimetric distortions due to relief.

One of the uses is multitemporal land use classification of Landsat MSS images. The classification yields about 8 classes, but misclassifications occur especially for settlements and the type of vegetation. This is due to the fact, that the pixel size is too large for European conditions and that certain chosen classes are not separable. A useful land use product results however, when a classification is superimposed on a 1:200 000 black and white topographic map base. In a multi-sensor application 2 Landsat MSS channels and 1 radar image of Seasat may be rectified to map data. The resultant combination shows settlements better due to the high reflective characteristics of radar for buildings.

Since 1982 the Landsat Thematic Mapper with 30 m pixel size is in operation. Now the important transportation network becomes visible. Image maps 1:100 000 may be compiled from TM-imagery. Classification into 8 classes becomes possible with higher reliability even for urban areas.

An optimal representation in image form is, however, not the classification itself, but a 1:50 000 thematic representation in which a gradient high frequency filtered image in green is jointly presented with a classification result in blue and red. The resulting product has a maplike appearance. It is even better to superimpose the classification or the image processed product onto a 1:50 000 map to obtain a useful product for map revision and for adding thematic content to a topographic map.

In 1983 the Fed. Rep. of Germany has operated the Metric Camera experiment on ESA's Spacelab-1 on NASA's Space Shuttle. 1050 space photographs have been taken throughout the globe with an aerial mapping camera RMK 30/23 operated from space. The resolution of this imagery was determined by edge gradient analysis. The MTF gave high contrast resolution of 40 lp/mm and low contrast resolution of 25 lp/mm for black and white and of 32 lp/mm for color infrared imges. This means that an 8 to 10 m pixel equivalent resolution has been reached, making the resolution suitable for what is interpretable for a 1:100 000 map. The advantage of space mapping is that a single space image covers several 1:100 000 map sheets.

Space photography is plottable on analytical plotters, like the Planicomp C100 with few modifications in the realtime program (earth curvature and projection errors to be corrected by a proper coordinate conversion). In the Analytical Plotter a DTM can be measured. An orthophoto may be produced on the basis of the DTM on the Orthocomp Z-2 orthoprinter. The orthophoto may be superimposed with computer generated contours (e.g. by the TASH program).

The orbital height of 250 m paired with a base height ratio of 1/3 for the normal angle camera permitted to generate 100 m contours from the Spacelab images. The orthophoto can be produced at the scale 1:100 000 corresponding to the resolution of the image. It corresponds to a map substitute which is much quicker to produce (about 5 times) and much less costly. Nevertheless certain features like minor roads, and under desert conditions, even major roads can often not be detected in the images at the original scale 1:820 000 without magnification and stereo interpretation. Therefore plotting of these features is necessary in an analytical plotter, as it has been done for an area in the Sudan. The orthophoto and the line plot have been combined. The product is a map substitute which now can be taken for field verification before printing.

The usefulness of the orthophoto under conditions of Developing Countries is demonstrated, when it is overlaid with existing map information 1:200 000, to show that the map has survey errors of sometimes more than 500 m in position. To make an analysis of the geometric accuracy of Spacelab images a strip over a well controlled area with a homogeneous triangulation network was taken over North Germany. 118 check points were used to determine orientation and precision by aerial triangulation. The control point coordinates for symmetrically identifiable features were taken from 1:5000 maps. With a σ_o of 6.2 µm a positional accuracy of σ_p = ± 7.7 m was reached and an elevation accuracy of σ_h ± 20.2 m.

When the test was repeated over a mountaineous strip in the border area of South Germany, Austria, Switzerland and Italy, the result was worse: σ_o ± 6.6 µm, position error of σ_p ± 16.2 m and elevation error of σ_h ± 32.5 m. This is due to the control coming from various maps and control systems, which have gaps: position in Germany is based on the 1:5000 map with a rather poor triangulation; in Switzerland, Austria and Italy it is based on the 1:50 000 maps having their own reference systems. Even the height systems (for Germany: tide gauge of Amsterdam, for Italy and Austria: tide gauge of Venice) are different in value and orientation. This accounts for the larger areas.

Another test was made for a Chinese strip, for which control point coordinates were placed at our disposal by the Wuhan Technical University for Surveys and Mapping (W.T.U.S.M.). The result for 74 control points was $\sigma_o = 39$ µm, σ_p = ± 48.6 m and σ_h = ± 47.4 m. This reflects the control point selection accuracy in difficult terrain, and second the precision of the 1:100 000 map over this region of Tibet.

Over North Germany another test was taken to compile a 1:100 000 map sheet by plotting in a Kern DSR 11 Analytical Plotter. The result was very success-ful. Almost all features to be represented in the map could be interpreted and plotted with the exception of minor unpaved paths and of individual buildings. For building representation an area symbol is neces-sary, and minor unpaved paths are to be determined in field verification to give a satisfactory map.

The Spacelab camera experiment of the Fed. Rep. Germany was to be reflown in August 1986 on Space Shuttle, but the disaster of Challenger in January 1986 interrupted further progress. Now the reflight is scheduled for 1991. For this reflight the camera has been equipped with image motion compensation. This image motion compensation not only permits to reduce image motion (at 1/1000 sec = 8 m, at 1/500 sec = 16 m), but it permits at longer exposures (1/250 sec or 1/100 sec) to use slower, but fine grain film. High altitude aerial photographic tests flown over the Montpellier area with the Spacelab camera confirm-ed that the resolution of 35 lp/mm could be raised to 80 lp/mm. In this way it is to be expected, that a pixel equivalent of 3 to 4 m can be reached, a value sufficient for 1:50 000 mapping.

Similar imagery has, in fact, already been flown in the 1984 NASA Space Shuttle Large Format Camera experiment, which yielded resolution pixel equivalents of 5 m. An experimental map 1:50 000 was in fact successfully compiled over the border area of Helmstedt, where only individual houses could not be shown. Instead a settled area symbol was used.

Not accessible to us for tests was the Soviet MKF-6 imagery. Even though not very useful for stereoscopic evaluations because of the very small base-height ratio, such imagery is claimed to have 16 m photographic resolution or a 6 m pixel equivalent. A comparison of this imagery over a GDR area with Landsat MSS at the same scale shows, that roads of only a few m width are identifiable in the MKF-6 imagery. A land use interpretation test for 1:50 000 land use mapping has been conducted by the Academy of Sciences of the German Democratic Republic. About 40 classes could be determined with MSK-4 aerial multispectral imagery. The test was repeated with MKF-6 spatial multispectral imagery, by which 15 classes could be obtained.

The first digital cartographic high resolution system is the French Spot-satellite, launched on Februar 22, 1986. It operates in a sun-synchroneous mode covering 2 strips of about 60 km width with 2 sensors. In vertical mode the sensors overlap by 3 km, but they are tiltable in steps of 5° until 27°. Each sensor can operate in panchromatic mode in which 6000 elements of 10 m ground pixel size can be taken. Or a multispectral mode is possible, in which 3000 elements of 20 m pixel size can be used in 3 spectral bands (green, red, infrared). The use of tiltable mirrors permits to cover images within a 950 km band. This is useful to increase the repetitivity of images over a given area from 28 days to 5 days. Furthermore it permits stereoimagery at varying base height ratios up to a value of 1. This makes Spot an ideal cartographic satellite sensor, if programmed accordingly.

Excellent multispectral imagery has been obtained worldwide with Spot. The panchromatic image with 10 m pixels shows very fine street details. We will discuss tomorrow the ways in which stereoscopic Spot imagery may be restituted, based on tests conducted at Hannover which is shown in a rectified Spot image here. Spot is capable of 1:100 000 topographic mapping and map revision with more speed and less cost than aerial photography.

4. Future Developments
4.1 Sensors
Sensors barely meet the requirements, they are currently determined by technological limits:
— high spectral resolution is limited by the sensitivity of the detectors. Nevertheless image spectrometers operating at low spatial resolution are being designed to operate in the visible and near infrared range in up to 256 spectral channels.
— High spatial resolution is limited by sensitivity; without image motion compensation an exposure of 1/1000 sec corresponding to about 8 m ground pixel is the limit for a satellite velocity of about 7.7 km/sec.

This may be surpassed by imge motion compensation operating either by moving the film with respect to the image being created. Linear arrays can operate in the same manner, if they are modified into area sensors consisting of consecutive linear arrays in which charges are passed from array to array.

The second limitation to obtain images of high resolution then becomes the data transmission rate, which is currently at 200 to 300 Mbit/s. If higher rates of data are accumulated, these can only be read out by a data reduction (e.g. run length coding).

Area detectors of 10 000 x 90 have been built in the USA for the purpose of image motion compensation. For a panchromatic limit of sensitivity of 10 m, 90 arrays would give a possibility to have a 0.11 m pixel size. At 7.7 km/s sensor motion this would, however, lead to a readout rate of 70 000 arrays per second times 10 000 elements per array = 700 M elan x 8 bit = 5600 Mbit/sec. To be able to read out the information at 300 Mbit/s there would be data reduction requirement by a factor of 20, which can only be reached by filtering of the image (gradient), by reduction to 1 bit (black & white) and run length coding, making the image only useful to detect certain high contrast objects. Furthermore the image size then only becomes 1.1 x 1.1 km on the ground, something which can be achieved by aerial photography much more cheaply over accessible terrain, making such systems of primary interest to military surveillance. Civilian systems, however, must produce images for a variety of purposes with full 8 bit range and at wide coverage, to be equally useful and competitive with aerial photographic techniques.

The future developments depend greatly on the following factors:
— *platforms:* whether the rocket based, single system launchers are used (Viking by USA, Ariane by ESA), whether Space Shuttle is used or whether polar platforms or space stations are developed;
— *data reception:* whether manned platforms permit to bring data back from space to the earth, or whether data have to be transmitted from unmanned serviceable or non-serviceable platforms. For transmission of data the USA has favoured the Telemetry Data Relay Satellite System TDRSS in which low orbiting satellites could transmit data to a chain of geostationary communication satellites, permitting to relay data continuously from any orbiting satellite to the USA. The Space Shuttle program delays, however, interrupted progress, since the second TDRSS blew up on Challenger, while many such satellites are required. For that reason the system of several ground base antennas is favoured for reliability and cost reasons again. These antennas receive data on HDDT's which must be converted to CCT's as an offline process.
— Commercialization is another big factor affecting the development of remote sensing systems. While the Landsat program from 1972 through 1985 has been a US Government subsidized program, the commercialization act in the US determined that remote sensing systems other than for meteorology should go commercial. This was a setback for the development of worldwide remote sensing. For most countries this decision came too early, as there is doubt that remote sensing can be made viable commercially in this decade at all. Remote sensing systems, since then, are in competition both as far as airborne and spaceborne platforms, and as far as international sensors are concerned.

A real cost comparison of satellite systems can only approximately be made, since many factors are not known or accountable. To plan for commercial remote sensing the cost of 4 operational (future) systems is considered for the purpose of thematic and topographic mapping: Landsat 6 and 7 and Spot 2, 3 and 4 will cost at least 500 M $. The systems will have the capacity to generate 150 000 images in 3 years, of these only about 1/3 may be sellable, that

is 50 000 images in 3 years. The price per image therefore becomes 10 000 $ per image. The current Landsat
    TM CCT price is, however, 4000 $ per image, and
the Spot CCT price 1500 $ per image. Therefore, under
these conditions, the Landsat program should be subsidized at least by 60 % and the Spot program by at
least 85 %.

    When, on the other hand, we place a camera on a
Space Shuttle manned module flight, which can generate
6500 images in 9 days on 10 films over meteorologically
controlled cloudfree areas, the cost per image for
2000 sellable images again becmoes 10 000 $, for
which only 1500 $ can reasonably be charged. Therefore
a 85 % subsidy is again required. The situation is
somewhat better for a space shuttle pallet operation,
in which 650 images (1 film) may be obtained in 9
days. If 500 of these imges are sellable, then the
cost per image becomes 4000 $, for which again 1500 $
may be charged, leaving the needed subsidy at 62 %.

    To compare the images, however, the difference in
coverage and resolution is required:
Landsat MSS covers 185 x 185 km with 80 m pixels, in 4
channels;
Landsat TM covers 185 x 185 km with 30 m pixels in 6
channels and with 120 m pixels in 1 channel;
the **Metric Camera** from Spacelab covers 190 x 190 km
with in the future 3 to 4 m pixel equivalents in 1 or
3 channels;
Spot covers 60 x 60 km with 10 m pixels in 1 channel
or with 20 m pixels in 3 channels;
high altitude photography covers 12 x 12 km with 0.5
m pixels in 1 or 3 channels.

    Comparing the current costs per image for Landsat
MSS, TM, Spot as CCT and on film with present and
future photographic imagery from space on the basis
of cost per 100 km or cost per 1 million pixel in one
or several channels one finds:
While Landsat TM has the most expensive at 4 500 $
per image, Spot becomes the most expensive per 100
km^2 at 45 $. But this is still cheaper than high alti-
tude aerial photography at 500 $ for 100 km^2. But
comparing cost per million pixels Spot is cheaper
than Landsat. A considerable price reduction for
Spot Imagery is obtained for photographic products,
which can be evaluated on analytical plotters very
economically. As compared to the cost of Landsat TM
and Spot costs per 100 km^2 space photographic systems
are considerably cheaper, and with respect to pixel

equivalents they are even more economical by an order
of 2.

Future developments are also expected in data
processing and data analysis of digital data:
While in the past the tendency was to obtain raw data
after HDDT to CCT conversion there will be an increasing tendency to obtain resampled corrected data by
polynomial fit to control, or as orthophoto by DTM
corrections; or even resampled geocoded data; and
finally as specialized feature extracted standard
product. Spotimage or CCRS work toward the tendency
to achieve high level products in a factory style
environment. It is still an open question as to whether
raw data evaluation by own facilities will be more
economical.

In data analysis everyone began to develop their
own raster based image processing systems with own
hard- and software. Since the past 5 years commercial
manufacturers such as Comtal, Gould-De Anza, I^2S,
Dipix, ContextVision, Sodetec and others provide
hard- and software for standard image processing
tasks.

Recently the raster- based remote sensing information is supplemented by vector-based information
coming from vector based systems such as Intergraph
and others (to be discussed with interactive graphics).

Software developed includes geometric transformation by resampling, grey level enhancement, filtering
by convolution or after Fourier transformation,
multispectral classification, operations with vector
information and the development of expert systems
currently in progress.

Data analysis procedures utilize these data
processing methods in a sequence in which the data
are processed according to certain rules based on
previous knowledge. The expert system groups these
operations, consisting of radiometric calibration,
geometric resampling, textural enhancement, multispectral classifications together aiding in the
processing rules to be decided.

There are various bodies concerned with remote
sensing:

- **UN organizations,** such as the UN Secretariate,
  FAO, ESCAP, ECA & COSPAR which are urged to avoid
  restrictive policies, otherwise a useful tool for
  human development is not available.
- **Governments,** particularly those offering space
  facilities can do their own policies, particularly

on commercialization of remote sensing. Particularly developing nations depend upon space oriented
nations for progress.
- **Scientific societies** such as IAF, IEE-IGARSS, ISPRS,
AARS, ACA, Earsel and SELPER as international or
regional organizations can do a lot to concentrate
and coordinate remote sensing activities. For this
purpose an international Joint Board of Remote
Sensing Activities (JOBRESA) has been created.
- Finally there will be more private organisations,
such as Eosat, Erim, Spotimage, which will promote
remote sensing data and products.

For optimal progress we depend on international
cooperation. There is no competition to civilian
remote sensing by military reconnaissance systems.

Bibliography

Konecny, G., "The Photogrammetric Camera Experiment on
Spacelab 1," Bildmessung und Luftbildwesen, Vol.52,
June 1984, pp. 195-200

Konecny,G., Kruck, E., and Lohmann, P., "Ein universeller
Ansatz für die geometrische Auswertung von CCD-Zeilenab-
tasteraufnahmen," Bildmessung und Luft bildwesen, Vol.
54, 1986, pp. 139-146

JEOS: A Low-Cost Approach to Earth Observation

Pierre Molette

MATRA Espace, Toulouse, France

ABSTRACT

A low cost domestic Earth Resource satellite has been designed, based on the use of a JANUS platform tailored to an Earth Observation mission over a limited area.

The JANUS Earth Observation Satellite (JEOS) will provide a limited coverage with real time transmission of image data, thus avoiding the need for on-board storage and simplifying operations.

The JEOS operates on a low earth, near polar sun-synchronous orbit and can be configured for either land or ocean applications.

The instrument for land application is a solid state push-broom camera which is composed of four optical lenses mounted on a highly stable optical bench. Each lens includes an optics system, reused from an on-going development, and two CCD linear arrays of detectors. The camera provides 4 registered channels in the visible and near IR bands. The whole optical bench is supported by a rotating mechanism which allows rotation of the optical axis in the accross-track direction. The JEOS typical performances for a 700 km altitude are : spatial resolution 30 m, swath width 120 km, off-track capability 325 km.

A similar instrument, with 8 narrow band channels in the visible and near IR, has been designed as an Advanced Ocean Color Monitor (AOCM) which would operate in two modes : high resolution (250 m) for coastal zones and medium resolution (500 m) for open ocean.

The payload data handling and transmission electronics, directly derived from the French SPOT Satellite, realize the processing, formatting, and transmission to the ground ; this allows reuse of the standard SPOT receive stations. The camera is only operated when the spacecraft is within the visibility of the ground station, and image data are directly transmitted to the ground station by the spacecraft X-band transmitter.

A typical operation scenario includes launch, transfer to the operational orbit, and routine operations, which mainly consist of payload scheduling.

1 - INTRODUCTION

A low cost domestic Earth Resource satellite has been designed, based on the use of a JANUS platform tailored to an Earth Observation mission over a limited area.

The concept of the JANUS multimission platform is to minimize the cost of the satellite by a maximum reuse of equipment from other programmes, and to reduce the launch costs by a piggy-back configuration optimised for ARIANE 4.

The application of the JANUS concept to an Earth Observation mission has the objective to provide a given country with a permanent monitoring of its earth resources by exploitation of spaceborn imagery. According to this objective, and to minimize the overall system and operational cost, the JANUS Earth Observation Satellite (JEOS) will provide a limited coverage with real time transmission of image data, thus avoiding the need for on-board storage and simplifying operations.

The JEOS operates on low earth, near polar sun-synchronous orbit. Launched in a piggy-back configuration on ARIANE 4, with a SPOT or ERS type spacecraft, it reaches its operational orbit after a drift orbit of a few weeks maximum. In its operational mode, the JEOS is 3-axis stabilised, earth pointed.

The JEOS can be configured for either land or ocean applications, with similar solid state push-broom cameras composed of optical lenses mounted on a highly stable optical bench. The camera is only operated when the spacecraft is within the visibility of the ground station, and image data are transmitted to the station via the payload data handling and X-band transmission electronics directly derived from the french SPOT satellite equipment.

2 - THE JANUS PLATFORM AND ITS APPLICATION TO EARTH OBSERVATION MISSIONS

2.1 - The JANUS piggy-back concept

JANUS is the generic name of a multimission spacecraft platform specifically designed to be launched as an additional piggy-back payload to ARIANE 4.

The JANUS concept takes full advantage of the flexibility of the new ARIANE 4 launcher which presents a variety of booster configurations that can be used in combination with short and long fairings and SPELDA's. The payload mass difference between two different configurations of boosters corresponds to the typical mass of a JANUS satellite, while its height of one meter is the difference in length between a long and a short fairing or SPELDA : a JANUS spacecraft can be piggy-backed to a main passenger nominally intended for single launch under a short fairing ; using instead a long fairing provides

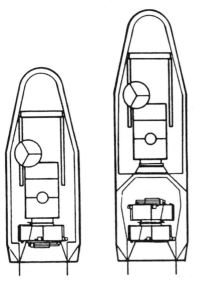

Fig. 1 Launch configuration with ERS.

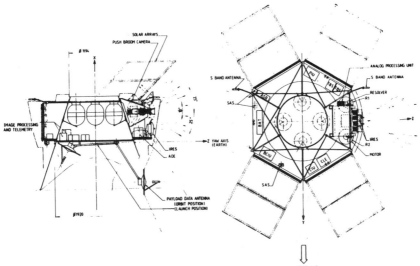

Fig. 2 JEOS spacecraft configuration.

volume for mounting JANUS without affecting the interface restraints to the main passenger. Several JANUS spacecraft can also be superimposed in a stack configuration, allowing to launch a cluster of satellites. Possible launch configurations with an ERS are shown in figure 1 in a single launch, or in a dual launch configuration with a stack of two JANUS.

2.2 - The JEOS spacecraft configuration

The JEOS platform is directly derived from JANUS and adapted to an Earth Observation mission. The resulting spacecraft is presented in figure 2, and the in-orbit configuration is shown in figure 3.

The spacecraft is made of a central conical structure that insures a direct interface between the ARIANE 4 Vehicle Equipment Bay and the main satellite. It so replaces the standard conical adaptors which normally carry the ARIANE payloads. The JANUS central conical structure is composed of:

- a JANUS 2624-1920 adaptor with a Φ1920 separation system,
- the JANUS 1920-1194 spacecraft structure. JANUS offers to the main spacecraft the same mechanical and electrical interfaces as the standard ARIANE configuration and the addition of JANUS in a launch configuration has therefore no impact on the main spacecraft.

Externally, the JEOS structures form an hexagonal cylinder, with four solar arrays hinged at its upper corners, and folded against the lateral walls during launch. In orbit the array wings are deployed with a cant angle which allows to optimize the power generation for the selected local time.

The Earth viewing compartment houses the payload camera and a few service equipment required for spacecraft pointing. All other service equipment are located in the 5

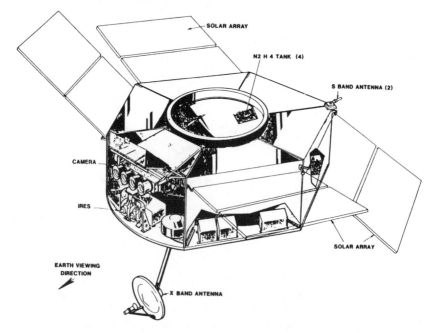

Fig. 3 JEOS in-orbit configuration.

remaining compartments. The hydrazine propellant tanks are located inside the central cone. At the bottom of the spacecraft, a large circular floor supported at the edges of the cone carries the payload telemetry units and antenna.

The orbital configuration is determined by the Earth viewing direction (as shown on figure 2) and the cone axis normal to the orbital plane.

The equipment supporting floors are located opposite to the sun direction to offer a simple passive thermal control : radiators and some electrical heaters, multilayer insulation blankets to minimize heat transfer through these surfaces.

2.3 - The JEOS electrical architecture

The electrical system of JANUS is derived from the one developped for the SATCOM International's Eurostar communication platform. Reusing equipment already qualified minimizes development risks and benefits from the reduced costs of a recurring production. Shown on figure 4 is the overall block diagram of the electrical system. Two main functions are performed by the electrical system :

- to distribute data to and from the various subsystems of the space- craft platform and payload.
- to supply electrical power to the same as required.

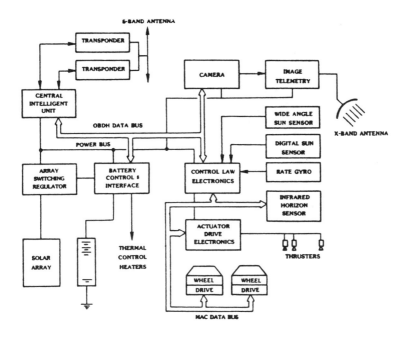

Fig. 4 JEOS electrical system block diagram.

i) The data system of JANUS consists primarily in a main processor called Central Intelligent Unit (CIU), coupled to the various subsystems (users) by a redundant standard OBDH digital data bus.

The CIU performs the on-board management functions:

- Satellite management including operation of the satellite in the various modes: transfer, attitude acquisition, operation, safe mode.
- Platform management : monitoring, failure passivation of all platform subsystems ; dedicated subsystem tasks such as battery charge/discharge management.
- Payload management : payload activation and monitoring, payload operation modes.
- OBDH bus management : command and telemetry interface with the remote terminal units, distribution of timing signals. The users of the data bus on the platform side are the Battery Controller and Interface Unit (BCIU) that is part of the power subsystem, and the Control Law Electronics (CLE) which is the main processing unit of the Attitude Control Subsystem. Coupling with the payload is via a specific equipment called the Payload Interface Unit (PIU) which is part of the payload itself.

The CIU also performs Telemetry-Telecommand related functions :

- Demodulation, validation and distribution of commands either direct from the ground or in time-tag mode.
- Generation of telemetry formats : real time format and report format which summarizes the housekeeping data out of ground visibility.

The housekeeping telemetry and commands are channelled through a redundant S-band transponder which operates in the 2200-2290 MHz band (Tx) and 2025-2110 MHz band (Rx).

ii) The power supply subsystem is designed to provide the required level of electrical power of the platform and the payload, taking into account that the peak power demand takes place during a relative short time of about 10 minutes when the payload takes pictures, and transmits data. The total power is 338 W during this period ; for the rest of the time, the power is only 186 W (excluding battery charging). The figure 5 shows a typical profile of solar array generated power compared to the power demand for a 10h local time and optimised solar configuration (+/-20° cant angle). The excess power is used to charge the battery for eclipse operation.

An Array Switching Regulator (ASR) controls the output of the solar array to supply a regulated voltage of 42.5 VDC +/- 1% in sunlight. During eclipse, current is supplied to the bus voltage then follows the battery discharge voltage, but never falls below 28 Volts.

2.4 - Attitude control

The choice of an attitude control system for JEOS has been guided primarily by the pointing accuracy and angular rates compatibles with a set of image quality criteria commensurate with a ground resolution of 30 m. Simplicity and reuse of existing equipments are other fundamental drivers to the control system design.

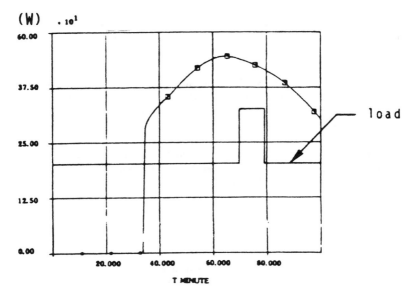

Fig. 5 Solar array output power.

The control system selected for JEOS consists in a fixed momentum wheel (FMW) oriented along the pitch axis of the satellite body, and a set of low thrust monoprellant (N2H4) thrusters providing control torques for wheel unloading and momentum vector orientation. Attitude measurement with respect to the local vertical is performed by an infrared horizon sensor developed for SPOT and providing angular error signals about pitch and roll axes.

The control is not activated during the time of image taking so as to maintain the quality of image. Conversely, the positions on the orbit when thruster actuation is performed can be selected by the microprocessor of the attitude control system. This mode of operation reults in the following performances:

- pointing accuracy : 0.2 - 0.4°
- angular rate accuracy : a few $10^{-3°}$/sec   } 3 axes

which are well compatible with the attitude requirements for a 30 m resolution.

In addition to maintaining the geocentric pointing of the satellite with small deviation in angles and angular rates during the normal imaging mode, the AOCS shall also provide control in other phases of the mission, namely:

- initial acquisition,
- orientation during orbit insertion and correction manoeuvres.

The central equipment of the attitude control system is the Control Law Electronics (CLE) which has been developed for the EUROSTAR platform, and is also used on HIPPAR-

COS. It interfaces with the spacecraft OBDH digital bus, processes the signals from the various sensors, and performs by software the implementation of the control algorithms in the various operation modes.

A second equipment of the AOCS is the Actuator Drive Electronics (ADE) which generates driving power to the thrusters upon reception of control signals from the CLE. Also developped for EUROSTAR, this equipment is modular in design, and capable to drive various types of actuators.

Interfacing of the various equipment of the attitude control system is performed in the most flexible way via the Modular Attitude Control System (MACS) digital data bus; Figure 6 displays the functional architecture of the AOCS and shows the interconnexion between the equipments through the MACS data bus.

3 - THE JEOS PAYLOAD

The satellite performances are strongly dependant on choice of the spatial resolution. In fact, irrespective of technological or operational limitations, the requirements may significantly vary from one country to another, according to the nature of ground patterns one intends to image. However, the selected resolution must also result from a compromise between performance and system complexity and costs. Because of the cost approach selected for the JANUS derived satellites, it is not reasonable to try to improve the ground resolution beyond a break point which happens to occur at about 30 m. Such a resolution is considered adequate with majority of applications, as illustrated by Landsat-4 and 5 and the indian IRS.

An essential feature of the JEOS imaging system is the off-track pointing capability (+/- 25°), also present on SPOT but not on the LANDSAT series. Two important

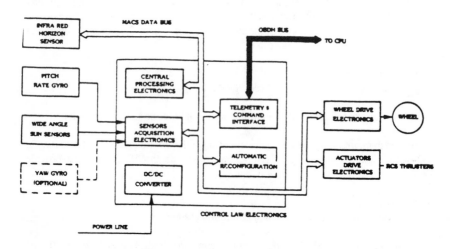

Fig. 6 JEOS AOCS block diagram.

capabilities derive from off-track pointing, namely :

- stereoscopic viewing, especially useful for mapping,
- dramatically improved time coverage of selected zones.

A final important property of any imaging system is the geometrical quality of the image. Two aspects are the residual distortions and the band-to-band registration. The following performances have been selected :

- absolute pointing accuracy ≤ 6 km on ground
- effect of angular rates ≤ 1 pixel over a 12x12 km2 frame (in fact a much better performance can be achieved),
- band-to-band registration ≤ 0.33 pixel.

3.1 - The optical imaging instrument (Figure 7)

The optical system of the push-broom imaging instrument consist in four identical refractive lenses mounted side-by-side on a common case. Each lens is dedicated to one spectral band, thereby eliminating the need for any dichroic beam splitter. The 324 mm focal length, f/4.5 lenses are identical to those designed and built by MATRA for the IRS (LISS-2) (see figure 8). The four spectral bands are : 0.45-0.52 μ m, 0.52-0.59 μm, 0.62-0.68 μm, 0,77-0,86 μ m. Two CCD arrays with 2048 elements each are staggered at the focal plane of each lense to achieve a total swath width of 120 km at nadir for a 750 km high orbit. The CCD's are thermally coupled to a radiator to maintain temperature low when the detectors are energized.

The lenses developed and qualified for the LISS optical system have some special features which make them practically unequalled in performances:

- Each lens is comprised of 8 elements. Each particular lens is separately optimized for its corresponding spectral band, so as to minimize chromatic aberrations. Measured MTF

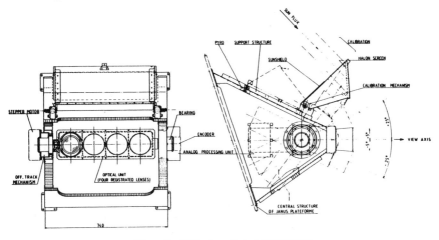

Fig. 7 JEOS imaging instrument.

Fig. 8 324 mm focal length lens.

is between 0.67 and 0.71 for each of the four spectral bands at 40 pl/mm for a CCD with 13 μ m pitch.

- For band-to-band registration stability, the mount is mechanically designed to avoid any lateral motion of the elements under mechanical or thermal constraints. On LISS, 0.5 micron of lateral displacement at image plane have been measured on the qualification model after vibrations and thermal cycling tests. This is to be compared with a value of 4 μ m at focal plane, required for a desired registration of 0.33 pixel.

- With a suitable selection of glasses and mount material, it has been possible to precisely athermalize the assembly. LISS achieves below 0.5 micron/°C, that means the defocus is unsignificant for temperature variations as large as +/- 15°C.

- Interference filters are placed in front of each lens to precisely define the required spectral bandwidth. The multilayer interference filters built by MATRA optics Division display outstanding performances for sharpness at cut-off frequency and off-band rejection ; moreover, the transmission factor within the usefull bandwidth is unusually good, maximizing the overall efficiency (see figure 9).

The photometric performances (radiometric resolution) of the JEOS imaging instrument are presented in figure 10.

The in-flight calibration can be performed by the sun illumination of the full aperture without complicating too much the design. The proposed calibration system consists in a fixed diffusing screen made of Halon, this material presenting nearly perfect lambertian properties. The position and orientation of the screen will be set so as to get sun illumination when the satellite is in visibility from the ground station. The off-track pointing mecanism will be activated to rotate the optical head assembly off the nominal pointing range so as to view the diffusing screen only. Baffles and shields will be suitably located to suppress stray light.

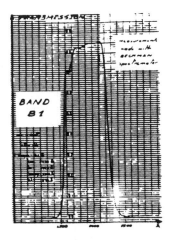

0.45 - 0.52 MICRON

Fig. 9 Interference filter characteristics.

SPECTRAL BAND	B1	B2	B3	B4	
SIGNAL CHARGE NS	145000	142000	215000	265000	
SHOT NOISE	380	376	. 463	514	IN NUMBER
DIGITALIZATION NOISE	621	621	621	621	OF
OTHER NOISE SOURCES ALLOCATION	100	100	100	100	PHOTOELECTRONS
TOTAL RMS NOISE NT	735	733	781	512	
NE Δ ρ $\theta S = S\sigma$ $\rho = 0.5$	0.5%	0.5%	0.35%	0.3%	

Fig. 10 Photometric resolution.

The image electronics is for most of its functions identical to that mounted on the SPOT HRV with a minimum of repackaging. The block-diagram of figure 11 shows the corresponding equipments and functions.

3.2 - Advanced Ocean Color Monitor (AOCM)

As an alternate payload to the imaging instrument, an instrument specifically dedicated to the observation of water surfaces and coastal zones could be of great use to countries which draw a significant part of their resources from the ocean. For these applications, the remote sensing bears primarily on the chlorophyll concentration which strongly influences the ocean colour, and is a good indicator of ocean productivity. In a recent past, the Coastal Zone Color Scanner (CZCS) on Nimbus 7 has been used with considerable success to map ocean colour and chlorophyll concentration over open ocean areas. The most recent push-broom techniques enable to obtain improved spatial resolution and better radiometric accuracy together with narrower spectral bands.

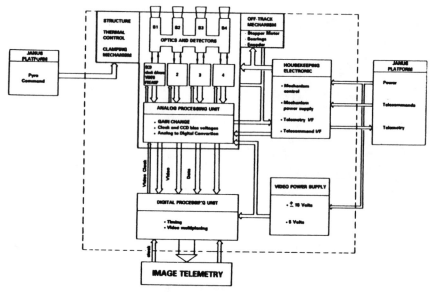

Fig. 11 JEOS instrument block diagram.

Chlorophyll is spectrally discriminable at short wavelength, where atmospheric effects - absorption and scattering - are important, while at the same time the reflectivity is low. Hence, the usefull signal is only a small percentage of the total upwelling radiance. It is therfore fundamental to correct for atmospheric effects, which in turns requires additional channels which are not found on land application instruments.

In its baseline design, the AOCM comprises 8 narrow band channels in the visible and near IR. The instrument would operate in two modes :

- high resolution (250 m) for coastal zone : NE ≤ 0.5%
- medium resolution (500 m) for open ocean : NE ≤ 0.025%

As for the imaging instrument, the AOCM is a modular concept with any one optical head fitted with one single CCD array. Along-track off setting of the viewing direction is provided for sunlight avoidance. As shown on figure 12, the AOCM is a similar concept to the imaging instrument.

3.3 - Payload telemetry

The function of the Payload Telemetry is to translate the video signal into a Radio Frequency signal and to transmit it to earth.

The 24.68 Mbit/sec formatted signal is sent to a QPSK modulator where it is transformed into an RF signal at 8.2 GHz. After amplification through the Travelling Wave Tube the signal is emitted to earth by means of the SPOT like X-band antenna, mounted at the end

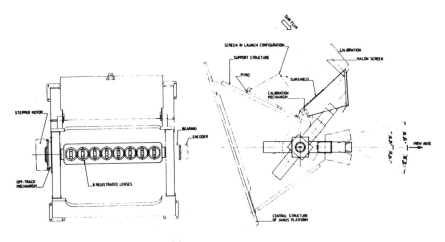

Fig. 12 Advanced ocean color monitor.

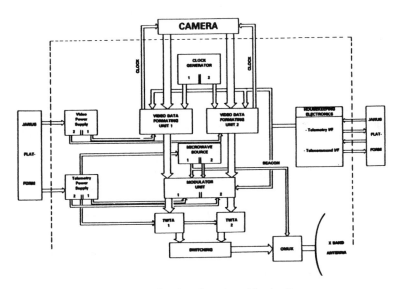

Fig. 13 Payload telemetry block diagram.

of a boom released after launch. All equipments are mounted on the bottom plate of JEOS so that heat can be easily dissipated to space.

Like for the test of the spacecraft, the equipments are directly reused or derived from existing hardware, i.e. the SPOT payload telemetry. This insures the compatibility of JEOS with the SPOT receiving and preprocessing ground stations.

The image telemetry block diagram is shown in figure 13.

4 - ORBIT SELECTION AND SPACECRAFT OPERATION

4.1 - Orbit selection

As a low Earth Observation mission, JEOS will be placed in a phased sunsynchronous circular orbit. The altitude of the JEOS orbit is determined by the ground coverage constraints and the instrument swath width, while the local time of the descending node is selected against sun illumination incidence (solar irradiance vs shadowing effects) and power budget of the satellite. Fixed panels being used, the efficiency of the arrays decreases when the descending node is shifted towards local noon. This is not actually a serious concern since optical imaging in the spectral range under consideration is best performed in the morning when atmospheric scattering and cloud cover tend to be minimum. The local time at descending node is selected to be 10.00 a.m.

The altitude is selected to allow observation of a given area at the same altitude and at the same viewing angle at regular intervals. On the other hand, complete area coverage requires a network of ground swaths compatible with the swath width of the camera. Since the swath width is about 120 km for an altitude of 750 km, one can compute the minimum number of ground tracks necessary to obtain the required coverage. As an example, for a country at around 20° latitude, an orbit with 315 revolutions per cycle for an orbital cycle of 22 days is best suited : its altitude is 780 km and inclination is 98.5° (the corresponding swath width is 125 km). This orbit is such that using the capacity of viewing +/-25° across track, any area of the country can be imaged within 3 days as shown in figure 14.

4.2 - Orbit acquisition

JEOS is intended to be launched in piggy-back to a SPOT or ERS type spacecraft. After separation of the main spacecraft, the Ariane third stage will orient the composite

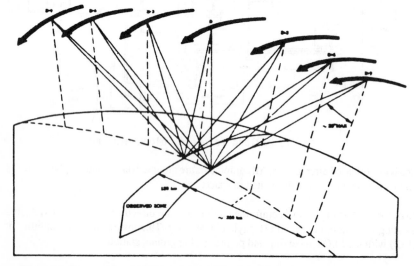

Fig. 14 Improved coverage capability with across-track pointing.

longitudinal axis normal to the orbit plane, spin up the composite to a low rate of a few rpm and initiate the separation of JEOS. The solar panels are then deployed, the automatic attitude acquisition takes place, the kinetic momentum is transferred to the wheel and earth is acquired with the IR earth sensor. This sequence is illustrated in figure 15.

Orbital manoeuvres have then to be performed to reach the operational orbit which differs from the initial insertion orbit by its altitude and the local time of its descending node. This is obtained by placing temporarily the satellite on a transfer orbit which is not sun-synchronous. Thus the line of nodes will drift slowly until the desired local time is attained.

The total duration of this phase is a compromise between minimizing the time from the launch to operation start, and minimizing the amount of propellant. For a 30 min change in local time, a duration of about one month is recommended. During this phase in-orbit verification and calibration of the payload can be performed.

4.3 - Routine phase

In the routine phase, when the spacecraft has reached its operational orbit, the operations to be performed on the satellite are essentially the payload scheduling and the orbit adjustments.

Every 24 hours the imaging programme is loaded in the CIU memories during a pass over the control ground station. This programme defining the exact date of viewing, the off-

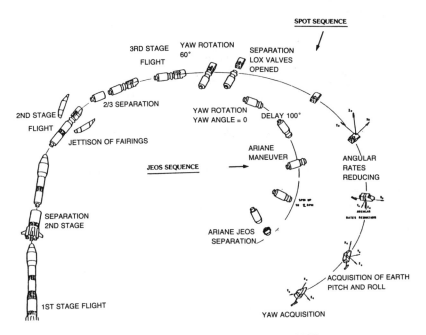

Fig. 15 JEOS launch sequence with SPOT.

nadir angles, the calibration factors, etc ... is then executed automatically by the spacecraft during the next passes over the country.

To maintain good imaging conditions and ground track repetitivity, orbit corrections are necessary. These maneuvers will be automatically executed by the spacecraft via time-tagged commands.

Image characteristics		
. spatial resolution	32 M	Sub-satellite point
. swath width	128 km	
. number of pixels/band	4000	2 CCD's at focal plane of each lens
. across-track offset	+ 25°	+ 360 km on ground
. spectral bands	450-520 NM	Fine tuning possible
	520-590 NM	to meet special requirements
	620-680 NM	
	770-860 NM	
. radiometric accuracy	NE$\Delta\rho$<0.5%	
. absolute calibration	< 5%	Design goal
. data rate	24.6 Mbit/sec	Quantization linear
Orbit parameters		
. altitude	780 km	Sun-synchronous
. local time descending node	10:00 a.m.	
. inclination	98.5°	
. period	100.57 MIN	
. orbit cycle	22 days	
. minimum revisit time	3 days	Using off-track pointing
Satellite		
. total weight at launch	590 kg	Including 130 kg monopropellant
. max electrical power	425 W	
. lifetime (nominal)	4 years	Computed reliability: 0.6
. TT & C	S-band	Standard ESA network
. image telemetry	X-band	Compatible with SPOT/ LANDSAT
. attitude & orbit control	Momentum biased system 25 NmSEC + 10 % Monopropellant (N_2H_4) reaction control Low thrust (0.5 N) and high thrust (14 N)	
. power subsystem	4 wings fixed solar arrays	
. data handling	Digital data bus (OBDH standard)	

Fig. 16 JEOS performance and characteristics.

5 - CONCLUSION

JEOS is a spacecraft for Earth Observation missions over limited areas, well suited for countries who want to acquire their own satellite for a more efficient management of their national territory and its natural resources. This approach enables the establishment and regular updating of inventories of limited resources (water, fauna, flora, soil ...) and the monitoring of the different media and their transformation processes (either natural or modified or induced by human activity), according to the needs and under the full control of the user. Main drivers of such an approach are :

- Availability of data not subject to shortage for reasons not under control of the user.
- Improved time coverage.
- Mission planning and operations under full control of the user.
- System specifications tailored to each using country particular needs and requirements.
- Property rights on data collected on national territory.

The collected information can be used for the purpose of :

- Prediction, detection and evaluation of certain evolutions of natural or artificial phenomena, some of which can be considered as detrimental (soil erosion, floods, drought, various pollution, land clearing, ...).
- Assisting certain economic activities with emphasis on development of certain regions, study of large infrastructures, management of large civil engineering installations such as dams or irrigation systems, agricultural and forestry work, mining and oil prospecting, navigation, fishing, ...

JEOS will meet all these objectives in a most economical way :

- by its concept of piggy-back launch,
- by the maximum reuse of existing equipment, which reduces development risk and schedule and provides confidence in mission performances..

Its special feature of off-track pointing capability makes it ideally suited for stereoscopic imaging which is of particular interest for cartography and geology and for quick revisit of any area in the country (within 3 days).

The main performances and characteristics of JEOS are summarized in the table of figure 16.

Second-Generation MOMS: A Low-Cost and Flexible Instrument with Stereoscopic and Multispectral Modes for Developing Countries

D. Meissner*

MBB-ERNO, Munich/Ottobrunn, Federal Republic of Germany

Abstract

This paper describes the present status of the MOMS (Modular Optoelectronic Multispectral Scanner) instruments, the space missions experiences and their results. Furthermore, future developments are summarized including short term missions focused on Shuttle-related activities e.g. German D2 mission as well as long duration missions as e.g. RADARSAT, POLAR PLATFORM and GEO-SPAS.

Introduction

With regards to MOMS' versatility, the concept of dedicated missions has been introduced with special emphasis on Shuttle-launched short remote sensing missions. When compared to LANDSAT and SPOT, MOMS can be adapted to different mission tasks suitable especially for the distinct needs of developing countries. The instrument can be operated in specific configurations, either flown on a Shuttle attached/detached space platform (with a mission duration of several days) or on a free flyer (with a mission duration of several months or years).

The modularity of MOMS combined with the various orbit configurations of the Shuttle makes it possible to cary out dedicated missions which may be regarded as optimally suited to provide complementary data to the operating or future remote sensing satellites.

The technical data of this MOMS 2nd generation instrument will be summarised.

The optimum solution for these dedicated MOMS missions, especially for developing countries, are small reusable plat-

*Project Manager, Space Systems Group.

416

forms like the spaceflight proven SPAS (Shuttle Pallet Satellite). This platform for remote sensing missions, designated GEO-SPAS also provides the additional chance of advanced sensor flight testing like X-band SAR or passive
microwave sounders. Mounted on this reusable platform, the
equipment can be flight-proven and returned to Earth after a
few days or a few months for adjustment, improvement or (in
worst case) for repair. The GEO-SPAS platform also allows
for easy on-board data storage. No ground station access
problem exists for remote areas to be covered.

Summary of MOMS-01 instrument

◆ First optoelectronic high resolution scanner in space
  → 2 spectral channels ←   → 20 m ground solution ←
Flown twice on-board the Shuttle / SPAS in June 1983 and
February 1984
◆ Developed and built by MBB
  - under contract of the German Aerospace Research Estab-
    lishment (DFVLR)
  - by order of the German Ministry of Research and Tech-
    nology (BMFT)
MOMS was made for tasks as
  - land surface thematic mapping
  - vegetation and sea surface monitoring
  - conventional photo interpretation and topographic map-
    ping
The ground coverage can be varied according to user needs by
  - changing the photometric layout and focal length

Fig. 1 Optical Module
of MOMS-01 (© MBB)

- combining several detector arrays (line extension) using
  the dual optics principle

Two successful system verifications under space conditions
demonstrate

- the technical capability for scan line extension by
  double lenses
- the ability of modules combination for two pixel-
  coincident channels

Fig. 1 shows the Optical Module of the MOMS-01 instrument

Summary of MOMS-02 instrument

MOMS-02, to be flown on German Spacelab D2 mission in 1991
includes (see Fig. 2):

◆ a Stereo Module, providing in situ panchromatic stereoscopic coverage with ≤ 10 m base and < 10 m height resolution. Main applications will cover topographic mapping
  up to 1:50,000 scales, evaluating of digital terrain
  models, orthophoto maps, and quantification of multispectral data under consideration of relief dependent parameters.

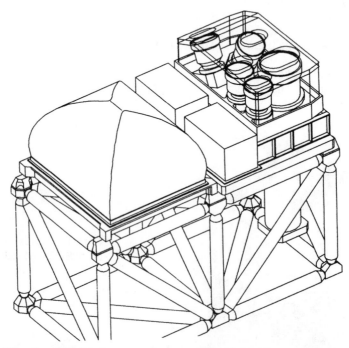

Fig. 2   MOMS-02 Integration on Unique Support Structure
         (USS) of Spacelab D2 Mission (© MBB)

◆ a Multispectral Module providing together with the Stereo
Module, in situ measurements with approx. 15 m ground
resolution. Main applications will cover optimized
geoscientific thematic mapping at small scales 1:100,000
on the basis of combined spectral and topographic data, 4
narrow band channels ($\Delta\lambda$ = 40 nm) for discreet phenome-
nas, large swath width for a high repetition rate, and
correlation of thematic information to geometric data for
all future geoscientific and commercial applications.
Future applications intend to carry out dedicated missions
which may be regarded as optimally suited to provide complementary data to Remote Sensing Satellites.
The proposed test sites for the D 2 mission are shown in
Fig. 3.

Summary of R-MOMS on RADARSAT

R-MOMS had been designed by MBB for the Canadian Radarsat
and includes:
- 4 spectral channels
- 400 km swath width
- 30 m ground pixel size
It can be easily redesigned for Polar Platform mission including a STEREO mode and the life-time can be extended for
a long mission period up to 7 years.
The mission objectives are to serve four groups of users
who desire information on
◆ navigation through ice-covered water
   - monitor ice and iceberg and predict location and type

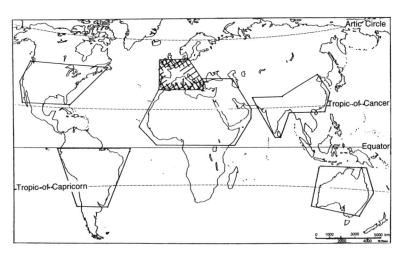

Fig. 3 Proposed Test Sites for MOMS-02 on D2 Mission

- support general offshore activities as oil, gas, mineral exploration
- analyse air/sea/ice interaction and climatology
◆ ocean surface data
  - ship route planning
  - management of off-shore oil drilling and production activities
  - fishing fleet and oil-slick monitoring
◆ land renewable resources
  - management of agriculture, forestry and hydrology resources
◆ non-renewable resources
  - update geological map of Canada and topographical maps
  - obtain a "world" stereo data set by radar imaging

The R-MOMS instrument is considered to be the primary optical sensor to give additional spectral information during daylight by imaging the earth surface.

In Table 1 the optical requirements to R-MOMS are shown, in Table 2 a summary of the technical design specification is given.

MOMS Missions on GEO-SPAS, Summary

◆ GEO-SPAS, a small reusable platform for dedicated MOMS missions, to be launched and retrieved by the Shuttle Orbiter
◆ Different mission types are feasible with GEO-SPAS like
  - short term missions of several days in an attached or free-flying mode
  - long term missions of 6 months or more for extensive instrument tests and operational applications

Table I Optical requirements of R-MOMS (©CCRS)

Swath Width	405 km
Ground Pixel Size: Band # 1	60 x 60 m²
Bands # 2, # 3, # 4	30 x 30 m²
Number of detector elements	13,500 (# 1: 6,750)
Size of detector elements	10.7 x 10.7 µm²
Focal length	359.2 mm (# 1: 179.6 mm)
Total field of view	22.8 deg
Instantaneous field of view	29.8 µrad (# 1: 59.6 µrad)
Scan line frequency (max)	216 Hz (# 1: 108 Hz)
Spectral band # 1	0.485 µm; Δ = 70 nm
Spectral band # 2	0.555 µm; Δ = 70 nm
Spectral band # 3	0.650 µm; Δ = 60 nm
Spectral band # 4	0.825 µm; Δ = 130 nm

Table 2 Technical design specification of R-MOMS (©MBB)

Optic Module, Size (L·W·H)	900 x 450 x 605 mm³
Weight	130 kg
Logic Box, Size (L·W·H)	420 x 225 x 200 mm³
Weight	14 kg
Power Box, Size (L·W·H)	380 x 225 x 200 mm³
Weight	16 kg
Total Power, Operating	345 W
Total Power, Stand-By	75 W
Input Voltage	41 ± 7 V
Output Voltages	+5 V, ± 15 V
Current Limiters	3 (switchable)
Telecommand, High Power	21
Telecommand, Memory Load	2 (16 bit serial)
Telemetry, analog	28
Telemetry, digital	25
Processor, normal and default mode	80 C 86

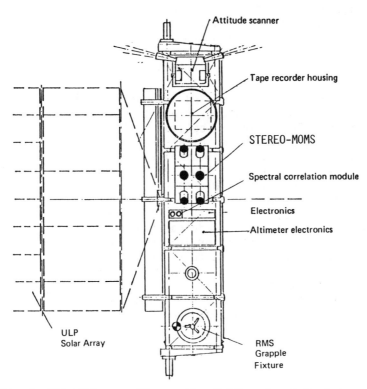

Fig. 4 GEO-SPAS, a small reusable platform for dedicated
shortterm commercial Remote Sensing missions (© MBB)

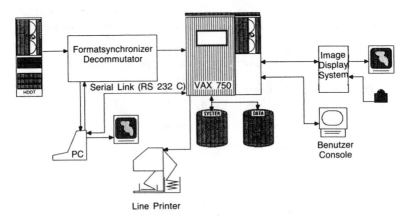

Fig. 5 Data Processing System for MOMS Instruments (© MBB)

◆ Best suitable for developing countries due to
    - low cost
    - opportunity for own missions and development tests
Fig. 4 gives an overview of the proposed GEO-SPAS platform
with MOMS instrumentation and support equipment

Conclusion

◆ Optimum Performance of 2nd Generation MOMS by
    - pacemaking and future-oriented optronic technology
    - best radiometric and optical abilities
    - adaptability of technical parameters as swath width,
      resolution, number of pixel etc.
◆ Adaptation to varying Requirements and Missions with
    - a modular and flexible concept
    - slightly modifications only for different missions as
      Shuttle-D2, Polar Orbiting Platform, Radarsat, Geo-SPAS
    - along-track STEREO mode capability
◆ Data Storage on board or Transmission to Ground
    - by on-board tape recorders or
    - by transmission via receiving station
◆ Data Processing with
    - reformatting to computer compatible tapes (CCT's)
    - processing of data according to user need and specifi-
      cation (see Fig. 5)

Typology of Lagoba: Ngooxoor's Landscapes (Saalum's Estuary) Using Data from SPOT 1 Satellite

Amadou Tahirou Diaw*
Universite de Dakar, Dakar, Senegal
and
Yves-Francois Thomas†
Ecole Normale Superieure, Montrouge, France

ABSTRACT

Saalum's estuary, 110Km South-East of Dakar in Senegal is a varied environment with a fast geomorphological evolution. A classification by clustering around mobile centers based on SPOT data gathered on May 9th, 1986 illustrates many categories of its landscape. A comparison between aerial missions of 1972, 1981 and SPOT imagery leads us also to the conclusion that high spatial resolution radiometers represent a new field of investigation for Earth Sciences.

INTRODUCTION

The advent of earth resources satellites enlarged considerably many geographical diagrams and improved our perception of landscapes. This trend has been amplified with the availability of data provided by satellites which could be branded as second generation ones (TM, MOMS, SPOT) with their approximate ground resolution of 10 to 30 meters.
For a century now, a huge amount of bathymetrics work has been conducted in the Saalum estuary which is a navigable way of paramount importance for the port of Kaolack (second city of Senegal).
Moreover, it is a varied and multifunctional environment with a fast geomorphological evolution. Owing to all those reasons, it has been utilized lately as an experimental

field for several projects: test for salted soils recovery
(1954-1962), production of partial vertical aerial views at
large scale (1972), a test zone for SPOT simulations in
West Africa (1981) and data assessment of SPOT satellite
(PEPS 1986).

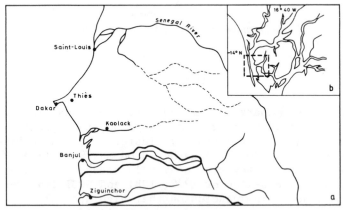

Fig. 1a   Localization of the Saalum coastal system (a) and
limits of the scene 022-322 and the studied area (b).

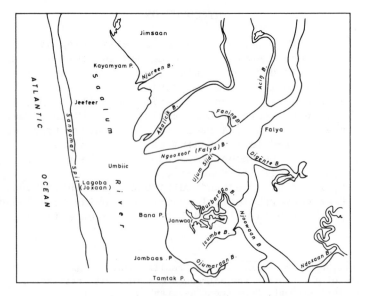

Fig. 1b   Toponymy and hydronymy of the Lagoba Ngooxoor's area.

CONDITIONS OF ACQUISITION AND METHODS OF STUDY

Acquisition and Processing Conditions

The processed image (Fig. 1a-b) was acquired by the receiving station of Kiruna (SWEDEN) and made available to us within the framework of a convention (PEPS N° 39 untitled "télédétection du littoral sénégalais. Evaluation des données du satellite SPOT") by the firm SPOT-IMAGE (Toulouse). The acquisition and processing conditions of the Saalum scene (60 km) are specified in table 1.

As for the hydroclimatic conditions, they are characterized by a clear sky, a little cloudy with light to moderate winds (average of 4m/s north-northwest direction, a barometric pressure equal to 1012 mb, a relative humidity of about 72% at 11:00 a.m., a sunburn of about 10.5, a good visibility (without any particular meteorological phenomenon), a temperature at ground level of 24°4 at 12:00 a.m. and a rising tide (1.5m at the Dakar station).

Methods of Study

An automatically plotted cartography has been made on an extract (512x640) of the Saalum scene; it is centered to the old mouth area (Lagoba or Joxaan) and the bolon (local name of tidal channel) of Ngooxoor. Information collection techniques are based here on a classification by clustering around mobile centers. The fundamental characteristics of this method can be diagrammed as follows: The classification is founded on the allocation of pixels to the center of divisions which are closer to them.

Table 1   Acquisition and processing conditions of the
Saalum scene

Mission	SPOT-1
Radiometer	HRV-2
Reference Grid	J = 322 and K =22
Orbit	009 03' 48"
Incidence	002 17' 00"
Spectral mode	Multiband
Preprocessing level	1B
Gains	XS1 = 5, XS2 = 6 and XS3 = 5
Latitude	014 01' 15"N
Longitude	016 33' 50"W
Solar rise	071 18' 00"
Solar azimuth	077 51' 00"
Satellite altitude	832 km
View intake date	05-09-1986
Time (U.T.)	11-46-03

Initially though the centers of gravity of the defined
divisions are determined; those centers in turn, represent
centers of clustering. The process is thus reiterated a
couple of times until the proportion of pixels whose
division changes during the iteration remains very low.

INTERPRETATION OF RESULTS

Hierachisation of Swamp Landscapes
        Tropical swamps which are highly influenced by tidal
dynamics, remain low stretches composed of fine grain size
with, as a main feature the adaptation of a particular
vegetation: mangrove. The low altitude of those soils, to
a great extent comparable with temperate swamps, but mostly
their topographic compartmentation (which is very low in
some cases, about centimeters) in relation with the
submersion rate of the tide remains at the basis of any
hierarchiation. But that hierarchiation, within the
tropical framework has long alienated itself and
systematically the diagram in vigor in temperate zones. To
that extent, if comparisons remain possible between
slikke-schorre and mangrove mudflat-tanne (Serer term
meaning barren flats) the existence of several local
particularities (vegetation, limited swamp extension,
sediments), hinders direct superposition, the assimilation
of forms of different climatic zones. In the absence of
studies in that sense, we are planning to carry out
research in the field. The first attempts represent a
synthesis conducted by M.D. THIAM[1] from whom we borrowed
the diagram of zoning of tropical swamps (Fig. 2).

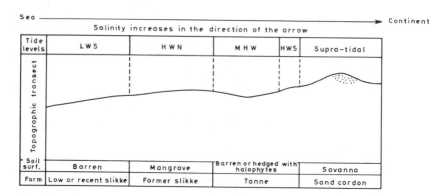

Fig. 2   Zonation within sheltered tropical coastal marshes.

Typology of Landscapes of the Studied Area

The typology of landscapes of the studied site is simple and it is related to the level of tides whose extension is determined by altimetric considerations of the soil. Consequently the tidal dynamics generates three types of environments.

Water Categories and Sand Banks (Fig. 3)

The distribution of water classes is related to depth variations and turbidity. The area drawn on the map is presented as a relatively high turbidity load environment, with a margin of 30 mg/1 along the mangrove or in turbulence areas. Bottoms, as a whole, remain weak with

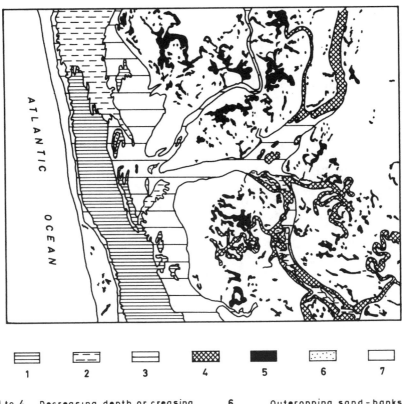

1 to 4 _ Decreasing depth or creasing turbidity

5   _ Brackish water areas

6   _ Outeropping sand-banks

7   _ Emerged surfaces

Fig. 3   Water categories of the Lagoba–Ngooxoor's area (method: clustering around mobile centers).

the existence of big sand banks that hinder navigation
considerably.

However, they play a crucial part in the shell-fish
gathering economy made by female oyster farmers of the
Saalum's islands.

The outlet of some Bolons (Njureen and Akalicik) is
marked by forms which can be assimilated to lateral levees
which F. VERGER and et al.[2] call "Pointes de Mangrove",
in comparison with "Pointes aux Herbes" developed in
temperate areas. Those levees are generally half curved
and stretched; they are dissymetric, with a concavity
directed to waters expulsion. Their shaping could be
mainly attributed to the ebb flow which contributes to the
mobilization of sedimentary stocks taken from mangrove mud
flats or suspended back from the bottoms.

The Mangrove - Tanne Complex

It corresponds to the intertidal zone and happens to
be, despite its apparent topographical homogeneity, a
highly interferent and intergrown environment, as regards
the morphological dynamics (formation process, warping up
problems, extension and regression process, state of those
processes etc.). Precise data on the topic are not
available and if much has been said on the unanimously
agreed on-role of the recent drought to explain a certain
number of evolutions, this should not hide realities which
appear to be more complex and which more detailed
observations alone could cover (role of water marks and
vegetation, frequence of submersions, sedimentation rate
etc.,). In any case, infographic map drawing obtained
(Fig. 4) offers an accurate spatial restitution of mangrove
formations with the individualization of the four classes,
which depend on the covering density. Mangrove groupings,
as P. NDIAYE[3] points out, have a low floristic complexity
and their configuration mainly defines two types:

- The first one is linear and closely follows the
edges of tidal channels; it is composed of dense Rhizophora
in the form of variable extension bands. The external
limit of those bands corresponds to what we defined as the
"contact line" between the maritime field and the vegetal
one (A.T. DIAW and et al.[4]).

- The second type of configuration corresponds to the
taxon composed of Avicennia, Conocarpus and Laguncularia.
It is located aside to Rhizophora, it is of variable
density and has no precise form.

Furthermore, those different formation have a specific
texture, which we approach through the geo-kinematic map
drawing of the Jimsaan area (Fig. 5a, b, c) by comparing

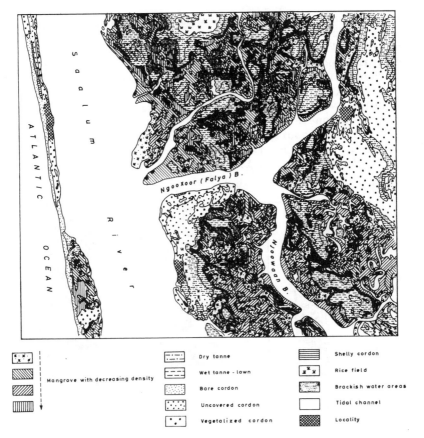

Fig. 4   Typology of Lagoba-Ngooxoor's landscapes
(Interpretation from the clustering around mobile centers).

the 1972 air mission (72 - SEN - 05/100/041 to 045), the
1981 SPOT simulation (81 SPOT - SEN 06/37 to 40) and the
1986 HRV data of SPOT 1 satellite. That procedure
represents a new field of investigation at a time of high
spatial resolution radiometers (TM, MOMS, SPOT). Recent
applications to the study of coastal lines (E. GUILLEMOT
and Y.F. THOMAS[5]) and to the study of the mobility of
dune fronts (A.T. DIAW & Y.F. THOMAS[6]) can show, if need
be, their profit worthiness.
      The examination of landscapes evolution in Jimsaan is
backed up by a textural analysis which can lead to the
following considerations:
      - The determination of mangrove texture is based on
various parameters among which the drainage organization
and/or less large scale conservation of species. That last

parameter is essential as regards the assessment of evolution and recovering grade. The utilization of those criteria make it possible to see for example a high textural opposition between the more massive, tighter and smoother Rhizophora formations of the high top, north of the bolon Akalicik (class 1) and that of smaller space extension, with thin texture which underlines the presence of tidal channels (class 2).

　　– The top south of the same bolon which is dominated by Avicennia, can be all in all put at the same level given its massivity as class 1. Its structure however remains

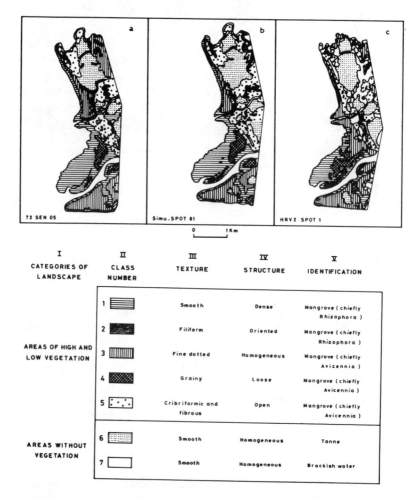

I	II	III	IV	V
CATEGORIES OF LANDSCAPE	CLASS NUMBER	TEXTURE	STRUCTURE	IDENTIFICATION
	1	Smooth	Dense	Mangrove (chiefly Rhizophora)
	2	Filiform	Oriented	Mangrove (chiefly Rhizophora)
AREAS OF HIGH AND LOW VEGETATION	3	Fine dotted	Homogeneous	Mangrove (chiefly Avicennia)
	4	Grainy	Loose	Mangrove (chiefly Avicennia)
	5	Cribriformic and fibrous	Open	Mangrove (chiefly Avicennia)
AREAS WITHOUT VEGETATION	6	Smooth	Homogeneous	Tanne
	7	Smooth	Homogeneous	Brackish water

Fig. 5a,b,c　Geokinematic mapping of Jimsaan's landscaping (1972 to 1986).

less tight with a semi homogeneous thin-dashed texture (class 3). This is different from class 4 which is also colonized by a scattered Avicennia with grainy texture and round "houppiers" arranged in species.

- The last observed class of mangrove (class 5) presents a whole cribriformic texture which shows an evolution towards the tanne. Vegetal formations of that class shows a fibrous arrangement on the air pictures, the "vegetal network" having a tendency to evanescence.

Here a confinement of vegetal species on more and more reduced areas can be pointed out, as well as an extension of barren areas (tanne) and brackish waters between 1972 and 1986.

Automatic map drawing (Fig. 4) also isolates precisely the edges of tanne, taxon with a complete and general uniformity. All in all they have concentric and annular forms. Their hypsometric zonation determination from the more or less long duration of the latest submersion by the tide, has been clearly established during previous works (A.T. DIAW and et al.[4]). In this study, we are more concerned with the definition of their space extension.

Supratidal Zone

It revolves around two points: "lawns" and sand cordons. Those units spectrally speaking, remains difficult to differenciate. However, they are easily landmarked on the point of view of space. Generally lawns are transitional areas, located along the tanne, with various grass cover. However, an itemized hierarchization was made by P. NDIAYE[3] who makes the following distinctions:

- The Sesuvium grouping around the tanne is the first herbaceous vegetation on highly salted areas. It is associated with halophytes like Philoxerus or Suaeda or other more or less tolerant species like Sporobolus.

- Grouping with Paspalum and Fimbristylis: those species can be associated as transitional elements linked with the first defined grouping. Their establishment is only determinant when unsalting processes are favoured. That is why they can be found on former rice rows with plants like Maytenus, Cassia and Argemone.

Topographically speaking, sand cordons represent the highest units. They are covered with vegetation on some parts; that vegetation, of a changing density, is composed of various facies (Dalbergia, Maytenus, Aphania, Parinari), be it natural or planted (coconut trees). In the Jonwaar, Falya or Jeefer sectors those units have naked areas resulting from fallowing or husbanding of those areas; that

fallowing, even selective - because the herb cover alone is
concerned - makes its that those areas, like tanne or
barren sands of the strand or dunes are here the most
strongly reflecting.

CONCLUSION

Validation studies, conducted on the Lagoba-Ngooxoor
area (Saalum's estuary, Senegal) show the interest of SPOT
imagery in the perception and understanding of geographic
facts or the elaboration of grading plans of coastal areas
through mapping. Results obtained here show undoubtedly an
improvement, brought about by high resolution radiometers
in precising taxons limits and the nature of represented
objects (tanne edges, water and mangrove categories, tide
extension). Consequently, they make it possible to have a
mapping which is close to reality, and at the same time
show the right track to geo-kinematic studies by mixing
available cartographic or photographic data with that
furnished by high resolution radiometers. That last
operation made on the Jimsaan site made it possible through
an exhibition of cartographic documentation to note changes
which occured in the landscape from 1972 to 1986.

REFERENCES

[1]Thiam, M.D., "Géomorphologie, Evolution et Sédimentologie des
Terrains salés du Sine Saloum," Doctorat Univ. Paris I, Dec. 1986,
pp. 16-21.

[2]Verger, F. et al., "Cartographie Automatique d'un Milieu
Littoral Tropical: Iles du Saolum (Sénégal), d'aprés les Données
Landsat," Rev. Photo Interprétation, Paris, No.6, Nov. Dec. 1978,
pp. 23-28.

[3]Ndiaye, P. "Méthode d'Inventaire, Analyse et Cartographie de la
Végétation. Exemple de l'Embouchure du Saloum (Sénégal)," Notes
de Biogéographie, Dakar, No.1, 1986, pp/ 32-41.

[4]Diaw, A.T. et al. "L'Exploitation des Données SPOT simulées en
Domaine Littoral Tropical. Les Iles du Saloum," Rapport TECASEN,
Dakar, No.1, Juin 1983, 95p.

[5]Guillemot, E., Thomas, Y.F., "Evolution de la Fléche Sableuse
de Los Torunos (Espagne)," Rev. Photo Interprétation, Paris, 1985,
pp. 11-15.

[6]Diaw, A.T., Thomas, Y.F., "Cartographie Satellitaire de Zones
de Déflation Eolienne; le Périmétre de Lompoul (Sénégal)," Actes
du Fi3G, Lyon, Juin 1987, pp. 496-508.

Chapter V.   Propulsion in Space

Propulsion for the Space Station

Vernon R. Larson* and Stephen A. Evans†
Rockwell International/Rocketdyne Division, Canoga Park, California

Abstract

Propulsion for the Space Station requires a unique, challenging combination of features. The propulsion requirements demand responsive capabilities for orbit control and countering on-board or externally induced forces, and precision control for station-keeping with minimum disturbances. The propulsion design must provide high performance to minimize propellant requirements, have extremely high reliability, provide an extended long-life capability, control emissions and outgasses that could impact in-space operations, and provide maximum safety.

Comprehensive evaluations have been performed investigating numerous design options. These efforts have included in-depth design and analytical evaluations, trade studies, and experimental test efforts.

The NASA Space Station program office has defined a baseline propulsion system based on hydrogen/oxygen thrusters, using an on-board water electrolysis system to provide the propellants. This system is combined with low thrust (100 millipound) resistojet propulsion thrusters that can operate using excess or waste fluids. With this combined on-board propulsion system, the need to transport and store hazardous propellants is eliminated, as is the need to collect, store, and return waste fluids to earth.

This paper is an update of IAF Paper No. 86-182, prepared by the authors and presented at the 37th IAF Congress between 4-11 October 1986.
*Director, Advanced Systems Programs.
†Program Manager, Space Station Propulsion Programs.

Prototype thrusters for the on-board propulsion have been fabricated and tested, with excellent performance and durability demonstrated. Features that enhance long life, maintainability, servicing, and health monitoring have been studied and are being incorporated.

A versatile propulsion test bed has been designed and an extensive demonstration program initiated, including system testing with the prototype hydrogen/oxygen thrusters, and with candidate water electrolysis units. Resistojet testing to demonstrate performance and life is now under way. To date, all results have been successful in demonstrating that the hydrogen/oxygen and resistojet systems will perform as planned on the Space Station.

Introduction

Since President Ronald Reagan committed the United States to the development of a permanently manned Space Station, the National Aeronautics and Space Administration has rigorously evaluated and analyzed candidate configurations for the Space Station, and for the subsystems that make up the Space Station. NASA has organized the Space Station program to challenge industry, in a highly competitive environment, to ensure the highest level of skills and expertise of all participants are fully employed. From these extensive efforts, a Space Station system design has emerged that has the capability and characteristics required for a near-term Initial Operating Capability (IOC), and also has the capabilities for growth to meet future needs.

The Space Station is illustrated in Fig. 1. In viewing this drawing, two principle features are most evident: the structural design and the power system design. The Space Station is characterized as a dual-keel design. The keels are the two main vertical structural members. The dual-keel design enhances on-board equipment and module placement, structural stiffness, and docking capabilities. The design also provides a gravity gradient that benefits spatial orientation stabilization. External payloads that will be oriented to space will be located primarily on the upper boom connecting the two keels, and earth-directed external payloads will be located primarily on the boom connecting the lower ends of the two keels.

The configuration shown in Fig. 1 would provide 87.5 kW of electrical power. The power system illustrated is a hybrid power system design that includes two photovoltaic modules and two solar dynamic power modules.

Fig. 1 Space station illustration

A candidate Space Station growth configuration is depicted in the drawing in Fig. 2; this hybrid power system
design employs additional dynamic modules, and would provide electrical power in the range of 300 kW. This design
employs an extension of the "cross structure," with additional solar dynamic power modules.

The on-board propulsion system for the Space Station
provides both the primary station keeping and attitude
control propulsion functions. In conjunction with the
other elements of the Space Station [i.e., the dual-keel
design; the on-board guidance, navigation, and control
system; and control moment gyros (CMG)], the on-board propulsion system must provide a very stable environment for
the users and operators of the Space Station. Propulsion
system basic functions include providing reboost thrust to
compensate for atmospheric drag, station keeping, and spatial attitude control. The propulsion system must also
accommodate and react to the routine rendezvous and berthing of the Orbiter and other spacecraft; provide a control
capability to "backup" the CMGs; provide the capability for
possible relocation; and provide propulsion for possible
contingency functions.

Based on the Space Station configuration studies,
studies of the on-board subsystems and the Space Station
operations, the NASA Space Station program office has defined a baseline on-board propulsion system that employs

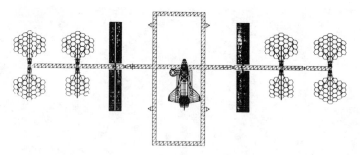

Fig. 2 Candidate growth space station

hydrogen/oxygen (H_2/O_2) thrusters using an on-board water electrolysis system to provide the propellants, combined with low thrust (i.e., 100 millipound thrust level) resistojet thrusters that can utilize a wide variety of excess or waste fluids. The major advantage of this system is that the need to transport propellants to orbit is basically eliminated, as is the return of waste fluids to earth. The savings in logistics (both costs and operations) is high, especially for the constant drag orbit mode of operation currently planned for the Space Station.

    This paper provides an introduction to the requirements demanded from the on-board propulsion system, summarizes the propulsion system design as it has evolved from the extensive comparison analyses and design studies performed, and defines the basic on-board propulsion system's features and characteristics. Design features that enhance life and reusability are highlighted; focusing on the propulsion thrusters--to provide a paper of manageable scope. Test status is then summarized.

Requirements of the On-Board
Propulsion System

    The basic functions and requirements of the Space Station's on-board propulsion system are summarized in Table 1, and include velocity corrections and spatial attitude control. The velocity correction requirements consist of atmospheric drag makeup (reboost), compensation for Orbiter docking, and such contingencies as possible debris avoidance and possible repositioning. The attitude control requirements include the capability to provide reboost attitude control, torques exceeding Control Moment Gyros (CMG) capacities, CMG backup, and CMG desaturation.

    Based on detailed propulsion requirements and design studies conducted as the Space Station configuration has

Table 1. On-Board Propulsion Requirements

- Velocity Correction
  - Drag Makeup
  - Reboost
  - Orbiter Docking
  - Contingencies
- Attitude Control
  - During Reboost
  - CMG Backup
  - CMG Desaturation
- Man-Rated
- Safe
- Fail Operational, Fail Safe, and Fail Restorable
- High Availability and Reliability
- Long Service Life
- Easily Maintained and Serviced
- Flexibility and Growth Capability
- Minimum Environmental Impact
  - Minimum Zero-g Impact
  - Minimum Impact of Exhausted Gases
  - Nonintrusive
- Low IOC Cost
- Low Life Cycle Cost
- Minimum Transport of Propellants
- Accommodate Initial Buildup of the Space Station
- Provide a Function—-Not Require an Operation

evolved, the design for on-board propulsion system has been optimized and developed to provide these requirements. The propulsion requirements have been significantly influenced by the Space Station's orbiting altitude, type of electrical power system (i.e., drag impact), the Space Station's configuration and its mass, and the design and operation of other on-board systems. The customer accommodations basically require a microgravity on the space station of less than 10^{-5} g.

The on-board propulsion system must be man-rated, have a maximum possible safety rating, provide high availability and reliability, provide a long, service-free life, and be easily maintained on-orbit. A propulsion system that minimizes transported propellants is highly desirable. Further, the on-board propulsion system must have sufficient capability and flexibility to provide the functions required for the Space Station, and for the growth Space Station (considering the increased drag area and mass of the 300 kW power growth space station). The station safety requirements specify a fail-operational/fail-safe/restorable system with "designed-in" safety. The latter would include damage containment. The basic design life for the Space Station's on-board propulsion is 10 years. The propulsion system must provide a minimum environmental impact both in terms of environment contamination (i.e., thruster exhaust gases and/or other gaseous venting). Low initial development cost, and low life cycle/ operating costs, are also key requirements.

The Space Station will operate in a "constant drag" mode at an altitude of 180 to 220 n mi, based on recent NASA plans. In this mode of operation, the Space Station will operate at an altitude, such that without reboost for a 90-day period, the orbit will not degrade to an altitude below minimum control capability. With this constant drag mode, the Space Station's operating altitude will vary depending on the atmospheric density in a given year and quarter. Drag deceleration at these conditions is limited to a maximum of 0.3 micro-g (3×10^{-7} g), which is equivalent to approximately 0.15 pound of drag force on the Space Station.

With this 180 to 220 n mi Space Station operating altitude (constant drag mode), the total impulse required from the on-board propulsion system is on the order of 60 million pound-seconds for a period of 10 years. A summary of the total impulse requirements is shown in Table 2. A 90-day nominal propulsive total impulse requirement for the IOC Space Station is on the order of 1 million pound-seconds. For the growth station, this increases to approximately 2 million pound-seconds.

A parametric graph of propulsion operating requirements for projected average daily total impulse requirements, showing hydrogen/oxygen (combustion) and resistojet operation, is depicted in Fig. 3. The specific impulse operating band shown for the resistojet thrusts encompasses

Table 2. Total Impulse Requirements

	Total Impulse, lbf-sec
● IOC Space Station	
● Daily Average	11,000
● 90-Day Period	1×10^6
● Growth (300 kW) Space Station	
● Daily Average	22,000
● 90-Day Period	2×10^6
● 10-Year Period	
● 1995 to 2004	60×10^6

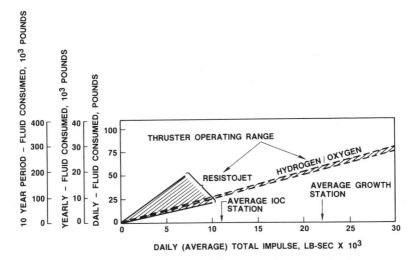

Fig. 3   Propulsion system on-board fluid consumption

a range of 150 to 500 seconds, depending on operating fluid
and temperature, as discussed later.   For the hydrogen/
oxygen thruster, this operating band is shown from 380 to
400 seconds specific impulse.   For average IOC station
total impulse requirements (11,000 pound-seconds), average
growth station requirements, the daily, yearly, and 10-year
period total fluid consumption requirements are shown. On
the average, approximately one-sixth of the daily requirements may be provided by the resistojet thrusters.

Another factor that impacts the on-board propulsion
system's design requirements are the station keeping and
attitude control capabilities needed as the Space Station
is "built-up" in orbit.   Attitude-control-only may be
adequate for the first element to be deployed, which is
the transverse boom.   However, it is possible that some
degree of propulsion system operational capability (including reboost to higher altitude) will be necessary beginning
with the second flight.   A design approach using modular
units that could be temporarily attached to the central
beam and then relocated to their operational position
appears favorable from the assembly standpoint.

Unique to the Space Station is the fact that personnel are available on-board for system repair and maintenance.   However, the goals of the Space Station's on-board
propulsion system are to provide a function--that of orbit
maintenance and altitude control--without requiring extensive maintenance or servicing (even though such is possible
with trained personnel on-board).

Space Station Propulsion System

Extensive evaluations have been conducted to define the on-board propulsion system design approach, propellants and design characteristics. The design approaches studied (Table 3) have included:

1. Monopropellant hydrazine (N_2H_4) system, with and without augmentation using electrically heated thrusters (resistojets)

2. Several design approaches employing earth-storable propellants (i.e., nitrogen tetroxide (N_2O_4) or mixed oxides of nitrogen (MON) as the oxidizer and hydrazine or related amines as the fuel),

3. A wide spectrum of hydrogen/oxygen propulsion systems, including subcritical and supercritical propellant storage, systems with and without accumulators, and pump- and pressure-fed systems

4. Resistojet systems (i.e., electrically heated thrusters) operating using a spectrum of working fluids as an adjunct to other propulsion systems.

Comparison studies considering such factors as development costs; operational costs; total life cycle costs (LCC); safety; environmental impact; reliability; life;

Table 3. On-Board Space Station Propulsion Candidates

- Monopropellant Hydrazine
- Earth Storage Bipropellants
  - Oxidizers: N_2O_4, MON (mixed oxides of nitrogen)
  - Fuel: N_2H_4, Amine Blend
- Hydrogen/Oxygen
  - Supercritical (cold, high pressure gas) Resupply
  - Subcritical (liquid resupply)
  - On-Board Water Electrolysis
- Warm Hydrogen
  - Supercritical Resupply
  - Subcritical Resupply
- Resistojets (used with any of above systems)
  - Hydrogen
  - Excess/Waste Fluids

maintenance and servicing; propellant supply; and transportation requirements were performed for these candidate systems along with variations and other design perturbations. (References 1 through 4 address some of the comprehensive comparison studies conducted to evaluate candidate propulsion options.) An example of these many evaluations and comparison studies is shown in Fig. 4. This example graph shows, for the Space Station constant drag mode of operation, that eliminating the need for transported propellants is very beneficial from an economic standpoint. In this example, transported hydrazine is shown as the comparison case. A similar trend in the results would be shown if transported hydrogen/ oxygen were used instead of transported hydrazine. In addition if the need for transported propellants is eliminated, logistics are greatly simplified.

From detailed Space Station propulsion requirements studies, and detailed studies of the characteristics, features and optimum design configurations for candidate propulsion systems, the basic result has been the selection of a hydrogen and oxygen propellant propulsion system, augmented by electrically heated resistojet thrusters. The hydrogen/oxygen system would operate principally using on-board, electrolysis produced hydrogen and oxygen and the resistojet thrusters are designed to maximize usage of excess and/or waste on-board fluids. This design approach provides many unique and highly attractive benefits in terms of: minimized environment contamination, enhanced safety, low operational cost, and ease of servicing. It

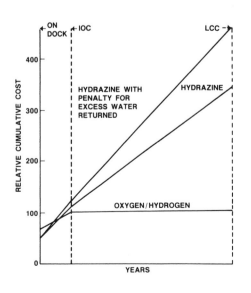

Fig. 4 On-dock, IOC, and life cycle cost comparison

minimizes (i.e., virtually eliminates) the requirement to store and transport hazardous propulsion fluids from the ground, and also minimizes the requirement to deorbit excess or waste fluids. Costs associated with propellant supply are greatly reduced.

Studies (e.g., Ref. 5) of the availability of hydrogen and oxygen provided by on-board water electrolysis, and the availability of waste and/or excess fluids for the resistojet, have indicated that most or all of the projected propellant needs can be met with these on-board sources (while also meeting all the other on-board water and oxygen needs). With this selected propulsion design approach, transport of propellant would only be required to provide on-board propulsion operation while the station was being "built-up." During buildup, the high pressure gas accumulators can be launched full. As the electrolysis units are brought up for the life support system, the accumulators will be refilled using these units. If necessary during early man-tended or initial-manned operation, some makeup water can be transported. Once full-manned operation is in place, no transport of propulsion fluids is required.

Sources of waste water include the Space Shuttle Orbiter, the environmental control and life support system (EC/LSS), the laboratories, and attached payloads. Reference 5 projects that excess water recovered from the EC/LSS will be on the order of 0.9 lb/man/day. Up to 1200 pounds of water can be scavenged from the Shuttle Orbiter on each mission. As the station grows, either a balance can be maintained between the availability of on-board excess/waste fluids and propulsion requirements or occasional makeup water can be transported. The availability of excess

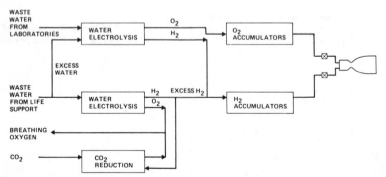

Fig. 5 Hydrogen/oxygen propulsion integrates naturally with fluid management

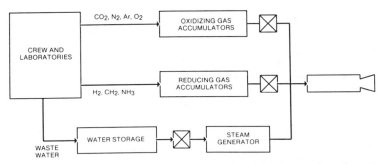

Fig. 6 Resistojets propulsion integrates with the on-board waste management system

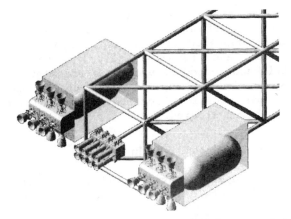

Fig. 7 Example hydrogen/oxygen and resistojet propulsion modules

fluids is dependent upon actual crew growth, realized laboratory usage, types of experiments, and amount of fluid recycling within the laboratories. Even with a postulated worst-case water resupply, life cycle costs are still significantly lower than other options (Ref. 5).

Propulsion Modules

A schematic diagram of the hydrogen/oxygen propulsion system is shown in Fig. 5. The resistojet propulsion system schematic is shown in Fig. 6. Although the schematics are shown separately for simplicity, there are obvious interconnections throughout the fluid management system that benefit both systems. Consonant with the selection of the propulsion system's propellants and the operation configuration has been the analysis and selection of the on-board packaging.

The baseline option employs four independent hydrogen/ oxygen propulsion modules, located on the Space Station's keels. Figure 7 illustrates two hydrogen/oxygen thruster modules and a single resistojet module. Each of the four independent modules would (as an example) contain nine hydrogen/oxygen thrusters, the high pressure (i.e., greater than 1000 psia) gaseous accumulators for the hydrogen and oxygen, and the associated valves and flow control devices. Hydrogen and oxygen lines run the gaseous propellant to the four modules from the central propellant production station. The three aft-facing thrusters provide principally reboost, the others provide attitude and reorientation control.

A single resistojet module (with eight thrusters) is located at the end of an aft-facing boom (Fig. 7). The module would be supplied from separate storage tanks for oxidizing gases and reducing gases. The resistojet thrusters will be oriented to provide a translation thrust, for the constant drag makeup. Location of hydrogen/oxygen thrusters at this location for reboost in addition to the resistojets is being studied.

Propulsion Thrusters

The gaseous hydrogen/oxygen and resistojet thruster assemblies for the Space Station propulsion are key ele-

Fig. 8 Prototype 25-lbf gaseous oxygen/gaseous hydrogen thruster

ments in the on-board propulsion system design, its operation, and its performance capabilities. The design of these
thrusters and their operation govern the life, availability, and reliability of the propulsion system, and in turn
maintenance and service requirements. In this section, the
hydrogen/oxygen bipropellant thrusters and the resistojet
thrusters are discussed.[6,7]

A key element in the design of these thrusters has
been the establishment of a design philosophy that emphasizes reusability rather than minimized weight (a goal
that is typically emphasized in thrusters for most rocket
powered systems). Since weight is not as critical as other
factors, the design philosophy and resulting design criteria can reflect a design approach that will maximize
life and safety, and thereby permit a design where margins
for these key criteria can be fully implemented. With such
designs, the need for on-orbit servicing and maintenance
are greatly reduced.

Hydrogen/Oxygen Thruster

The design goal for the thrusters is to provide ample
thruster design life for 10 years of nominal Space Station
operation. Based on this 10-year Space Station design life
(without replacement) and four hydrogen/oxygen modules--
each with three rear-facing thrusters--a single rear-facing
thruster will be designed to have an operating life equivalent to 15 million pound-seconds of total impulse, on the
average. (In each hydrogen/oxygen module, two of the three
rear facing thrusters are considered redundant.) At a nominal thrust level of 25 pounds for a hydrogen/oxygen
thruster, this total impulse corresponds to a total average
10-year operating life of 165 hours. Cyclic life requirements are projected to be on the order of several thousand

Table 4. Hydrogen/Oxygen Thruster Design Configuration

● Hydrogen/Oxygen Propellant

● Fuel-Cooled, Regenerative, Channel Wall Design
  ● Copper Alloy Wall Material
  ● Electrodeposited Closeout

● Coaxial Injector

● Independent Fuel and Oxidizer Valves

● Integrated Spark Igniter

● Operates at Mixture Ratios 3:1 to 8:1

● Operates Throttled 2:1

full thermal cycles. Pulse life will be on the order of 10,000 pulses. These are not severe requirements for a hydrogen/oxygen thruster designed to operate at relatively low temperatures, and designed with ample structural and thermal margins.

A prototype 25-pound gaseous hydrogen/gaseous oxygen thruster designed for the Space Station is shown in Fig. 8 with the features summarized in Table 4. The thruster incorporates a regeneratively cooled thrust chamber to a nozzle exit area ratio of 30:1, a coaxial injector, spark igniter, and close-coupled propellant valves. This thruster has been tested for over 24 hours at mixture ratios from 3:1 to 8:1. The longest single firing was 6.1 hours while the shortest pulse was 0.3 second. Over 10,000 pulses were demonstrated during pulse operation.

The regeneratively cooled thrust chamber is based on very conservative thermal margins. The main chamber and nozzle liner, stainless steel manifolds, and an electro-deposited nickel closeout of the liner coolant channels. The fabrication technique has been proven in production (Space Shuttle) and experimental engines.

The thruster incorporates separate, identical coaxial solenoid fuel and oxidizer valves. The thruster is designed for a combustion chamber pressure of 100 psia. The 30:1 nozzle area ratio permits the chamber to be tested in existing vacuum test facilities. The design permits addition of a radiation-cooled nozzle skirt to provide higher performance. Total weight of this prototype thruster is approximately 8 pounds. Basic design characteristics are summarized in Table 5.

Resistojet Thruster

The resistojet thruster for the on-board propulsion system is designed to operate using waste gases/fluids from the Environmental Control and Life Support System (ECLSS), the Materials Technology Laboratory from attached payloads, the Shuttle Orbiter, and in the future excess propellants from orbit transfer vehicles. The gases/waste fluids that the resistojet can use include water, carbon dioxide, methane, hydrogen, nitrogen, oxygen, and inert gases.

Carbon dioxide, methane, mixtures of carbon dioxide and methane, and water vapor will be available from the ECLSS. Gaseous oxygen and hydrogen will be available from the water electrolysis system(s) or the ECLSS. Oxygen and hydrogen also may be available from the transfer stage propulsion supplies if a hydrogen/oxygen transfer system is used. The laboratory modules will require disposal of various gases, potentially including nitrogen, helium, and argon.

In addition to operating with different fluids, the resistojet thrusters must be capable of relatively large flow throttling and electrical power throttling. Projected values are on the order of 3 to 1. For example, when large quantities of "excess" fluids are available, these can be used to generate thrust without regard to efficiency; in this situation the resistojets would operate at a lowered electrical power input so as not to exceed thrust requirements. The resistojet thrusters also must be able to operate at less than maximum flow when low levels of excess fluids are available but when some thrust is required. In this case, the electrical power would be adjusted to ensure that maximum (i.e., long-life) operating temperatures are not exceeded. In addition, the resistojet thruster must be able to provide a (warmed flow) venting of waste gases capability, with the warm flow utilized to prevent exhaust condensation.

A resistojet thruster prototype is shown in Fig. 9 with features summarized in Table 6. Thrust for this prototype unit is 0.1 pound. Projected nominal Is values for various candidate fluids are shown in Table 7. Basic thruster characteristics are in Table 8. The high temperature achieved by the resistojet working fluid/exhaust gases, combined with the high area ratio nozzle will

Table 5.   Hydrogen/Oxygen Thruster Characteristics

Design Characteristics	Design Goals	Demonstrated
Thrust, pounds	23 at = 30:1 25 at = 100:1	12.5 to 25
Chamber Pressure, psia	100	50 to 110
Mixture Ratio, o/f	3:1 to 8:1	3:1 to 8.2:1
Area Ratio	30:1 Regen Cooled	30:1
Specific Impulse (at MR =4.1 and = 30:1)	400	405
Minimum Pulse Duration, milliseconds	30	30
Minimum Impulse Bit, lbf-sec	Less than 0.5	Under 0.5
Life	Meet 10 years Space Station Life	24.1 hours; 2M lb-sec
Pulse Capability	Over 1 million	Over 10,500
Weight, pounds	--	8.25

Fig. 9 Prototype Resistojet

Table 6. Resistojet thruster Design Configuration

- Concentric Configuration
- Geometrically Centered Platinum Conductor/Heater
- Annular Insulation
- Platinum Sheath
- Commercially Available Heater Materials
- Designed to Operate on Many Fluids: CO_2, CH_4, CO_2/CH_4 Mixtures, Water Vapor, Gaseous Oxygen, Gaseous Hydrogen, Nitrogen, Helium, Argon, and Air
- Operating Design Temperature Consistent with Commercial Heater with 14,000 hours Demonstrated Life

Table 7. Preliminary Resistojet Performance

Fluid	Specific Impulse, seconds
Argon	135
CO_2	130
Helium	425
Hydrogen	500
Nitrogen	160
Steam	200

accelerate the discharged fluids to hypersonic velocity, which will minimize backflow contamination.

The prototype thruster is designed for an operating life (goal) of 10,000 hours. The thruster incorporates a platinum sheathed heater that is derived from commercially demonstrated long-life heaters. The thruster is a coaxial configuration with a geometrically centered platinum conductor/heater, surrounded by an insulator. This assembly is enclosed in a platinum sheath. The design has a projected thermal efficiency of 92%. The design goal has not been to realize maximum performance, but (as with the hydrogen/oxygen thrusters) to achieve long life and reusability over the 10-year period and beyond.

Design Features for Long Life with Minimized Maintenance and Servicing

Design for minimized maintenance and servicing begins with a design for long life and high relability. Ample structural and thermal margins, a strong data base, and verification testing enhance these characteristics. The selection of the design configuration, materials, operating temperature levels, and structural safety factors all provide a means for ensuring long design life and high reliability, and thereby avoid the need for maintenance or servicing. Redundancy in the design can further reduce maintenance and servicing needs. Design features that provide ease of servicing and simple maintenance operations benefit on-board operations.

Table 8. Resistojet Thruster Characteristics

Thrust, pounds	0.1 nominal
	(0.03 to 0.1 range)
Chamber Pressure, psia	50
Area Ratio	over 1000:1
Maximum Design Fluid Temperature, C (F)	1000 (1800)
Maximum Design Chamber/Heater Wall Operating Temperature, C (F)	1400 (2500)
Design Life	10 years
	Space Station Life
Firing Duration (minimum), hours	10,000
Total Impulse, lbf-sec	2×10^6
Maximum Specific Power, watts/millipound of thrust	5.5 (CO_2)
	15.6 (H_2)

Table 9. Design Features to Minimize Maintenance
and Provide Servicing Ease

==

- Redundancy
  - Thrusters, Valves, Tanks
  - Thruster Elements/Components

- Design Factors for Long Life
  - Design Configuration and Approach
  - Thermal Margins
  - Safety Factors

- Orbital Replaceable Unit (ORU) Approach

- Accessibility - Simple/Quick Disconnects

- Minimized Impact of Vented Fluids

- Provision Attachments/Interfaces for
  Replacement Unit

==

These design features are enhanced by a design approach that emphasizes health/status monitoring of operating conditions that might lead to an anomaly. Health/status monitoring would measure key operating conditions; this would ensure that possible detrimental conditions are recognized early before hardware degradation would occur. With such monitoring (1) the anomaly could be corrected, service is performed, or (3) shutdown and transfer of operation to a redundant or alternate node of operation until maintenance is performed could be accomplished. Even with this attention to long life and high reliability, and the built-in design characteristics that ensure realization, the capability for on-orbit operations (e.g., maintenance, servicing, monitoring, etc.) is necessary.

The processing and operations requirements include the need to access, service, and maintain all elements of the on-board propulsion system. The on-board system monitoring, maintenance, and servicing design (and operational approach) are all directed to ensure maximum utilization of condition-monitoring fault detection to provide early detection of any possible anomaly, and easy replacement at lowest Orbital Replacement Unit (ORU) component element if such a requirement exists. For the Space Station's on-board propulsion, these and other features either already have been incorporated, or are in the technology/evaluation phase to be adopted as defined desirable.

Some of the design features that enhance low maintenance and minimum servicing are summarized in Table 9. For a rocket thruster, a generic listing of life limiting categories are summarized in Table 10. These categories

Table 10. Generic Life Limiting Categories for Thrusters

I. Duration/Run Time Factors
   1. Erosion/Reaction of Nozzle Coatings
   2. Erosion/Reaction of Chamber Nozzle
      Material by Propellants or Exhaust
      Products
   3. Hydrogen Embrittlement
   4. Erosion of Catalysts
   5. Erosion of Overheated Metal

II. Cyclic Operation Factors
    1. Cyclic Yield Stresses Caused by Thermal
       Cycling with High Thermal Gradients at
       High Temperature Levels
    2. Erosion/Degradation of Valve Seats

III. External Environmental Factors
     1. Oxidation/Reaction of Material
     2. Space Effects
        o Radiation
        o Thermal Cyclic Due to Solar Influx
        o Local Atmosphere Impact (e.g., MON-
          Atomic Oxygen)

provide a basis for design emphasis, evaluation, and verification testing.

The on-board hydrogen/oxygen propulsion system
thrusters are (representatively) grouped in modules, with
nine thrusters to a module—three for each axis (Fig. 7).
This provides a high degree of redundancy, since ample
thrust is provided by operation of only one of the module
thrusters in any axis. With a design life of 165 hours
operation and 50,000 full thermal cycles for the hydrogen/
oxygen thrusters and 10,000 hours for the resistojet
thrusters, an individual thrusters must be serviced (i.e.,
replaced) only on an average of once every 10 years.

An overview of the basic thruster elements and design
approaches that can be provided to enhance operation,
life, and reliability are listed in Table 11. As this
summary illustrates, key factors in achieving a design
that will require minimum maintenance and servicing can be
grouped into several system and component design areas.

For long life and high reliability, both the hydrogen/
oxygen and resistojet thrusters employ ample thermal and
structural factors, and well characterized proven materials. The copper-alloy chamber wall material in the hydro-
gen/oxygen thruster is well proven and its long cyclic life
design characteristics are fully test verified. The design
wall temperatures for the hydrogen/oxygen thruster are conservative, and have been determined (in the testing con-

Table 11.   Thruster Elements and Design Approach for Long–Life Design with Minimized Maintenance and Servicing

Thruster Components	Active/Key Elements	Potential Degradations	Potential Impacts	Counters/Design Attributes
• Valve	• Actuator	• Electrical • Structural • Thermal	• Degraded Activation	• Dual Actuators • Redundant Actuator Coils • Monitor
	• Mechanical Closure	• Structural • Thermal	• Leakage/Failure	• Dual Seat Valves • Seat Design • Thermal Margin • Monitor
• Injector	• Injector Orifice	• Clogging/Orifice Restriction • Erosion	• Degraded Performance • Localized Wall Heating	• Filters • Orifice Element Design • Monitor
	• Thermal/Face Cooling	• Overheating	• Burnthrough • Degraded Performance	• Filters • Thermal Margin • Monitor
• Chamber Body (Combustor and Nozzle)	• Inner Wall (Contain Gases	• Overheating	• Burnthrough • Degraded Performance • Failure	• Thermal Margin • Monitor
	• Cooling Passage	• Clogged	• Burnthrough • Degraded Performance • Failure	• Filter • Passage Design • Monitor
	• Outer Wall (Structural)	• Thermal Failure • Structural • Delamination Performance	• Design Failure • Leakage • Degraded	• Structural Margin • Thermal Margin • Monitor
• Other Elements				
• Valve-Injector Body Attachment	• Structural	• Structural	• Leakage • Design Failure	• Structural Margin • Thermal Margin • Monitor
• Actuation Signals	• Electrical	• Signal Loss • Signal Degraded	• Failure to Operate • Fail Open	• Monitor • Fail-off (Closed) Design
• Fluid (Propellant) Supply	• Impurities • Loss of Pressure/ Flow	• Clogging • Degraded Flow	• Degraded Performance • Overheating • Failure to Operate	• Filters • Thermal Margin • Monitor

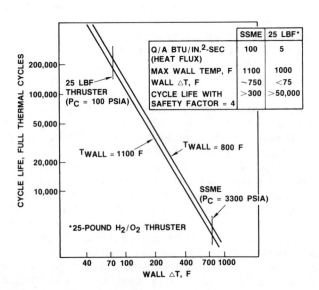

	SSME	25 LBF*
Q/A BTU/IN.²-SEC (HEAT FLUX)	100	5
MAX WALL TEMP, F	1100	1000
WALL ΔT, F	~750	<75
CYCLE LIFE WITH SAFETY FACTOR = 4	>300	>50,000

Fig. 10   Gaseous oxygen/gaseous hydrogen thruster estimated cycle life (*25–pound hydrogen/oxygen thruster)

ducted) to be below 600 C (1100 F), with a gradient across
the wall of less than 40 C (100 F). This combination provides an outstanding thermal cyclic life: in excess of
50,000 thermal cycles (Fig. 10), which exceeds requirements
of the Space Station (i.e., pulses are not full thermal
cycles).

The resistojet design has been configured and key
design parameters (and materials) selected to ensure long
life and minimum maintenance. For the resistojet thruster,
a maximum heater temperature of 1400 C (2500 F) was established based on long life considerations (this results in
a gas temperature of about 1000 C (1800 F). Wall thickness
has been selected to realize the 10,000-hour design life
at 1400 C (2500 F) with all anticipated thermochemical and
thermophysical effects included, i.e., creep, chemical
interaction, etc. The resistojet's heater is based on the
use of commercially available materials. (For example, commercially produced heating element assemblies operate at
temperatures of 1400 C (2500 F) in oxidizing environments
for periods in excess of 10,000 hours; heater lifetimes of
14,000 hours are typical). Key elements of the resistojet

Fig. 11 Space station propulsion test bed

design have a safety factor of better than four under worst operating conditions (Ref. 7).

Redundancy--in addition to multiple thruster units--can also be provided in the component elements of the design. Dual-seat valves provide redundancy in event of seat leakage or failure. Redundant electrical valve actuators (i.e., solenoid or torque motors) can provide redundancy in event of an electrical failure.

As illustrated in Table 11, monitoring, to provide controlled operation and/or immediately reduced operating level (i.e., reduced temperature level) or shut down, is a capability that can be employed to counter many possible degradations. For example, to determine when servicing is needed, on-board, integrated measurements of key thruster parameters can determine if operating conditions are deteriorating. For example, in the regeneratively-cooled hydrogen/oxygen thruster, measurement of the chamber back-wall temperature (at three locations), combined with upstream and downstream propellant pressure measurements, and chamber pressure measurement can determine if such conditions as:

1. Overheating of the chamber wall

2. Coolant channel full or partial blockage

3. Injector orifice obstruction or overheating

4. Internal leakage

Should any of these conditions become evident, the thruster operation can be controlled (i.e., reduced mixture ratio, increased coolant flow) and lower temperature operation initiated.

Valve actuation voltage, actuation time, and flowrate measurement through the valve (resistance) can be used to assess valve condition.

Should a thruster become unfit for further operation, either the thruster or the entire module can be shut down and replaced. The most probable is the simple replacement of the thruster. To enhance such removal and replacement, features for accessibility and ease of replacement are being assessed at present for incorporation into the propulsion module design. The use of robotic systems to provide for thruster removal/replacement without an Extra Vehicular Activity (EVA) is one such approach.

To minimize servicing time, when servicing is desirable, it may be beneficial to provide in the propulsion module the capability for an independent hookup for a replacement thruster. Such a design approach could allow

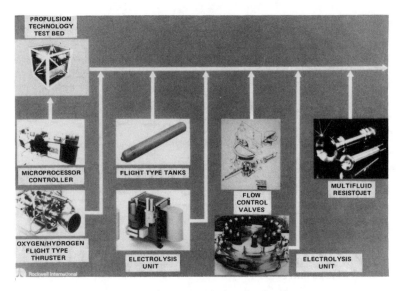

Fig. 12 Space station propulsion technology program

a replacement thruster to be quickly added, with the inoperative thruster removed when convenient. Another feature for thruster replacement is the incorporation of quick-disconnects. Several concepts have been designed and tested.

The use of hydrogen/oxygen gaseous propellants minimizes the impact of opening a propellant feedline during servicing. Valving at the thruster modules will prevent large emissions of either hydrogen or oxygen or the fluids to be used by the resistojet thrusters. In the highly unlikely event of a line failure, the use of the nontoxic, noncorrosive hydrogen/oxygen propellants greatly enhances on-board safety, and valving design provides isolation of the leak as well as alternate flow paths for continued operation.

Test Bed

To provide the means for an ambitious, energetic test program and to permit testing to be conducted on a timely basis, under NASA sponsorship[4] Rocketdyne has designed and fabricated a propulsion test bed (Fig. 11) for use by the NASA/ Marshall Space Flight Center for extensive duty cycle testing. This test bed is designed for hydrogen/ oxygen propellants. It represents a complete propulsion module. The design and packaging permit the test bed to be

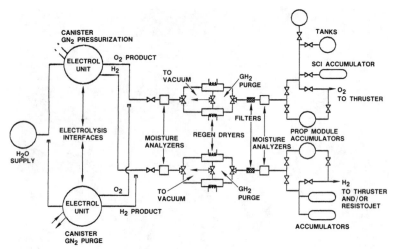

Fig. 13 Schematic of space station test bed with dual-test electrolysis unit

placed in a vacuum chamber and the entire module operated at approximately 10^{-5} torr conditions.

The test module permits growth propulsion systems to be readily accommodated. Testing is now underway. Plans call for incorporation of two types of electrolysis units and a resistojet module as shown in Fig. 12 and 13. Figure 12 depicts this planned integration of subsystem into the propulsion test bed. Figure 13 illustrates through the schematic diagram, the integration of the electrolysis units and flight type tanks into the test bed for "end-to-end" system testing.

Test Results

An ambitious program is currently being conducted to test and demonstrate the life, reliability, performance, and operating characteristics, as well as maintenance and serviceability for the on-board thrusters. For the hydrogen/oxygen prototype thruster this program is well underway. For the resistojet this program is now in the early phases. However, the resistojet design is backed by over 100 resistojet type thrusters that have already flown demonstrating extensive space operation, and by the design and selection of materials and operating temperatures that are conservative and well within demonstrated material/design capabilities.

The hydrogen/oxygen thruster was first tested at mixture ratios of 3:1 to 4:1. Over 1 million lbf-sec of impulse were demonstrated at these mixture ratios, with vacuum specific impulse values of 415 to 405 seconds established. Over 10,500 pulse mode operations have been performed. Minimum impulse bits much lower than required for Space Station have repeatedly been demonstrated.

When a propellant supply provided by water electrolysis was determined advantageous, the thruster was modified and further testing performed. Propellant supply by water electrolysis requires that the thruster be capable of operating at the stoichiometric mixture ratio of 8:1. An additional 1 million lbf-sec of impulse has been performed at this mixture ratio, with a vacuum specific impulse of 360 seconds established. The modified thruster will still operate with very ample margins in a hydrogen-rich coolant mode at the lower mixture ratios as was demonstrated by stepping from mixture ratio 8:1 down to 5:1 during a single extended firing.

A total operating time of 24 hours and a total impulse of 2 million lbf-sec have been accumulated on the hydrogen/oxygen thruster shown in Fig. 8. Testing also has been conducted by the NASA/Lewis Research Center on two other thrusters, one of similar design and the other of a radiation/film cooled design. These tests have been supportive of the technical readiness of low thrust oxygen/hydrogen engines. The test results indicate that all major technology issues for long-life, gaseous hydrogen/gaseous oxygen thrusters for Space Station application have been resolved. Test results are highlighted in Table 12. The Space Station's on-board propulsion system as represented by the test bed, with test bed thrusters installed, has been tested for full duration (300 seconds) at a mixture ratio of 8:1; specific impulse demonstrated was in the 360-second range and a total impulse of 7500 pound-seconds was accumulated. The electrolysis units are being installed

Table 12.   Prototype Hydrogen/Oxygen Thruster Test Summary*

Mixture Ratio, o/f	Total Duration, seconds	Maximum Single Run Time, hours	Total Pulses (Pulse in Mode), 0.1 second
3:1	32,148		10,200
4:1	12,618	0.5	
5:1	208		
6:1	477		
7:1	425		
8:1	40,442	6.1	
Total	86,318		

*Rocketdyne Design

in the test bed and an ambitious test program of the end-
to-end propulsion system (i.e., from water electrolysis to
propulsion thrust) is planned. This testing will include
pulsing tests, multiple thruster tests, mixture ratio
excursion tests, throttling tests, and a variety of propellant generation duty cycles.

Conclusions

Outstanding progress is being made for the system
design and technology verification for a safe, reliable,
simple on-board propulsion system for the Space Station.
The selected design meets the defined requirements, is
highly cost effective, and provides an outstanding degree
of synergism with other on-board systems. Design features
that enhance long life, safety, and minimize maintenance
and servicing have been implemented and others continued
to be evaluated for potential incorporation. Tests underway have confirmed the outstanding characteristics and
features of the prototype hydrogen/oxygen thrusters.
Further testing, both with propulsion components and with
the NASA Space Station propulsion test bed will provide a
comprehensive data base.

Acknowledgments

The authors wish to acknowledge and credit the many
NASA and industry personnel who are very effectively contributing to the success of the Space Station propulsion
program and whose work has gone into the evolution of the
design of the system. Information from these people
presented in various NASA coordination meetings has been
used throughout without formal reference but is
acknowledged here.

References

[1]Wilkinson, C. L., Brennan, S. M., and Valgora, M. E., "Space
Station Propulsion Options," AIAA-85-1155, July 1985.

[2]Shoji, J. M., Meisl, C. J., Glass, J. F., Tu, W., Ebert, S. J.,
Evans, S. A., Jones, L., and Campbell, H., "Oxygen/Hydrogen Space
Station Propulsion System Concept Definition for IOC," AIAA-86-
1561, June 1986.

[3]Graetch, Joe, "Design Drivers of the Space Station Propulsion
System," presented at the AIAA 24th Aerospace Sciences Meeting,
Reno, Nevada, January 1986.

[4]Briley, G. L. and Norman, A. M., "Space Station Propulsion Test
Bed - A Complete System," AIAA-86-1402, June 1986.

[5]Graetch, Joseph E. and Unterberg, Walter, "Fluid Independence of the Space Station," AIAA-86-2309, Sept. 1986.

[6]Heckert, B. J., Yu, T. I., Allums, S. L., and Carrasquillo, E. A., presented at the 1986 JANNAF Propulsion Meeting, New Orleans, Louisiana, August 1987.

[7]Pugmire, T. K., Cann, G. L., Heckert, B., and Sovey, J. S., "A 10,000 Hour Life Multipropellant Engine," AIAA-86-1403, June 1986.

Space Shuttle RCS Plumes:
Radiation and Temperature Measurement
on Mission 51–G

Turki Al-Saud* and I-Dee Chang[†]
Stanford University, Stanford, CA

Abstract

An experiment was conducted on Mission 51-G to understand the phenomena responsible for the visible radiation from thruster plumes. The TV camera on the elbow of the Space-Shuttle Remote Manipulator System (RMS), with three color filters, was used for the measurement.

Results show that the visible radiation is independent of the environment (contrary to previous suggestions), and is mainly thermal radiation from glowing carbon particles produced by thermal cracking of monomethylhydrazine (MMH) fuel, due to fuel-rich conditions influenced primarily by film cooling rather than pulse-mode operation. This is similar to the well-known problem with hydrocarbon fuels.

Simultaneous measurement of plume temperature and absorption coefficient is made possible, using different filters. Results from different firings are used to avoid the lag problem. The measured temperature is higher than that calculated for the same chamber-to-exit pressure ratio, but agrees well with the calculation for lower-pressure ratios. The stagnation pressure calculated from the measured temperature also agrees well with a prior full-scale ground test.

Introduction

The Reaction Control System (RCS) is used for maneuvering the Space Shuttle. There are two types, primary and vernier. This experiment applies to the primary engine, which has 870-lb thrust and uses MMH and N_2O_4 as fuel and oxidizer, respectively. Visible radiation from Space-Shuttle RCS engines was observed, but not measured, in previous flights. However, the work associated with the Space-Shuttle

---

*Ph.D. Candidate, Department of Aeronautics and Astronautics.
†Professor, Department of Aeronautics and Astronautics.

surface glow and its enhancement by thruster firings raised the possibility of a parallel mechanism for the visible radiation of thruster plumes similar to that responsible for surface glow; namely, the interaction of the plume with the space environment[1]. On the other hand, there is an interesting similarity between this phenomenon and the visible radiation from rocket and aircraft engines using hydrocarbon fuels, which have been studied for many years and attributed to the formation of soot particles (primarily carbon) not expected under equilibrium conditions[2].

Table 1 shows the principal equilibrium products of the engines[3]. The radiation of these gases is emitted at wavelengths in the infrared region of the electromagnetic spectrum. On the other hand, the emission spectrum from solid particles is continuous.

If the influence of the environment is determined to be negligible (and we think it is, in this case), then the presence of solid particles is expected to be the source of this radiation.

As a first attempt to understand this problem, within the constraints of the the Space-Shuttle schedule, we utilized the existing television system installed in the Space Shuttle to make our measurements.

Formulation of the Problem

Formulation of our physical model is based on thermal radiation emitted from carbon particles in the plume. The physical situation and coordinate system is shown in Fig. 1. Our basic assumptions are the following:

1. The gas and particles are in local thermodynamic equilibrium (LTE).

2. Scattering and absorption are small, compared to emission.

3. The plume is circularly symmetric.

4. The plume is optically thin in the visible spectrum.

The equation of radiative transfer along a line-of-sight at x in the direction of y is:

$$\frac{dI_\lambda(x,y)}{dy} = K_\lambda(x,y)I_{b\lambda}(x,y) \tag{1}$$

Table 1 Principal Equilibrium Products

Species	Mole Fraction
H_2O	0.3176
N_2	0.3026
H_2	0.1693
CO	0.1345
CO_2	0.03404
H	0.02276
O_2	0.0009

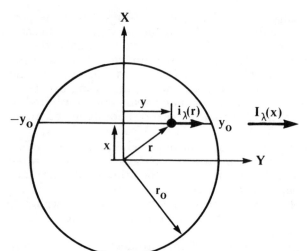

Fig. 1 Physical Model.

Here $I_\lambda(x, y)$ is the spectral radiance at a point along the line, K_λ is the absorption coefficient, and $I_{b\lambda}$ is the black body spectral radiance given by Wein's formula[2]:

$$I_{b\lambda} = 2c_1\lambda^{-5}exp(-c_2/\lambda T) \qquad (2)$$

The integral of spectral radiance along the physical dimension $-y_o$ to y_o becomes:

$$I_\lambda(x) = \int_{-y_o}^{y_o} K_\lambda(x, y)I_{b\lambda}(x, y)dy \qquad (3)$$

Expressing the integral in r as

$$I_\lambda(x) = 2\int_x^{r_o} \frac{K_\lambda(r)I_{b\lambda}(r)r}{(r^2 - x^2)^{1/2}}dr \qquad (4)$$

the measured lateral radiance becomes

$$I_m(x) = 2\int_x^{r_o} \int_{\lambda 1}^{\lambda 2} \frac{R(\lambda)K_\lambda(r)I_{b\lambda}(r)d\lambda}{(r^2 - x^2)^{1/2}}rdr \qquad (5)$$

The total response of the system (optics and detector) is $R(\lambda)$ and the absorption coefficient can be expressed as

$$K_\lambda = \frac{\alpha(r)}{\lambda^q} \qquad (6)$$

where $q = 1.39$, as given by Hottel and Broughton[4], and α is a constant that depends on particle concentration, which is a function of position, to be determined from the experiment.

The relation between lateral and radial radiance distributions is expressed as

$$I_m(x) = 2\int_x^{r_o} \frac{i_m(r)r}{(r^2 - x^2)^{1/2}}dr \tag{7}$$

where, from Eq. (5),

$$i_m(r) = \int_{\lambda 1}^{\lambda 2} R(\lambda)K_\lambda(r)I_{b\lambda}(r)d\lambda \tag{8}$$

and the solution to the Abel integral in Eq. (7) is:

$$i_m(r) = -1/\pi \int_r^{r_o} \frac{I'_m(x)}{(x^2 - r^2)^{1/2}}dx \tag{9}$$

Abel transformation was obtained numerically, using the method given by Bockasten[5].

The detector output signal is linear with radiance and therefore, by using a reference signal for a known radiance value, we can determine the radiance at any other point from the relation:

$$\frac{S}{S_{ref}} = \frac{I_m}{I_{ref}} \tag{10}$$

where S is the output signal.

From the measured lateral radiance distribution, we can find the radial radiance from Eq. (9). Using this result with two filters, we can numerically solve Eq. (8) for the temperature and the constant in the absorption coefficient expression as functions of the radial distance.

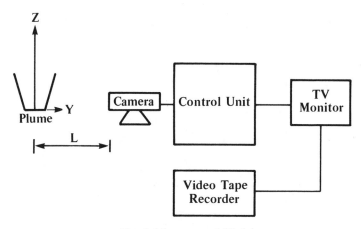

Fig. 2 Measurement Model.

Table 2   Camera Characteristics

Gamma	1
Responsivity	160 $\mu A/Lm$
Gain	1600
Dark Current	7 nA
F-No.	1.4
Focal Length	18 mm
Target Area	122.88 mm^2
Solid Angle	0.39 sr
Distance to plume (L)	14.2 m

Experiment

The measurement model is shown in Fig. 2. The RMS Elbow Television Camera of the Space Shuttle was used for the measurement. It is a modified version of an RCA 4804H Silicon-Intensified Target (SIT) Vidicon. It has three color filters (red, blue, and green) alternating every 1/60 th of a second.

Detailed description of the system was given by Freedman[6]. Camera characteristics are given in Table 2. The total response of the system was obtained by multiplying the response of the detector with lens and filter transmittances, and is shown in Fig. 3.

For fast operation time, the maximum signal just before saturation corresponds to a radiance value of $6 \times 10^{-7} W/cm^2/sr$. This we took as a reference value with respect to which other points are measured.

Twelve separate and one simultaneous engine firings of varying duration from 0.16 to 1.2 seconds were measured in different vehicle

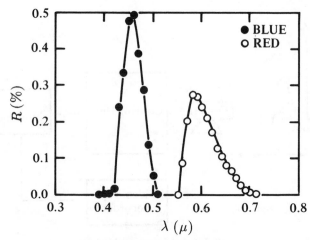

Fig. 3   Total Camera Response.

orientations at nighttime and at an altitude of about 190 nautical miles.

The camera was focused between the centerlines of L4L and L4U engine plumes. The distance from the plumes to the camera is about 14.2 m. The measured region extends from 45 cm to 276 cm along the centerline. The data was recorded on videotapes on board the Space Shuttle. The videotapes were subsequently digitized by using a Quantex Image Analyzer (QX-7) and a Beta Cam Tape Player. Field by field, high-quality images were obtained without flicker between the fields. The image of each field was stored in the computer and smoothed, using a 3 x 3 averaging filter. Lateral radiance profiles were measured along the centerline and at three positions across, then plotted directly onto a laser printer.

Results

The first important result we found, by conducting the experiment in different vehicle orientations, was that the radiance values remained constant and, consequently, the environment has a negligible effect on plume radiation. This result has been supported by recent observation of thruster-plume visible radiation at nighttime during a ground test[7].

From the analysis of video images it became apparent that camera lag prohibits the measurement of a single firing with sequential color filters, since only the first filter is unaffected. We therefore used the results of four firings, two starting with red and two starting with blue.

Two firings for each filter showed equal results at a plume radiance corresponding to 10.21 atm chamber pressure. The lateral

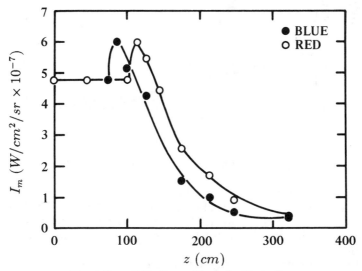

Fig. 4 Lateral Radiance Along the Centerline.

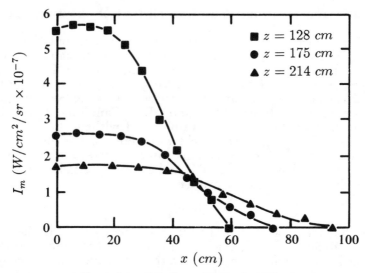

Fig. 5 Lateral Radiance Profile, Red Filter.

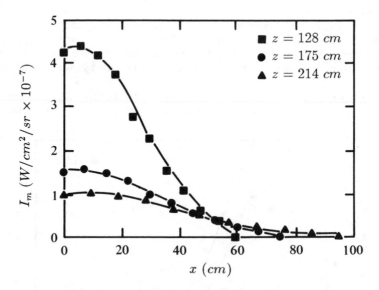

Fig. 6 Lateral Radiance Profile, Blue Filter.

radiance distribution along the plume centerline is given in Fig. 4. The peak corresponds to the reference maximum-radiance value. The region to the left of the peak is the saturated region and has a lower value because the signal was cut by the camera system as it responded to overload.

The noise was generally not a problem because of the broad filters used. It only became a problem when the radiance was low

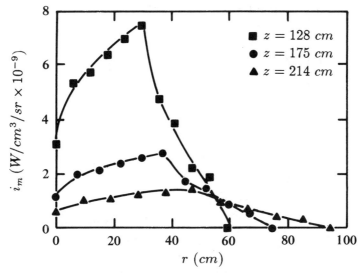

Fig. 7 Radial Radiance Profile, Red Filter.

enough to be close to $6 \times 10^{-8} W/cm^2/sr$, which corresponds to the dark current of the camera. The effect of noise can be seen as the two curves in Fig. 4 approach each other.

We have taken the lateral radiance profile at three positions, far from both the saturation and the noise region. Profiles from both filters are given in Figs. 5-6. There is clearly a dip in the radiance along the centerline, and this can also be seen in the plume photograph.

The values close to the plume particle boundary, where noise is a problem, were obtained by continuing the slope of the profile from the point above the dark current value to the outer boundary, where the radiance should vanish. The radial radiance profile, obtained by Abel transformation, is shown in Figs. 7-8. It has an off-axis peak that diminishes further along the axis. The temperature was evaluated from the ratios of the radial radiance for both filters, as shown in Fig. 9. The resulting temperature profile is given in Fig. 10. It has a maximum at the centerline and drops fast toward the boundary.

The constant in the absorption coefficient expression was obtained as a function of radius. The coefficient was calculated at wavelength equal to 0.633 μ, as shown in Fig. 11. It also has an off-axis

Table 3 Theoretical Exit Temperature, (K^o)

O/F	$P_c/P_e = 50$	$P_c/P_e = 150$	$P_c/P_e = 700$
0.8	768	572	374
1.6	1376	1075	743
2.2	1720	1379	991

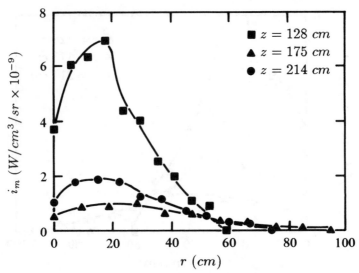

Fig. 8 Radial Radiance Profile, Blue Filter.

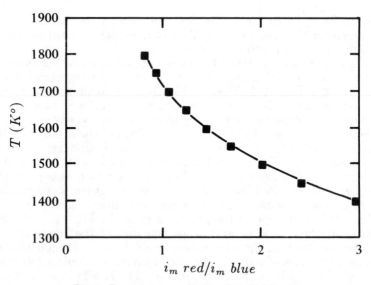

Fig. 9 Temperature vs Ratios of Radial Radiance.

peak and is very low at the centerline, indicating lower particle concentration.

The temperature at the centerline, measured at $z = 128$ cm, is higher for high chamber-to-exit pressure ratios (P_c/P_e) than the exit temperature calculated by the Ramp program[3]. However, it is closer to the calculation for lower pressure ratios as shown in Table 3, assum-

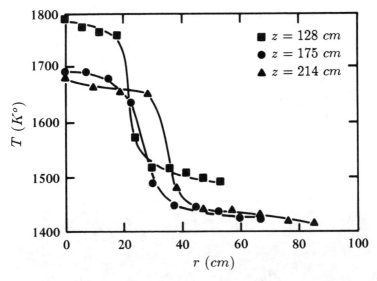

Fig. 10 Temperature Profile.

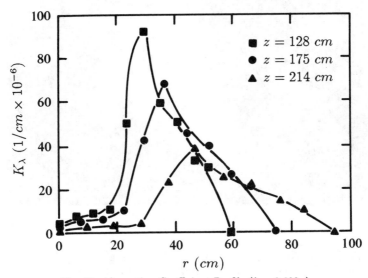

Fig. 11 Absorption Coefficient Profile ($\lambda = 0.633\mu$).

ing the temperature drop from the exit to the point in question to be small. We calculated the radial distribution of the stagnation pressure (P_s) behind a normal shock at $z = 128$ cm [8], by using the measured temperature distribution plus the density and velocity profiles given by Alred[9], for $\gamma = 1.3$ and $c_p = 0.4$. The results were compared to the ground test done at 111.8 cm from the exit plane along with

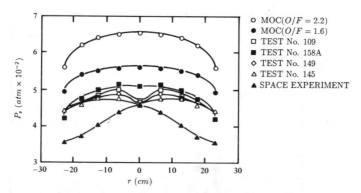

Fig. 12   Radial Distribution of Stagnation Pressure.

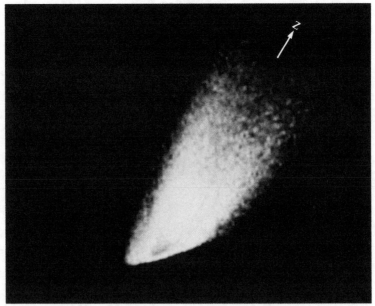

Fig. 13   Plume Photograph.

calculations by the method of characteristics given by Kanipe[10]. All of these results were plotted in Fig. 12. Our result agrees with the test at the centerline, but is lower for other points. The disagreement between the off-axis points is perhaps influenced by the theoretical density and velocity profiles used, which are of a source type. In fact, the real density profile should have a maximum at an off-axis point, as clearly seen in the absorption coefficient (Fig. 11) and the plume photograph (Fig. 13). This would explain the drop of the stagnation pressure at the centerline in the ground test as due to the drop in density. In general, the agreement is better than the method of characteristics.

Table 4 Theoretical Carbon Mole Fraction

O/F	$P_c/P_e = 50$	$P_c/P_e = 150$	$P_c/P_e = 700$
0.8	0	0.01296	0.04937
1.6	0	0	0
2.2	0	0	0

These results strongly suggest the presence of carbon particles in RCS plumes, and the probable mechanism of carbon formation seems to be the thermal decomposition of MMH fuel. This is true especially near the chamber walls, where low oxidizer to fuel (O/F) ratio exists (below 0.8)[11], due to film cooling. Table 4 gives the mole fraction of carbon at different O/F ratios[3].

Also, the reported presence of methane gas in relation to thruster firings[12] seems to be another indication of the decomposition process.

Conclusions

We have concluded from the experiment that the visible radiation from the RCS plumes is predominantly thermal and the effect of the environment interaction is negligible. The probable emitter is solid carbon resulting from MMH decomposition, and is probably caused by the film-cooling process.

Simultaneous measurement of temperature and absorption coefficient is possible using the television camera employed in the experiment, but separate firings with different filters are needed in doing so because of the lag problem, which prohibits measurement of a single firing with sequential color filters. The temperature measured is higher than that calculated for the same exit pressure, but agrees with the calculation for higher exit pressure values. Our results agree well with data from the stagnation pressure test.

Acknowledgments

We would like to acknowledge the following people: the NASA Discovery 51-G Flight Crew (and especially Sultan Al-Saud, who implemented the experiment in space flight), Dr. John Alred, Charles Chassay, W. F. Eichelman, Dr. David Kanipe, Walt Karakulko, and Dr. L. J. Leger of NASA JSC. Also, Dr. A. Dabbagh of UPM, Saudi Arabia, and our Stanford colleagues Prof. Donald Baganoff, Prof. Sidney Self, and Dr. Roger Williamson; plus Mill Packard, Quantex Corporation; Larry Freedman, RCA; and Dr. Emil Lawton, Caltech.

References

[1]Green, B. D., Marinelli, W. J., and Rawlins, W. T., "Spectral Identification/Elimination of Molecular Species in Spacecraft Glow," *Proceedings of*

Second Workshop on Spacecraft Glow, NASA Conference Publication 2391, May 6-7, 1985, pp. 82-97.

[2]Siegel, R. and Howell, J., *Thermal Radiation Heat Transfer*, Hemisphere, 1981, pp. 452-456, pp. 658-670.

[3]Smith, S. D., "High Altitude Chemically Reacting Gas Particle Mixtures - Volume III Computer Code User's and Applications Manual," LMSC-HREC TRD 867-400-III, 1984.

[4]Hottel, H. C. and Broughton, F. P., "Determination of True Temperature and Total Radiation from Luminous Gas Flames," *Industrial and Engineering Chemistry*, Vol. 4, 1933, pp. 166-175.

[5]Bockasten, K., "Transformation of Observed Radiances Into Radial Distribution of the Emission of a Plasma," *Journal of the Optical Society of America*, Vol. 51, No. 9, Sept. 1961, pp. 943-947.

[6]Freedman, L. A., "The Space Shuttle Closed Circuit Television System," *Proceedings of the IEEE National Aerospace and Electronics Conference*, May 19-21, 1981, pp.23-30.

[7]Karakulko, W., Private Communication, NASA JSC, Houston, Texas, Sept. 1986.

[8]Liepmann, H. W. and Roshko, A., *Elements of Gas Dynamics*, Wiley, New York, 1957, p. 149

[9]Alred, J. W., "Flowfield Description for the Reaction Control System of the Space Shuttle Orbiter," *AIAA 18th Thermophysics Conference*, June 1-3, 1983.

[10]Kanipe, David B., "High Altitude RCS Plume Test," unpublished NASA Johnson Space Center Memorandum EX33/7907-90, July 16, 1979.

[11]Haynes, B. S. and Wagner, H. G., "Soot Formation," *Progress in Energy and Combustion Science*, Vol. 7, 1981, pp. 229-273.

[12]Green, B. D., Caledonia, G. E., and Wilkerson, T. D., "The Shuttle Environment: Gases, Particulates and Glow," *Journal of Spacecraft and Rockets*, Vol. 22, No. 5, Sept.-Oct., 1985, pp. 500-511.

Light Weight Electrostatic Generator for Aerospace and Other Uses: Cylindrical Parametric Generator

F. Cap *
University of Innsbruck, Innsbruck, Austria

Abstract

Using a time-varying capacity a new principle to convert mechanic energy into electric energy is described. Arbitrary as well as sinusoidal time varying high-voltages can
be produced. The generator consists of rotating slotted
hollow cylinders. On the basis of the solution of the boundary problem of these cylinders the parametric pump function
of the device is derived. Using a nonlinear mechanism stability of the device is guaranteed.

1. Introduction

In this paper a light weight generator is introduced.
It may be of interest for application in cars, boats, helicopters, airplanes, portable generators for laser telemetry
and especially for space stations. Whatever the primary energy source of a space-platform will be (thermal solar, nuclear etc), for higher power (tens of kwatts) a device will
be necessary to convert heat and mechanical energy into electromagnetic energy. For aerospace uses (space-platform,
spacecraft) and for airplanes, helicopters, maybe also for
cars and other purposes like laser installations, etc., low
weight is important. In this paper a nonlinearly self-sta-
bilizing EMP-safe electrostatic generator of low weight is
introduced. Patents have been granted in several countries,
e.g., the USA, Austria, Israel, Canada, Argentina, etc.;
patents in other countries are pending.

The parametric electrostatic generator is especially
suited for high-voltage high frequency low ac current production. Its principle consists of a capacity varying periodically with time. To obtain a sinusoidally varying ac-

  * Professor of Physics.

voltage an inductance has to be added. A disc operating
laboratory model has been built and delivered several watts
at 1500 volts and 400 cycles. This is a frequency preferred
in aerospace electronics. In this paper a new construction
will be presented consisting of 2 coaxial hollow cylinders.
The outer cylinder is fixed in space (stator) and has an inner radius r_2. The inner cylinder has an inner radius r_1
and its wall thickness is d. The space gap between the 2
cylinders has a thickness a and is filled with air, vacuum
or SF_6 under pressure. Both cylinders carry alternatively
conducting and nonconducting meridional slotted sectors.

2. The boundary value problem of two rotating slotted cylinders

Since the relevant distances are quite small effects of
retardation may be neglected and the solution for the electrostatic potential U is described by Laplace's equation.
In cylindrical coordinates r,ϕ,z it is separable into U
= R(r),$\Phi(\phi)$,Z(z) and the ordinary differential equations
are[1]

$$R'' + \frac{1}{r}R' + (1^2 - m^2/r^2)R = 0 \qquad (1)$$

$$\Phi'' + m^2\Phi = 0, \qquad Z'' - 1^2Z = 0 \qquad (2)$$

The solutions of these equations are

$$R_1(r) = AJ_p(1r) + BY_p(1r), \qquad \Phi_m = C\cdot\sin(m\phi) + D\cos(m\phi) \qquad (3)$$

$$Z_1 = E\exp(1z) + F\exp(-1z)$$

Here J_p, Y_p are Bessel functions. If the length L of the
cylinder is assumed to be relatively long (L $\gg$ r_1, r_2) then
the dependence of the potential U(r,ϕ,z) on z is negligible.
Then instead of (3) we may use the solution of (1) - (2) for
1 = 0 which reads

$$R(r) = Ar^1 + Br^{-1}, \qquad \Phi = C\sin(1\phi) + D\cos(1\phi) \qquad (4)$$

In order to produce a time varying capacitance the inner cylinder is assumed to rotate relative to the standing outer
cylinder. Then the boundary condition on the inner cylinder
is time-dependent. Let us assume that both hollow cylinders
carry 2p conducting and nonconducting sectors (slotted or
filled with dielectric). If n is the number of revolutions
per second of the inner cylinder and if at the time t = 0 a

conducting sector of the inner cylinder faces a conducting
sector of the outer cylinder then the maximum C_{max} of the
capacitance is obtained. After the time span $\tau/2p$ has passed, a conducting sector of the inner cylinder faces a nonconducting sector of the outer cylinder and the minimum C_{min}
of the capacitance is obtained. After the full operational
period τ/p again C_{max} is reached. We thus see that the relative twisting angle δ is given by

$$\delta = 2\pi t/\tau \quad \text{for} \quad 0 \leq t \leq \tau/2p$$

$$\delta = -2\pi t/\tau + 2\pi/p \quad \text{for} \quad \tau/2p \leq t \leq \tau/p \tag{5}$$

If the central angle of one of the 2p sectors be β, then we
have $\beta p = \pi$ for geometric reasons. Let V_O be the potential
on the conducting sectors of the inner cylinder and $-V_O$ the
potential on the standing outer cylinder then the symmetric
boundary conditions read

inner cylinder

$$U(r=r_1+d,\phi) = 0 \quad \text{for} \quad -\pi/p+\delta \leq \phi \leq \delta$$

$$U(r=r_1+d,\phi) = V_O \quad \text{for} \quad \delta \leq \phi \leq \pi/p+\delta$$

outer cylinder (6)

$$U(r=r_2,\phi) = 0 \quad \text{for} \quad -\pi/p \leq \phi \leq 0$$

$$U(r=r_2,\phi) = -V_O \quad \text{for} \quad 0 \leq \phi \leq \pi/p$$

We write the general solution (4) in the form

$$U(r,\phi) = \sum_{l}\left\{(A_l r^{pl}+B_l r^{-pl})\sin(pl\phi)+(C_l r^{pl}+D r^{-pl})\cos(pl\phi)\right\} \tag{7}$$

We now define a_l, c_l, $\overline{a_l}$, $\overline{c_l}$ by:
outer cylinder $(r = r_2)$

$$\sum_{l=1}(A_l r_2^{pl} + B_l r_2^{-pl})\sin(pl\phi) = \sum_{l=1}c_l \sin(pl\phi) \tag{8}$$

$$\sum_{l=0}(C_l r_2^{pl} + D_l r_2^{-pl})\cos(pl\phi) = \sum_{l=0}a_l \cos(pl\phi) \tag{9}$$

inner cylinder $(r = r_1 + d = b)$

$$\sum_{l=1}(A_1 b^{pl} + B_1 b^{-pl})\sin(pl\phi) = \sum_{l=1}\overline{c_1}\sin(pl\phi) \qquad (10)$$

$$\sum_{l=0}(C_1 b^{pl} + D_1 b^{-pl})\cos(pl\phi) = \sum_{l=0}\overline{a_1}\cos(pl\phi) \qquad (11)$$

Inserting the boundary conditions (6) and integrating the Euler-Fourier formulae we obtain

$$a_1 = 0, \quad c_1 = -2V_o/\pi l$$
$$\overline{a_1} = -(2V_o/\pi l)\sin(pl\delta), \quad \overline{c_1} = +(2V_o/\pi l)\cos(pl\delta) \qquad (12)$$

We do not consider a_o and $\overline{a_o}$, because they are of no interest since later on we have to differentiate.
    Defining $(b/r_2) = x$ and inserting into (8) - (11) we receive

$$A_1 = -\frac{2V_o}{\pi l}\frac{1+x^{pl}\cos(pl\delta)}{1 - x^{2pl}}r_2^{-pl}, \quad B_1 = +\frac{2V_o}{\pi l}\frac{x^{pl}+\cos(pl\delta)}{1-x^{2pl}}b^{pl} \qquad (13)$$

$$C_1 = +\frac{2V_o}{\pi l}\frac{x^{pl}\sin(pl\delta)}{1 - x^{2pl}}r_2^{-pl}, \quad D_1 = -\frac{2V_o}{\pi l}\frac{x^{pl}\sin(pl\delta)}{1 - x^{2pl}}b^{pl} \qquad (14)$$

We thus have the following solution of the boundary value problem

$$U(r,\phi) = \frac{2V_o}{\pi}\sum_l \frac{1}{l(1 - x^{2pl})}\left\{\left\langle\left[-1-x^{pl}\cos(pl\delta)\right]\sin(pl\phi) + \right.\right.$$

$$\left. + \left[x^{pl}\sin(pl\delta)\right]\cos(pl\phi)\right\rangle\cdot\left(\frac{r}{r_2}\right)^{pl} + \qquad (15)$$

$$\left. + \left\langle\left[x^{pl}+\cos(pl\delta)\right]\sin(pl\phi) - \left[x^{pl}\sin(pl\delta)\right]\cos(pl\phi)\right\rangle\middle/\left(\frac{b}{r}\right)^{pl}\right\}$$

For $t = 0$, $\delta = 0$ we obtain the electrostatic potential at time $t = 0$ which will give C_{max} and for $t = \tau/2p$, $\delta = \pi/p = \beta$ we should have the minimum capacitance C_{min}.

3. The variation of the capacitance with time

Multiplication of (15) with $\varepsilon_o\varepsilon_r$ (dielectric constant), deriving $-\partial U/\partial r$ at $r = b = r_1 + d$ (inner cylinder) and integrations

$$p\int_{\delta}^{\pi/p+\delta} \cdots \cdots r d\phi$$

along the conducting surfaces of the inner cylinder of length L as well as division by the relative potential difference $2V_o$ yields the capacitance

$$C(t) = \frac{2\varepsilon_o\varepsilon_r Lp}{\pi} \sum_{l=1,3}^{N} \frac{1}{l}\frac{1}{1-x^{2pl}}\left[x^{2pl} + 2x^{pl}\cos(pl\delta) + 1\right] \quad (16)$$

In deriving (16) the integrals

$$p\int_{\delta}^{\pi/p+\delta}\sin(pl\phi)d\phi = \frac{2}{l}\cos(pl\delta), \qquad \int_{\delta}^{\pi/p+\delta}\cos(pl\phi)d\phi = -\frac{2}{l}\sin(pl\delta)$$

have been used. Using (5) and the addition theorem for $\cos(-pl2\pi t)/\tau + 2\pi pl/p)$ we may write (16) in the form

$$C(t) = \frac{2\varepsilon_o\varepsilon_r Lp}{\pi} \sum_{l=1,3..}^{N} \frac{1}{l}\frac{1}{1-x^{2pl}}\left[x^{2pl} + 2x^{pl}\cos\left(\frac{pl2\pi t}{\tau}\right)+1\right] \quad (17)$$

The maximum and minimum values of the capacitance may be calculated from (7). For the maximum C_{max} we have $t = 0$, $\delta = 0$ and

$$C_{max} = \frac{2\varepsilon_o\varepsilon_r LP}{\pi} \sum_{l=1,3..}^{N} \frac{1}{l}\frac{x^{2pl} + 2x^{pl} + 1}{1-x^{2pl}} \quad (18)$$

and for the minimum $t = \tau/2p$, $\delta = \pi/p$ we get

$$C_{min} = \frac{2\varepsilon_o\varepsilon_r Lp}{\pi} \sum_{l=1,3..}^{N} \frac{1}{l}\frac{x^{2pl} - 2x^{pl} + 1}{1-x^{2pl}} \quad (19)$$

Using these expressions we may define

$$C_o = \frac{C_{max} + C_{min}}{2} \qquad \Delta C = \frac{C_{max} - C_{min}}{2} \qquad (20)$$

Then we may write (17) in the form

$$C(t) = C_o + \frac{2\varepsilon_o \varepsilon_r Lp}{\pi} \sum_{l=1,3..}^{N} \frac{1}{1} \frac{1}{1 - x^{2pl}} \cdot 2x^{pl} \cos\left(\frac{pl2\pi t}{\tau}\right) \simeq$$

$$\simeq C_o + \Delta C \cdot \cos(p2\pi t/\tau) \qquad (21)$$

Here we have neglected all terms $l > 1$. As a careful analytical investigation has shown[2] the neglection of all terms
$l \geq 1$ does not modify significally all the results.

4. Operation without an inductance

When a time varying capacitance described by (21) is
connected to a consumer with ohmic resistance R_o, the circuit is described by

$$R_o \frac{dQ}{dt} + \frac{Q}{C(t)} = 0 \qquad (22)$$

where $Q(t)$ is the charge of the capacitor. The solution

$$Q = Q_o \exp\left(-\frac{1}{R_o}\int_o^t \frac{dt'}{C_o + \Delta C \cos(2\pi p t'/\tau)}\right)$$

describes the forced periodic charging and discharging of a
capacitor. The integral is given by

$$\int_o^t \frac{1}{C_o + \Delta C \cos(2\pi p t'/\tau)} dt' =$$

$$= \frac{1}{\frac{2\pi p}{\tau}\sqrt{C_o^2 - (\Delta C)^2}} \operatorname{arctg} \frac{\sqrt{C_o^2 - (\Delta C)^2} \operatorname{tg}\left(\frac{\pi p t}{\tau}\right)}{C_o + \Delta C} \qquad (23)$$

In this mode of operation no sinusoidally varying voltage
is produced. Such voltage may however be of use for special
engineering purposes. More generally speaking, the choice
of the function C(t) describing the time-variation of the
capacitance may result in any given arbitrary form for Q(t),
$\dot{Q}(t)$ and the voltage U(t) = Q(t)/C(t). The differential
equation (22) is called non-autonomous because the independent variable t appears in it explicitly. It is also called
parametric, since a parameter, namely the capacitance C,
varies periodically with the independent variable.

5. Parametric resonance and production of sinusoidal ac

In order to produce a sinusoidal ac voltage we need to
add an inductance L_o to the circuit. Then (22) takes the
form

$$L_o \frac{d^2 Q}{dt^2} + R_o \frac{dQ}{dt} + \frac{Q}{C(t)} = 0 \qquad (24)$$

This Weigand differential equation possesses then and only
then an <u>unstable</u> solution when the parametric resonance condition is satisfied[2]. This condition reads in our case

$$2\pi p/\tau = 2/\sqrt{L_o C_o} = 2\omega_o \qquad (25)$$

The <u>main idea</u> of the generator consists now[3] in that
1. it should work in the unstable regime,
2. the voltage amplitude will be limited by nonlinear terms
3. that a Poincaré limiting circle transforms the Weigand
   functions, - i.e. the solutions of (24) into stable sinusoidal periodic functions,
4. to obtain nonlinear self-stabilization and nuclear EMP-
   safety.
In order to achieve this goal, we replace the inductance L_o
by $L_o \cdot f(Q')$. We thus assume that the inductance becomes a
function of the current Q'. As a first approximation to
solve the circuit equation, now reading

$$L_o C(t) \cdot \left[f'(Q')Q'Q'' + fQ'' \right] + C(t)R_o Q' + Q = 0 \qquad (26)$$

we make the ansatz[2]

$$Q(t) = r\cos(\omega_o t) + s\sin(\omega_o t) \qquad (27)$$

and

$$f(Q') = 1 + \alpha Q'^2 \qquad (28)$$

Neglecting again higher frequencies, using the relations of the type

$$\sin^2(\omega_o t)\cos(\omega_o t) \simeq \tfrac{1}{4}\cos(\omega_o t); \quad \cos^2(\omega_o t)\sin(\omega_o t) \simeq \tfrac{1}{4}\sin(\omega_o t)$$

$$\sin^3(\omega_o t) \simeq \tfrac{3}{4}\sin(\omega_o t); \quad \cos^3(\omega_o t) \simeq \tfrac{3}{4}\cos(\omega_o t)$$

$$\cos(2\omega_o t)\cdot\sin(\omega_o t) \simeq -\tfrac{1}{2}\sin(\omega_o t); \quad \cos(2\omega_o t)\cos(\omega_o t) \simeq \tfrac{1}{2}\cos(\omega_o t)$$

$$\cos(2\omega_o t)\sin^3(\omega_o t) \simeq -\tfrac{1}{2}\sin(\omega_o t)$$

$$\cos(2\omega_o t)\cos^3(\omega_o t) \simeq \tfrac{1}{2}\cos(\omega_o t); \quad \cos(2\omega_o t)\sin^2(\omega_o t)\cos(\omega_o t) \ne 0$$

$$\cos(2\omega_o t)\cos^2(\omega_o t)\sin(\omega_o t) \simeq 0$$

and inserting (21), (25), (28) and (7) into (26) one obtains[3] homogeneous linear equations for r and s. The coefficients contain $A^2 = r^2 + s^2 =$ const. The condition that the system has a non-trivial solution is given by the vanishing of the determinant of the coefficients. This allows to calculate[4]

$$r^2 + s^2 = A^2 = \frac{4L_o C_o}{3\alpha}\sqrt{\left(\frac{\Delta C}{2C_o}\right)^2 - \frac{R^2 C_o}{L_o}} \tag{29}$$

Since the amplitude A^2 must be real, we have

$$\frac{\Delta C}{2C_o} \ge R_o\sqrt{\frac{C_o}{L_o}} \tag{30}$$

This is a threshold condition expressing that the power pumped into the generator mechanically by varying $C(t)$ must be greater or equal than the power dissipated in the resistance R_o. The sinusoidal voltage

$$U(t) = U_o\cos(\omega_o t + \bar{\delta}) \tag{31}$$

has a limited amplitude U_o. $\bar{\delta}$ is the phase difference between voltage and current. It depends on R_o and the parameters of the load. The voltage is connected through $U = Q/C$

with r and s which are limited by (29). In designing a generator one should not select U_o greater than about half of the breakdown voltage of the medium in the gap between the cylinders. If there is a load connected in series to L_o and having the parameters R_1 and L_1, then one has to replace $L_o \rightarrow L_o + L_1$, $R_o \rightarrow R_o + R_1$ in all formulae.

6. The power produced by the generator and its weight

The electromagnetic power P is given by $U \cdot I$, where $I = Q'$. Due to $Q = UC$ we may write for one period $\tau_o = 2\pi/\omega_o$ of the current

$$P = \frac{1}{\tau_o} \int_0^{\tau_o} UI\,dt = \frac{1}{\tau_o} \int_0^{\tau_o} \left(U\frac{dU}{dt}C + U^2\frac{dC}{dt} \right) dt \qquad (32)$$

With U given by (31) and C given by (21) the first integral vanishes and the second integral yields

$$P = \frac{1}{\tau_o}U_o^2 \pi \Delta C \cdot \sin(2\overline{\delta}) = \frac{p}{2\tau}U_o^2 \pi \Delta C \cdot \sin(2\overline{\delta}) \qquad (33)$$

Inserting (20), (18), (19) we obtain

$$P = \frac{2\epsilon_o \epsilon_r L p^2 U_o^2}{\tau} \cdot \sin(2\delta) \sum_{l=1,3..}^{N} \frac{1}{1} \frac{x^{pl}}{1 - x^{2pl}} \text{ Watts} \qquad (34)$$

We see that the power produced is proportional to the frequency ω_o, to the squares of the number of slotted nonconducting sectors and of the peak voltage. For a gap wide a $= 5 \cdot 10^{-4}$ m we may assume

	U_o volts	E = U/a
Air	10^3	20 kV/cm
Baylectrol	$9 \cdot 10^3$	180 kV/cm
Vacuum	$5 \cdot 10^4$	1000 kV/cm

Reasonable dimensions of a generator could be

inner radius of inner cylinder	$r_1 = 0,1$ m
wall thickness of the inner cylinder	$d = 2 \cdot 10^{-3}$ m
gap width between cylinders	$a = 25 \cdot 10^{-5}$ m
number of nonconducting sectors	$p = 40$

We then obtain x = $(r_1 + d)/(r_1 + d + a)$ = 0,9976 and

$$C_{max}/L = 6,08 \text{ nF/m} \qquad C_{min}/L = 0,45 \text{ nF/m}$$

$$\Delta C = 2,815 \text{ nF/m} \tag{35}$$

where $\varepsilon_o = 8,859 \cdot 10^{-12}$ As/Vm, $\varepsilon_r = 1$ (vacuum).
    For vacuum $U_o = 5 \cdot 10^4$ volts and according to (33) the power per unit of length becomes for $\tau = 10^{-2}$ (6000 revolutions/min)

$$P/L = 44 \sin(2\bar{\delta}) \text{ kW/m} \tag{36}$$

at a frequency $p/2\tau$ = 2000 cycles. This seems to be sufficient for a space-platform. According to publications[5], the ISF (industrial space facility by Westinghouse, 1990) will have a solar array power of 16 kwatts only.
    The weight of the generator consists of
1. the weight of the rotating inner cylinder + bearing + axis
2. the weight of the outer cylinder
3. the weight of an inductance if necessary

If made of aluminum ($\rho = 2,7 \cdot 10^3$ kg/m³) the inner cylinder has a weight of $[(r_1 + d)^2 - r_1^2] \cdot \pi \rho$ = 3,4 kg/m per unit of length L and the outer cylinder of inner radius $r_2 = r_1 + d + a$ and of smaller thickness (d/2) since it does not move has a weight of 1,7 kg/m. For C_o as given by (20), (35) and a frequency of about 1000 – 2000 cycles the necessary inductance has to have some Henry and will have a weight of 2 – 3 kg. Thus the total weight will be about 8 kg/m and for a generator of L = 1 m length one has a <u>specific power</u> of about 10 kW/kg which is far better than the specific power (1 kW/kg) of magnetic generators now in use.
    The power given by (33) has been calculated with the use of the approximate expression (21) which neglected higher harmonics in the pump function C(t). We may however use the exact formula (17). Then the result for the power produced is given by

$$P = \frac{2\varepsilon_o \varepsilon_r L p^2 U_o^2}{\tau} \sin(2\bar{\delta}) \frac{x^p}{1 - x^{2p}} \text{ Watts} \tag{37}$$

This is due to the fact that integrals of the type (32) vanish for $l \neq 1$. The comparison between (34) and (37) gives a measure for the neglection of higher harmonics. We obtain for x = 0,9976, p = 40

$$\sum_{l=1,3..}^{500} \frac{1}{l} \frac{x^{pl}}{1-x^{2pl}} : \frac{x^p}{1-x^{2p}} = 6,2445 : 5,1941 = 1,2022 \quad (38)$$

Thus the contribution of higher harmonics may be estimated to be about 20 %.

The power produced by a generator without inductance pumped by

$$C(t) = C_o + \Delta C \cdot g(t) \quad (39)$$

can also be calculated. Here g(t) is an arbitrarily given periodic function of time (which could be expanded into a Fourier series like (17) or (21)). According to (21) the instantaneous power is given by

$$P_i(t) = U(t) \cdot I(t) = \frac{Q(t)}{C(t)} \cdot \frac{dQ(t)}{dt} = \frac{1}{2C(t)} \frac{dQ^2}{dt} \quad (40)$$

Since Q(t) is given by

$$Q(t) = Q_o \exp\left(-\frac{1}{R_o}\int_0^t \frac{dt'}{C_o + \Delta Cg(t')}\right) \quad (41)$$

and having (23) in mind, we see that the power is again determined by C_o, ΔC, p and τ. The function $Q(t)/C(t)$ according to (41), (23) is plotted in Fig. 1.

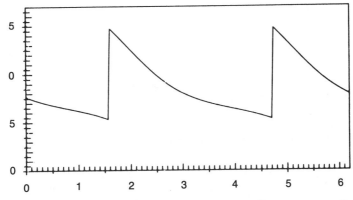

Fig. 1 $Q(t)/C(t)$ from (41), (23) for $Q_o = 10^{-8}$, $C_o = 12 \cdot 10^{-9}$, $\Delta C = 0,9 \cdot 10^{-9}$, p = 40.

7. Losses and Efficiency

If the generator is operating with SF_6 or Baylectrol as the medium in the gap, then there are friction losses to be considered. The surface of the inner rotating cylinder is given by $2(r_1 + d)\pi L$. On this surface a frictional resistance is acting. The strain is given by $\eta dv/dr$. For a laminar flow in the gap of width a we may assume

$$\eta\frac{dv}{dr} \approx \eta\frac{v}{a} = \eta\frac{\omega(r_1 + d)}{a} \qquad (42)$$

so that the friction per unit of the surface is given by

$$\eta\frac{(r_1 + d)}{a}\cdot\omega\cdot(r_1 + d)$$

since the lever arm is $(r_1 + d)$. The total power = = torque·angular velocity is then

$$R = 2\pi(r_1 + d)^3\omega^2\eta L/a \text{ watts} \qquad (43)$$

The viscosity η is measured in $kg\ m^{-1}s^{-1} = Nsm^{-2} = Pas$ (Pascalsec). For air we have $\eta = 1{,}7\cdot 10^{-5}$, SF_6 has $3{,}6\cdot 10^{-4}$ at a pressure of 0,4 MPa and Baylectrol 4900 has $6\cdot 10^{-3}$. For vacuum $\eta \sim 0$. Since we may assume a loss in the bearings of 1 – 2 % of the total power, the efficiency η_w is given by

$$\eta_w \tilde{=} \frac{pU_o^2\Delta C/2\tau}{1{,}02pU_o^2\Delta C/2\tau + 2\pi(r_1 + d)^3(2\pi)^2\eta/a\tau^2} \qquad (44)$$

The friction losses in the vacuum can be neglected so that η_w = 98 % in vacuum, probably about 90 %. For Baylectrol the friction losses of a generator with $r_1 = 0{,}1$ m, $\tau = 10^{-2}$, $L = 1$, $d = 2\cdot 10^{-3}$ m, $a = 25\cdot 10^{-5}$ m are given by (43) and are some 63 kW (air: 0,18). One thus sees that the efficiency may be as low as 50 – 70 % if the gap is filled with liquid medium. Low efficiencies may be of interest only if the small weight is more important. On the other hand, the generator is very well suited to work in an artificial vacuum or in the vacuum of cosmic space.

8. The generator under load

If the resistances R_0 and R_1 are connected in <u>series</u>, then the power P_0, P_1 consumed in the two resistances is given by

$$P_1/P_0 = R_1/R_0 \tag{45}$$

If the resistances are connected in <u>parallel</u>, then

$$P_1/P_0 = R_0/R_1 \tag{46}$$

Let us assume that R_0 be the resistance of the inductance L_0 and R_1 the resistance of the load. Then we have the conditions $R_1 > R_0$ for series connection and $R_0 > R_1$ for connection in parallel. According to (30) we should have $(\Delta C/2C_0) \geq R(\sqrt{C_0/L_0})$ for the total resistance R. For series connection we thus have the condition

$$\frac{\Delta C}{2C_0}\sqrt{\frac{L_0}{C_0}} - R_0 \geq R_1 > R_0 \tag{47}$$

whereas for connection in parallel

$$\frac{\Delta C}{2C_0}\sqrt{\frac{L_0}{C_0}} \geq \frac{R_0 R_1}{R_1 + R_0} \quad \text{and} \quad R_0 > R_1 \tag{48}$$

This agrees with other calculations.[3]
     The phase $\overline{\delta}$ between voltage and current is difficult to be calculated since it is doubtful, if the usual formula

$$tg(\overline{\delta}) = (\omega L - 1/\omega C)/R \tag{49}$$

for connection in series of R, C, L and

$$tg(\overline{\delta}) = R(\omega C - 1/\omega L) \tag{50}$$

for connection in parallel of R, C and L are still valid due to the nonlinear parametric differential equation for current and voltage. Using (25) both equations (49) and (50) would yield $\overline{\delta} = 0$. Experiments have however shown[3] that this is not the case.

References

[1]Moon, P., Spencer, D., "Field Theory Handbook", Springer, New York 1971, p. 13.

[2]Schmidt, G., Dum, R., Cap, F., "Globale rheonichtlineare Schwingungen. Anwendung auf parametrische Stromgeneratoren", Z.f.Angew. Math.u.Mechanik (ZAMM), Berlin 1987 (in press).

[3]Cap, F., "Ein elektrostatischer Stromgenerator", Elektrotechnik und Maschinenbau (EuM), Springer, Vienna, Vol, 102, Nr. 12, 1985.

[4]Schmidt, G., "Parametererregte Schwingungen", Deutscher Verlag der Wissenschaften, Berlin 1975.

[5]SATELLITE NEWS, October 6, 1986, p. 7.

Chapter VI. Lunar Activities

Lunar-Based Energy and Power Systems

D. R. Criswell*
University of California at San Diego, La Jolla, California
and
R. D. Waldron†
Rockwell International, Downey, California

ABSTRACT

Systems deployed from Earth can power a small, lunar research base. Key characteristics of such power systems are reviewed. Solar power systems could be manufactured on the moon and expanded by using lunar resources. Lunar manufacturing can lead to much larger opportunities. A functionally superior version of the space solar power satellite system can be produced on the moon from local materials. This lunar power system (LPS) can provide new motivations for a return to the moon and encourage others to consider larger possibilities for industry in space and on the moon.

Introduction

Electrical energy and power needs of both unmanned and manned lunar missions have been studied since the early days of the space program. Several pre-Apollo studies assumed availability of liquid water from hydrated minerals, and even the availability of hydrocarbons.[1-5] Major studies such as MIMOSA[6-7] focused on nuclear thermal sources (radioisotope or reactor) that yielded 5 to 100 kWe. Potassium or mercury Rankine cycles, Brayton cycle, or thermoelectric direct conversion were considered for permanent bases. Nuclear energy was favored over solar energy due to the perceived difficulties in storing sufficient solar energy during approximately 350 hours of lunar night.[8]

Lunar-Base Power Requirements and Growth

The energy or power utilization of the space station or future lunar bases will more closely resemble terrestrial utility operations rather than the total-energy operations of space probes. New figures-of merit are needed to compare alternative energy or power concepts for permanent bases in space.[9]

Power requirements for life support, scientific observations, experiments, communications, robotic controls, etc. are relatively modest (tens of kilowatts) for most early phases of lunar base activities. Power for lunar surface transportation will depend on the traffic model but will require portable or stored energy systems for most applications. Larger scale base power requirements (hundreds of kilowatts or more) will arise primarily from industrial activities in three types of applications with different time scales.

1. Materials processing such as in chemical processing for refined materials, propellant production, or hot forming for manufacturing will require large average power continuously.

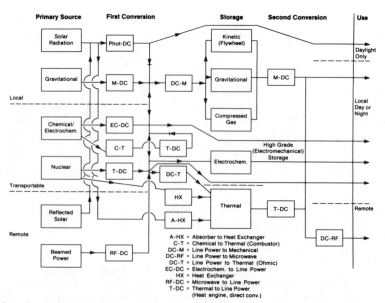

Fig 1   Potentially useful subsystem options for a lunar base.

2. Surface powered propulsion for inbound or outbound traffic such as for mass drivers, lunar dust braking, or suborbital material transport will require very large power peaks of minutes duration. Strategies such as use of lunar dust ejection and suborbital capture can reduce peak power.[10]

3. Power transmission to remote locations, including Earth, can be continuous and at large power levels.

Figure 1 depicts the major components of potential power systems grouped into primary sources of power, means to covert that power into alternative forms, systems to store fractions of the energy, and systems to provide power to the final user. The power sources may be divided into those locally available, transport deliverable, or remote (i.e., beamed power). Only local solar power is restricted to daylight operation.

Conversion and storage options are divided into those that can be manufactured primarily from available materials and those requiring significant inputs of materials or components from Earth. The former class is generally preferable in meeting long range power demands.

Engineering Characteristics of Energy and Power Systems

Figure 2 is an overview of comparative specific energy (joules/kg) of available sources. Also shown is the specific power (watts/kg) of conversion and storage devices that use thermal, kinetic, gravitational, thermo-electrochemical, or nuclear energy means. Specific energy increases orders of magnitude going from gravitational energies associated with practical differences in lunar elevation to kinetic, chemical, and nuclear means. Material stresses limit flywheels. Thermal specific energies modestly exceed flywheel energies but are subject to higher conversion losses because of thermodynamic losses.

Heat engines generally need fluid means for removal of heat and structures to radiate heat to space. These additions lower the device-specific power or energy. Area-type power converters such as photovoltaic arrays and antenna systems for beaming power can generally be thermally self-regulating without separate heat radiators.

The specific power for thin-film solar converters is greater than for any other practical system for supplying lunar daytime power. Photovoltaic systems

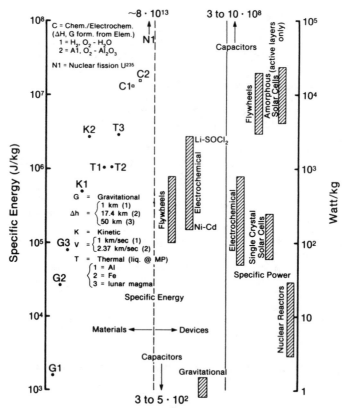

Fig. 2 Comparative specific energy and power of materials, devices, and systems.

can be made primarily from lunar materials, which can have far reaching implications for availability of multimegawatt power at a lunar base with minimal import mass requirements. Orbital mirrors can have a larger specific power, for reflected sunlight, than very thin solar cells. Useful orbital mirrors will be large.

Selection of a storage technology is strongly affected by several fundamental considerations: rates of charging and discharging; duration of storage cycle; magnitude of stored energy; material changes in the storage medium; reliability; and safety. Costs will be strongly affected by complexity and size of equipment, operations near physical limits, need for special materials, heavy maintenance, and avoidance of catastrophic failure. The lunar system is characterized by a very long cycle time of 300 to 700 hours, many decades of total operation, moderately large total energy storage, and the need for very high safety and reliability. Ignoring conversion losses the total energy storage over one lunar night for 300 kWe is 3.8 E11 Joules or 1.1 E4 kW-hrs.

This is approximately equivalent to the energy in:
- 100 tons of chemical explosive (4 E6 Joules/kg);
- the cooling of 1,000 tons of molten, lunar like materials through 200 degrees Kelvin assuming a specific heat of 2 kJ/kg-degree Kelvin;
- shifting 300,000 tons of lunar rock one kilometer vertically;
- a 23-meter radius chamber (48,000 cubic meters) of oxygen compressed to 70 atmospheres;
- 2,000 tons of nickel-iron batteries; or
- 1,000 tons of flywheels made of lunar glass, assuming a working strength of 100,000 psi.

It seems likely that thermal power will be stored by means of heated lunar materials stored in chambers under the lunar soil. Lunar materials are available in virtually unlimited quantities on the moon. Because of their particular physical properties and the lack of a lunar atmosphere, common lunar dust provides excellent insulation against loss of energy from high-temperature reservoirs.

Specific energy-to-power ratios or characteristic times may be associated with mass-to-mass rate ratios (Tons per Tons/Hr) for capital equipment, chemical inventory, power plant, radiator and other components of production processes. Such times, if based on a common process mass rate or throughput, are additive. The total time represents the so called "payback time" required to operate the process. Table 1 lists representative power-to-mass rates (megawatts per Ton/Hr) and characteristic times (Hr) for a number of industrial and aerospace processes and operations.

The chemical refining estimates are based on studies of a lunar processing plant and its associated power system. Lunar materials refining using an HF acid leach process has been subject to a detailed sizing analysis.[11] The total equivalent time for reagents, vessels, heat exchangers, photovoltaic power, and space radiators amounts to 78 hours, of which 30 hours represent installed power plant. To operate a 1 ton/hour plant would require transport of 30 tons of solar collectors and 48 tons of remaining equipment. Operating only in daylight would yield about 4,000 tons per year of refined lunar materials. Installation of lunar manufactured energy storage systems could approximately double the annual output. Industrial plants on the moon could produce energy systems, propellants, machines of production, and many other products that would

Table 1. APPROXIMATE MANUFACTURING SCALES[a]

Function versus Specific

-Mass T/T/Hr)	-Power (MW/T/Hr)	-Consumables (T/T/Hr)
Excavation or Beneficiation[b]		
0.1-1	E-4	<E-7
Cold Forming		
10	0.1	Small
Mass Drivers		
50	1	0.01
Cold Assembly		
100	0.1	Small
Welding and Hot Forming		
1-10	E2 to 3	Small
Chemical Refining		
100	10	0.01
Lunar Rockets		
150	10-20	>2
Micro Parts		
3000	50	0.1
Earth Rockets		
>2000	>100	20

[a] From NASA, General Dynamics, Massachusetts Institute of Technology, and the Lunar and Planetary Institute.
[b] Magnetic, electrostatic, or physical processes used to concentrate a selected fraction of a population of grains.

sharply reduce the need to bring equipment and supplies from Earth and open new opportunities for establishing mankind beyond Earth.

Space Solar Power Satellites and the Moon

Dr. Peter Glaser[12] first proposed construction of giant solar power satellites (SPSs) in geosynchronous orbit (GEO) about Earth. Each SPS would beam 5 to 10 gigawatts (GW) of microwave power to a few large receivers (rectennas) on Earth. Two thousand GWe of electric power is presently produced worldwide. Approximately 30 million dollars was spent by NASA, the Department of Energy (DoE) and aerospace companies in the 1970s to study the SPS concept.[13] It was not pursued. SPS construction would have required operations on Earth and in space thousands of times larger than all 1980s space programs, a major 10 to 20 year research and development effort, and a long period of high investments.

Following the suggestion of O'Neill[14] studies were conducted at the Lunar and Planetary Institute,[15] Massachusetts Institute of Technology,[16] and General Dynamics-Convair,[17] which revealed that over 90% of an SPS could be made of lunar materials by factories deployed from Earth to the moon and geosynchronous orbit. MIT researchers concluded that most of the factories could be made from lunar materials. Waldron et al.[18] concluded that raw lunar materials could be processed to provide a wide suite of materials appropriate to construction and logistics support. The General Dynamics study found the lunar approach competitive with launch of SPS components from Earth.

Concept of Lunar Power System

Lunar Power Systems may provide a method for efficiently acquiring high-quality, solar-derived energy to create NET NEW WEALTH[19,20] both on and off Earth. Waldron and Criswell[21] provide an extended discussion of the concept, the scale of a demonstration base, and growth of a demonstration base to a large lunar power system. The moon always displays the same face toward Earth. Small fractions of the lunar surface on the east and west limbs (visible edges) of the moon can be transformed into solar collectors and transmitters (Fig. 3). Power in the form of many converging microwave beams can be transmitted from segmented, oversized antennas at one or the other of the stations to much smaller receivers (rectennas) on the moonward side of Earth.

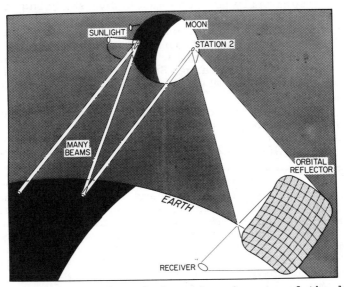

Fig. 3 Schematic view of the major elements of the lunar power system. showing: two lunar bases, mirror orbiting the moon, microwave reflector orbiting Earth, and rectennas on Earth receiving microwave power beams directly and via the reflector.

Each field or station of segmented antennas would have a projected area of 10 to 100km as seen from the receiver. This assures that receivers on Earth can be in the near field of a station or that a beam of 10-centimeter microwaves could be focused on a receiver that is significantly smaller in diameter than the transmitter station.

Power would be received directly from the moon approximately eight hours per day. This is possible except for a two- to three-day period about new moon. Because the lunar power can be supplied in a predictable manner with few interruptions due to weather, terrestrial power storage facilities can be accurately planned both for maximum capacity and intensity of use. A rectenna could be one tenth or less the area of a terrestrial photovoltaic array of equal average output. The simplest lunar power system can evolve to a more efficient configuration.

Microwave reflectors (0.1 - 1 km diameter) can be placed in high-inclination orbits about Earth and well inside geosynchronous altitude to distribute converged lunar power beams about Earth. Each reflector would have a mass per unit area of 1% to 10% that of an SPS, could be made of composite materials provided from Earth or possibly lunar derived glasses, and should be less expensive per unit of area than an SPS. Fine bore sight alignments of beams reflected from the orbital mirrors can be done electronically on the moon by shifting the center of the field of active transmitters servicing a given rectenna. NASA, Department of Defense, and private companies maintain major developmental programs directly applicable to orbital reflectors (e.g., Conference on Large Space Antennas Systems Technology-1984; December 4-6, 1984, NASA Langley Research Center).[22]

Very lightweight solar mirrors, actually heavy solar sails, can be placed in orbit about the moon to illuminate lunar power bases during the 300-hours-long lunar night.[23,24] Life times of decades to hundreds of years seem achievable.[25]

A very large transmitting aperture on the moon can project convergent power beams from the moon that are tightly constrained. Stray power would result primarily by scattering from the atmosphere and rectenna. The large lunar aperture is synthesized from many appropriately positioned, individual screen-wire reflectors illuminated by low power, solid state diodes. Hundreds, perhaps thousands, of power beams could be transmitted.

The N reflectors could resemble passive billboards. The microwave reflective billboards can be built on the moon of foamed glass posts and glass

fibers coated with lunar iron or aluminum. The order
of several 10s of SPS mass of lunar-derived glass
components would be adequate to make reflectors that
could handle several hundred times more reflected
power than could be produced by 10 SPS. They would
be arranged in an elliptical area near the lunar limb
so as to apparently overlap into a filled, circular
transmitting aperture 10 km to 100 km across when
viewed from the receivers.   Cost of the aperture
would be divided over many power beams.   Segmented
lunar reflectors can decouple the problems of beam
geometry from the details of microwave sources, power
levels, and rectenna sizes on Earth.

Antenna engineers are not usually concerned with
the development of optics-like microwave beams which
can be converged far from the aperture even though
the engineering and physics are very well
understood.[26-28] Segmented reflectors have been used
in receiving antennas. Reciprocity obviously applies
to their use in transmitters.[29,30]

Very large, segmented antennas on the moon seem
possible because of the lack of atmosphere and
because the moon is a body with very little internal
energy.   Horizontal and vertical ground motions
evoked over large areas by deep moonquakes are the
order of 1 to 10s of nanometers.[31]

The moon does not propagate coherent seismic
waves as does Earth. Solid body tides of the entire
moon are likely to be the greatest distortions.
These tides are regular, predictable in their
distortion, and their effects on the reflectors can
be eliminated electronically.

It is conceivable that the stray microwave power
of the LPS beaming 10,000 GWe could be less than a
few billionths of a watt per square centimeter (few
nanowatts/cm^2) at Earth. This is 10 to 100 less
intense than the human body radiates as incoherent
power in the microwave region.[32] If such a low level
of incoherent stray power can be attained, then there
is no biological risk from exposure to incoherent
microwaves outside the main beams.

Costs predictions are difficult.   Only
preliminary estimates are available. Kerwin and
Arndt,[33] while agreeing LPS can be considered as a
lunar and space power supply, maintain the early
lunar power system would not be competitive with a
satellite solar power system for the supply of power
to Earth.

Waldron and Criswell[21] disagree with the latter
conclusion and provide arguments that even early
lunar power facilities can be competitive with SPS.
They have estimated the costs of a LPS sequentially

broadcasting power directly to three rectennas positioned 120 degrees apart along the equator of Earth. It was found the costs would be approximately the same as a reference SPS delivering the same annual energy. For SPS it has been estimated that five gigawatts of capacity requires 12 billion dollars and would provide power at approximately 5 cents per kilowatt hour.

Due to the lack of atmosphere and clouds, lunar photovoltaic cells will be regularly exposed to 2 to 5 times more flux than cells on Earth. Thus, LPS can use significantly lower efficiency cells than can terrestrial photovoltaic systems. The rectenna constitutes approximately 20% of the costs of an SPS system. The three rectennas for LPS constitute 60 to 70% of LPS costs.. Introduction of orbital reflectors would permit continual beaming of power to all rectennas (factor of 3) and at a less oblique angle on average than directly from the moon (factor of 1.5). In addition, each rectenna aperture could be almost uniformly illuminated (factor of 2-3). An overall factor of 9 to 12 more power can be provided by the mature LPS to the rectennas than can the early LPS. It appears the mature LPS can introduce approximately 2 to 3 times more power into a rectenna than can SPS of the reference design. A factor of 9 to 12 reduction in costs of power seems possible.

Earth power systems now deliver 50 MWe/km2 of land.[34] Microwave heating of animals would be negligible at this intensity. Alternatively, power would be cheaper at higher beam intensities such as might be possible by directing several different beams to a given rectenna in a remote area. The full Lunar Power System offers more ways to efficiently feed power to rectennas than does an SPS, especially at high latitudes and to small rectennas.

The terrestrial rectennas to collect microwaves from SPS can be considered to be composed of concrete supports and the electrical equipment (antennas, rectifiers, collection wires, etc) to collect the power. The concrete stands were estimated in earlier DoE and NASA studies to have a mass the order of 200,000 tons per GW, about the same as a coal-fired plant but of much simpler construction. The electrical equipment was estimated at 500 tons per GW.

LPS looks very attractive for the supply of power to Earth because it requires small quantities of mass to provide useful electric energy virtually anywhere in the world to simple, rugged, and relatively low-mass receivers. In addition, it uses components in the environments appropriate to them. LPS could supply power that will decrease in cost over time,

require a minimal incremental capital investment for
terrestrial equipment, and introduce new net wealth
into the biosphere without the pollution or the
depletion of other resources.

The electrical equipment to receive beamed power
in space would be much more rugged and simpler than
solar cells. Receivers in space would be composed of
only the electrical equipment components of the
rectenna. Space rectennas could have a specific
power of more than 2000 watts/kg. This compares well
with other systems in Figure 2. The mature LPS could
provide convergent power beams to space-borne
receivers out to approximately 80,000,000 km from the
moon at intensities greater than that of local
sunlight.

Construction of a Lunar Power System

Figure 4 depicts one early concept for LPS
construction. The operations are large compared to
lunar science bases discussed at this time. However,
they are very small compared to shipping lunar
materials to space for construction of SPS.[17]
Massive construction and assembly operations in space
are not needed. See Waldron and Criswell[21] and Table
1 for scaling parameters used to estimate the
equipment masses and power of a demonstration base.
A lunar power base could support much of the research
and development projects that have been discussed for
an early lunar science base and do so at marginal
rather than full costs for the research.

An astronaut is inspecting the operation of an
automatic tractor about the size of the rovers used

Fig. 4 Astronaut viewing an automatic machine preparing the
lunar soil and emplacing solar power convertors. Previously
established power plots are in the background.

by the Apollo astronauts. The tractor is plowing up the top 10 centimeters of the lunar soil. Lunar soil has about 0.4% to 1% by weight of tiny blebs of metallic iron left over from meteoritic bombardment of the surface.[35,36] This iron can be extracted with a permanent magnet attached to the front of the tractor. The tractor can extract per hour approximately its own mass in iron. Iron and adhered glass are taken to a solar furnace and melted. The iron is then formed into rolls of wire. Molten glass is formed into support sections and then iron is coated onto glass fibers, which are made into the segmented microwave reflectors (not shown in this drawing). Excess molten glass is placed in a refractory hopper on the tractor and sprayed onto the soil surfaces of north-south aligned ridges to form hard surfaces. Iron wires to collect the solar electricity are placed under the soil. Ridge surfaces are sloped so the geometry of exposure of solar conversion devices approximately levels the electrical output of a given power plot over the course of the 300-hour lunar day.

Many alternative approaches are possible. Very thin film photoconversion devices could be manufactured in sheet form, placed on the sloped surfaces and connected to the iron pickup wires. Steady developments are resulting in amorphous silicon solar cells of large area (10s square centimeters) for terrestrial applications.[37,38] Such developments in thin-film technology encourage the thought that devices could be vapor deposited directly on substrates of lunar-derived glass. Recent industrial studies[39,40] indicate that the production of complex integrated circuits in space will be reasonable in the next decade due to the needs of manufactures to produce ultra small and ultra clean components in vacuum on Earth.

Electric power is collected by the buried wires within the small areas indicated by the three small rectangular power plots just beyond the astronaut. Locally collected power energizes sets of solid state microwave transmitters indicated by the dots at the center of each power plot. A subset of transmission elements in front of each of the many billboards is coordinated in phase by local absolute clocks or fiber-optics networks laid underground between the many power plots. Each coordinated subset of transmission elements produces at least one power beam. Adjustments of beam pointing may be accommodated by providing extra solid state transmitter sets, small movements of complete transmitters, and special shaping of the lunar reflectors.

Scale of Example and Growth

Waldron and Criswell[21] estimated that a mass of machinery and associated habitats for twenty people of approximately 600 tons would be adequate to emplace 1 megawatt of power conversion and transmission capacity every 24 hours over the course of a year. The choice of emplacement rate is arbitrary. Much lower initial emplacement rates seem likely. There have been no studies to determine the smallest reasonable size. Initially approximately 50 kg of components for the power system are needed from Earth per megawatt of beamed power. This ratio should approach zero as experience in gained producing and building power systems on the moon from local materials. An SPS would require 10,000 kg of mass from Earth per megawatt. A unit of installed capacity returns its electrical energy of construction in a few 10s of days. This includes the energy (5MW-Hr) to uplift the 50 kg of terrestrial components but assumes the uplift energy for the 600 tons of equipment is amortized over many emplacement operations to a small value per unit of production.

In two years a four-shuttle fleet, assuming 5 flights per year per shuttle, could take to orbit about Earth the facilities, transfer rockets, and people adequate to establish the demonstration base. This assumes oxygen derived from lunar materials is provided in low earth orbit. Efficient acquisition of lunar oxygen would significantly reduce uplift requirements from Earth. Availability of heavy lift vehicles derived from the shuttle and Apollo systems could place the necessary components in to orbit about Earth in less than a year.

LPS provides several advantages in the demonstration phase. Power can be used locally to support production of oxygen for life-support or transportation, to gather materials for shielding, make structure, or to conduct more energy intensive research than would otherwise be done. Beaming of power back to Earth for transmission tests can be done early on. Early in the LPS demonstration the reflectors can be installed far apart, like components of an interferometer, to produce a very narrow central test beam. The initial test beam would contain only a fraction of the radiated power but will be useful in testing power transmission and switching and reception, control of stray power levels, support of space facilities, and for research. As more reflectors and associated power plots are added the beam will approach 100% confinement of transmitted power.

As with factories in space designed to make an SPS,[16] most of the machines used to build lunar power components from lunar materials can themselves be built of lunar materials. Assuming 90% of an installation unit can be made from lunar-derived materials, we approximately estimate the order of 60 people using somewhat less than 1,000 tons of manufacturing facilities, tools, and habitats could manufacture one installation unit in 30 days.[21] One installation unit is the complete collection of machines, much as depicted in Fig. 5, required to initially install 1 megawatt of power transmission capacity on the moon every twenty four hours. The installation rate of power can grow exponentially.

By the standards of terrestrial experience such fast rates of power production and growth of emplacement capacity seem fantastic. However, it results from the high specific power of the LPS components and because the moon provides the correct

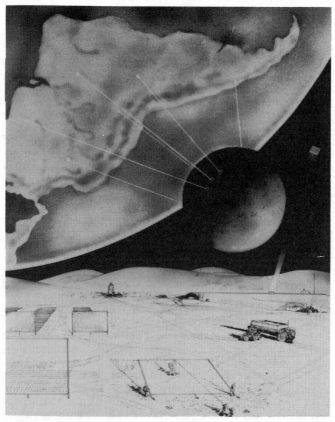

Fig. 5 Artist's montage of a demonstration lunar base (lower half) with a few reflectors in place. Power is shown being beamed from the moon to Earth (top half).

materials and environments for large area
photoconversion and microwave systems. Criswell[41]
summarized the relevant scales of similar terrestrial
power systems. All other systems are more massive,
require far more area and maintenance, consume fuels,
waste process heat, or yield ash. It is possible in
principal to design exponentiating manufacturing
systems to build solar power systems for use on
Earth. However, due to the massiveness and area of
such systems compared to LPS they would be far more
intrusive (large area), require a far larger scale of
manufacturing, require much more repair and
replacement activities, and have a much longer energy
payback time.[41] In addition, either energy storage
of indeterminate duration, backup power sources, or
worldwide transmission and retransmission of energy
would be required. It appears LPS could provide
dependable power that follows the immediate power
needs of any particular area. The power would enter
the biosphere only once, not produce pollutants, and
provide net energy to Earth.

Few if any factors (such as launch costs) would
limit the downward trending costs of producing LPS as
industrial production experience is acquired. The
mass, complexity, and costs of LPS components needed
from Earth can be steadily decreased. Industrial
learning-curve experience will be fully applicable to
LPS production and to the manufacturing of tools and
systems of production. Lunar products can decrease
the costs of logistics from Earth and create new
industrial operations on the moon. Key factors are
the development and application of new skills by
people on the moon and by the many more people back
on Earth who can work remotely on the moon.[42]

Immediate Activities

LPS research and development can start
immediately. Compared to SPS, LPS is not highly
integrated physically. Realistic demonstrations and
systems analyses can be readily conducted on Earth in
a progressive, certain manner. INDUSTRIAL engineers
can focus on minimizing cost, accelerating growth,
and using lunar materials to support major phases of
space logistics. LPS can support an open-ended, far
larger scientific program than did Apollo. The
Office of Technology Assessment (OTA)[43] maintains the
space program needs wider goals. A suitably designed
space-station program can create support facilities
in low earth orbit (LEO) and provide needed equipment
(transfer rockets, habitats, etc.) for the lunar
effort at reasonable additional expense compared to a

space station designed only for use in LEO.[44] LPS can
spawn new industries in cis-lunar space.

We believe all the research and development
necessary to understand, adequately cost out, and
perform a demonstration of the LUNAR POWER SYSTEM can
be done expeditiously on Earth prior to a return to
the moon. Modest systems studies of the concept,
which takes into account the unique resources and
environment of the moon, could be done immediately.
We expect they would justify definitive experiments
and analyses leading to a lunar power program.

References

[1]Boring, E., "Lunar Surface Solar Array Characteristics,"
Intersociety Energy Conversion Engineering Conference Record,
1968, p.571.

[2]Boretz, J. E. and Miller, J.L., "Large Solar Arrays for
Multimission Lunar Surface Exploration," Intersociety Energy
Conversion Engineering Conference Record, 1968, p.581.

[3]Kendrick, J. B., "Energy Conservation Methods Applied to
Lunar Surface Operations," Intersociety Energy Conversion
Engineering Conference Record, 1968, p.641.

[4]Pringle, J. K., "Lunar Gravity as a Power Source,"
Proceedings of the 6th Annual Meeting of the Working Group on
Extraterrestrial Resources, 1967, p. 117.

[5]Woodcock, G. R. and Brewer, J. B., "Reactor Power for Lunar
Exploration," Proceedings of the 6th Annual Meeting of the
Working Group on Extraterrestrial Resources, 1968, p.139.

[6]National Aeronautics and Space Administration. Mission Modes
and Systems Analysis for Lunar Exploration (MIMOSA) [three
technical volumes, three analysis volumes, summary report
(pp. 100)], LMSC-A847940, Contract No. NAS8 - 20262 (MSFC),
April 1967.

[7]Hollax, E., "Power Systems for Permanent Lunar Stations,"
Proceedings of the XXth International Astronautics Congress,
Pergamon, New York, 1972, p.409.

[8]Miller, P. R., "High Power, Long Life Electrical Generation
Systems for Lunar Base Missions," Proceedings of the 7th
Annual Meeting of Working Group on Extraterrestrial Resources,
1969, p.11.

[9]Criswell, D. R. and Waldron, R. D., "Lunar-Based Power
Systems" [a shorter version of the present paper],
Proceedings of the International Astronautics Federation,
Innsbruck, Austria, Oct. 9, 1986, p. 8.

[10]Waldron, R. D., "Useful Properties and Novel Methods for
Transportation of Lunar Soil in or to Earth Orbit," Space

Manufacturing 4, <u>Proceeding of the 6th Princeton Conference on Space Manufacturing</u>, American Astronautical Society, 1981, p.331.

[11]Waldron, R. D., "Total Separation and Refinement of Lunar Soils by the HF Acid Leach Process," Space Manufacturing 5, <u>Proceedings of the 7th Princeton Conference on Space Industrialization</u>, American Institute of Aeronautics and Astronautics, NY, 1985, p.132.

[12]Glaser, P., "Solar Power from Satellites," <u>Physics Today</u>, Vol. 30, No. 2, Feb. 1977, pp. 30-38, and letters to the editor, Vol. 30, No. 7, July 1977, pp. 9-13, 66-69.

[13]Kraft, C. C., "The Solar Power Satellite Concept - The Past Decade and the Next Decade," AIAA 79-0534, 15th Annual Meeting, Washington, D.C. Feb. 6-8, 1979, p. 17.

[14]O'Neill, G. K., "Space Colonies and Energy Supply to the Earth," <u>Science</u>, Vol. 190, No. 4218, Dec. 5,1975, pp.943-947.

[15]Lunar and Planetary Institute. "Extraterrestrial Materials Processing and Construction" (D.R.Criswell - Principal Investigator, R.D. Waldron - co-Principal Investigator), NASA Contract NSR 09-051-001 Mod. No. 24, (available on microfiche) National Technical Information Service, 1980, p.500; 1978 p. 450.

[16]Miller, R., "Extraterrestrial Materials Processing and Construction of Large Space Structures," NASA Contract NAS8-32935, NASA CR-161293, Space Systems Laboratory, Dept. Aeronautics and Astronautics, Massachusetts Institute of Technology, Cambridge, Mass.   Three Volumes, Dept. of Aerospace Sciences, Massachusetts Institute of Technology, Boston, 1979.

[17]Bock, E., "Lunar Resources Utilization for Space Construction," 3 volumes, General Dynamics-Convair, San Diego, NASA Contract NAS 9-15560, 1979.

[18]Waldron, R. D., Erstfeld, T. E., and Criswell, D. R., "The Role of Chemical Engineering in Space Manufacturing," <u>Chemical Engineering</u>, Feb. 12, 1979, pp. 81-93.

[19]Criswell, D. R., Waldron, R. D., and Aldrin, Buzz, "Lunar Power System," presented at conference on "Space Development: The Next Frontier," National Center for Policy Analysis, University of Dallas, June 1985.

[20]Mueller, G., "The 21st Century in Space," <u>Aerospace America</u>, Jan. 1984, pp. 84-88.

[21]Waldron, R. D. and Criswell, D. R., "Concept of the Lunar Power System," <u>Space Solar Power Reviews</u>, Vol 5, No. 1, Pergamon, New York, 1985, pp. 53-75.

[22]Browning, D. L., "Large Space Structures, <u>Space Industrialization</u>, " edited by B.J. O'Leary, CRC Press, Boca Raton, FL, 1982, Vol. II, pp.55-123.

[23]National Aeronautics and Space Administration, Space Resources and Space Settlements, SP-428, U.S. Government Printing Office, Washington, D.C., 1979, p.288.

[24]Office of Technology Assessment, Congress of the United States. Solar Power Satellites, Library of Congress Number 81-600129, U.S. Government Printing Office, Washington, D.C., 1981, p. 298.

[25]Fink, D., Biersack, J. P., and Stadele, M., "On the Performance and Lifetime of Solar Mirror Foils in Space," Space Solar Power Reviews, Vol. 5., No. 1, Pergamon, New York, 1985, pp. 91-100.

[26]Stone, J. M., Radiation and Optics, McGraw-Hill, New York, (see Sect. 7-16) 1963, p. 544.

[27]Steinberg, B. D., Principles of Aperture and Array System Design (Including Random and Adaptive Arrays), Wiley, New York, 1976, p. 356.

[28]Brookner, E., "Phased Array Radar," Scientific American, Feb., 1985, pp. 94-102.

[29]Schell, A. C., Franchie, P. R., Goggins, W. B., and Forbes, G. R., "An Experimental Evaluation of Multiplate Antenna Properties," IEEE Transactions on Antennas and Propagation, Vol. AP-14, No. 5, Sept. 1966, pp. 543-549.

[30]Schell, A. C., "The Multiplate Antenna," IEEE Transactions on Antennas and Propagation, Vol. AP-14, No. 5, Sept. 1966, pp.550-560.

[31]Nakamura, Y., "A1 Moonquakes: Source Distribution and Mechanism," Proceedings of the Lunar Planetary Science Conference 9, Vol. 3, Geochimica Et Cosmochimica Acta, Pergamon, New York, 1978, pp.3589-3607.

[32]Frey, A. H., "From the Laboratory to the Courtroom: Science, Scientists, and the Regulatory Process," edited by Steneck, N. H., Risk/Benefit Analysis-The Microwave Case, San Francisco Press, Inc., CA, 1982, pp. 197-228.

[33]Kerwin, E. M. and Arndt, G. D., "Cost Comparisons of Solar Power Satellites versus Lunar Based Power Systems," Space Solar Power Review, Vol. 5., No. 1, 1985, pp. 39-52.

[34]Smil, V., "On Energy and Land," American Scientist, Vol. 72, Jan.-Feb. 1984, pp. 15-21.

[35]Pearce, G. W., Strangway, D. W., and Gose, W.A., "Magnetic Properties of Apollo Samples and Implications for Regolith Formation," Proceedings of the Lunar Conference 5, Geochimica et Cosmochimica Acta, Vol. 3, Pergammon, New York, 1974, pp. 2815-2826.

[36]Goldstein, J. I., Axon, H. J., and Yen, C. F. "Metallic Particles in the Apollo 14 Lunar Soils," Proceedings of the

Lunar Conference 3, vol. 1, Geochimica et Cosmochimica Acta, MIT Press, Boston, 1972, pp. 1037-1064.

[37]Blieden, H., "Continuous Production of Thin Film Amorphous Silicon Solar Cells," Proceedings of the SPIE International Society of Optical Engineering, Vol. 428, 1983, pp. 66-71.

[38]Hanak, J. J., Fulton, C., Myatt, A., Nath, P., and Woodyard, J. R., "Ultralight Amorphous Silicon Alloy Photovoltaic Modules for Space and Terrestrial Applications," American Chemical Society, 8412-0986-3/86/0869-329, Washington, D.C., 1986.

[39]Hallett, R. and Kugath, D. "Space Station Automation Study - Automation Requirements Derived from Space Manufacturing Concepts," NASA Contract NAS 5-25182, Vol. II - Tech. Report., General Electric Corp., Space Systems Division, Philadelphia PA, Nov. 27, 1984, p. 117.

[40]National Aeronautics and Space Administration, Advanced Automation for Space Missions, CP-2255, U.S. Government Printing Office, Washington, D.C., 1982, p. 393.

[41]Criswell, D. R., "Cis-lunar Industrialization and Higher Human Options," Space Solar Power Reviews, Vol.5, No. 1, 1985, pp. 5-38.

[42]California Space Institute, "Automation and Robotics for the National Space Program," by the NASA Space Station Automation and Robotics Panel, April 1985. Study mandated by the Congress of the United States and Conducted for NASA by the California Space Institute of the University of California, Steering Committee - Dr. R. A. Frosch, Dr. C. A. Rosen, Prof. J. R. Arnold, and Dr. D. R. Criswell.

[43]Office of Technology Assessment, Congress of the United States, Civilian Space Stations and the U.S. Future in Space, Library of Congress Number 84-601136, U.S. Government Printing Office, Washington, D.C., 1984, p. 234.

[44]Chapman, P.K., Csigi, K. I., Glaser, P. E., Thomas, R. G., Criswell, D. R., and Woodcock, G. A., "Assessment of Alternative Space Base Locations for Scientific Research and Development," report to National Science Foundation NSF-BOA-PRA-8400692, Little, Cambridge, MA, Dec. 30, 1983, p. 34.

Concrete for Lunar Base Construction

T. D. Lin*
*Construction Technology Laboratories,
The Portland Cement Association, Skokie, Illinois*

Abstract

Prior to the establishment of lunar scientific and
industrial projects envisioned by the National Aeronautics
and Space Administration, suitably shielded structures to
house facilities and personnel must be built on the Moon.
One potential material for the construction is concrete.
Concrete is a versatile building material, capable of
withstanding the effects of extreme temperatures, solar
wind, radiation, cosmic rays and micrometeorites. This
paper examines data published by NASA on lunar soils and
rocks for use as concrete aggregate and as possible raw
materials for producing cement and water, and investigates
the technical and economic feasibility of constructing
self-growth lunar bases. A hypothetical 3 story, 210 ft
(64m) diameter concrete cylindrical building was analyzed
for conditions of a vacuum environment, lunar gravity,
lunar temperature variations, and 1 atmosphere internal
pressure. The advantages of concrete lunar bases are
subsequently discussed.

Introduction

A project proposed within the National Aeronautics and
Space Administration to build permanent lunar bases after
the turn of the century has drawn tremendous interest from
scientific and engineering communities across the nation.
Lunar bases will enable mankind to extend civilization
from Earth to the Moon. Ample solar energy, low
gravitational force, and abundant minerals on the Moon
will provide excellent conditions for the development of

scientific and industrial space activities. Prior to the
establishment of these activities, suitably shielded lunar
structures must be built to house facilities and to
protect personnel from the effects of solar wind,
radiation, cosmic rays, and micrometeorites.

As a material capable of withstanding these effects,
concrete is proposed for construction and can be produced
largely from lunar materials. The discussion covers the
process of making cement from lunar material, concrete
mixing in a lunar environment, physical properties of
concrete at lunar surface temperatures and structural
designs suitable to lunar conditions.

Cementitious Materials

Cements used in construction on Earth are made
basically with raw materials such as limestone, clay and
iron ore. A burning process transforms the raw materials
into primarily calcium-silicate pebbles called clinker.
The clinker is then ground into micron sized particles,
known as cement. A wide variety of cements are used in
construction. The chemical compositions of these cements
can be quite diverse, but by far the greatest amount of
concrete used today is made with portland cements. A
typical portland cement[1] consists of about 65% calcium
oxide (CaO), 23% silica (SiO_2), 4% alumina (Al_2O_3) and
small percentages of other inorganic compounds. Among
these constituents, calcium oxide is the most important in
the cement manufacture.

Other types of cements produced with lower calcium
oxide content are available, e.g., slag cement, expansive
cement, alumina cement, and low calcium silicate cement.
High alumina cement has 36% calcium oxide while low
calcium silicate cement has only 30%.[2] Theoretically, a
cementitious material can be made with any proportion of
$CaO:SiO_2:Al_2O_3$ that falls within the calcium-silica-
alumina phase diagram.

Lunar Materials

Information of Apollo lunar soils and rocks indicates
that most lunar materials consist of sufficient amount of
silicate, alumina and calcium oxide for possible
production of cementitious material. Table 1 shows the
chemical compositions of some selected lunar samples.[3-5]
It appears that the content of calcium oxide in lunar
material is relatively low in comparison with other major

cement ingredients, our discussion, therefore, will center
around the calcium oxides.

A review of available literature on Apollo lunar
samples reveals that a typical mare soil has a CaO content
of nearly 12% by weight, highland soil 17%, basalt rocks
14% and anorthosite rocks, a calcium rich plagioclase in
the feldspar group almost 19%. A rock type with 19% CaO
content is a good candidate for lunar cement production.
Lunar sample 60015, a coherent, shock-melted anorthosite
rock, is an example. The rock is approximately 12x10x10 cm
and is largely coated with a vesicular glass up to 1 cm
thick as shown in Fig. 1. The glass layer has been
interpreted as a quenched liquid derived from melting the
surface layer of the anorthosite rock. Quenched glass

Table 1.   Chemical Compositions of Selected Lunar Samples

Element	Mare Soil (10002)	Highland Soil (67700)	Basalt Rock (60335)	Anorthosite Rock (60015)	Glass (60095)
	Major Elements, wt %				
SiO_2	42.16	44.77	46.00	44.00	44.87
Al_2O_3	13.60	28.48	24.90	36.00	25.48
CaO	11.94	16.87	14.30	19.00	14.52
FeO	15.34	4.17	4.70	0.35	5.75
MgO	7.76	4.92	8.10	0.30	8.11
TiO_2	7.75	0.44	0.61	0.02	0.51
Cr_2O_3	0.30	0.00	0.13	0.01	0.14
MnO	0.20	0.06	0.07	0.01	0.07
Na_2O	0.47	0.52	0.57	0.04	0.28

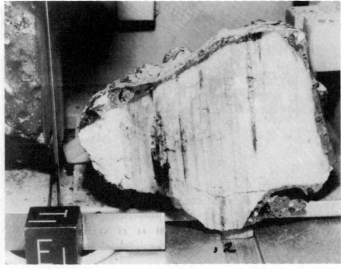

Fig. 1 Pristine cataclastic anorthosite glass-coated sample 60015

generally is amorphous substance, and represents a
potential cementitious material if ground to fine particle
size.

Glasses are common in lunar soils. Table 2 shows the
averaged glass contents in lunar samples brought back by
the Apollo missions. Note that samples taken from Shorty
Crater rims have glass content as high as 92.3%. The
chemical compositions of glass could possibly be similar
to glass sample No. 60095 shown in Table 1.

Process Methods

Figure 2 shows condensation temperatures of various
elements in basalt rocks.[6] Interestingly, all
cementitious elements including Ca, Al, Si, Mg and Fe have
condensation temperatures above 1400 K, at least 200
degrees higher than those of noncementitious elements.

Table 2.   Average Glass Content of Lunar
Samples

Mission	Average Glass Content, %
Apollo 11	6.6
Apollo 12	18.0
Apollo 14	12.2
Apollo 15	29.4
Apollo 16	10.6
Apollo 17	31.1

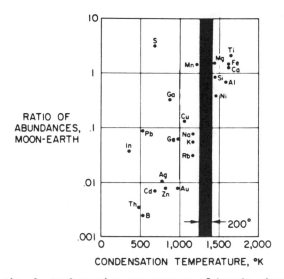

Fig. 2 Condensation temperature of basalt minerals

This unique physical property may enable us to separate cementitious elements from noncementitious ones in the process of cement manufacture.

However, a temperature of 3000 K or higher will be needed for the elemental evaporation in the process. This may cause some degree of difficulty in finding suitable material for the containment use.

Figure 3 shows residual fractions of multicomponent melt consisting FeO, MgO, SiO_2, CaO and Al_2O_3 of solar elemental abundances during the evaporation process at temperatures up to 2000°C.[7] The complete evaporation of FeO in Stage I may be utilized for metallic iron beneficiation. The remaining residues in Stage IV have high concentrations of CaO and Al_2O_3, and a small amount of SiO_2. Calculated $CaO:Al_2O_3:SiO_2$ proportions of the combined residues at lines A and B of Fig. 3 fall in the stoichiometric range of commercial high alumina cement.[8]

Aggregates

Aggregates generally occupy about 75% by weight of the concrete and greatly influence concrete properties. Aggregates, according to American Standard for Testing and Materials (ASTM), are not generally classified by

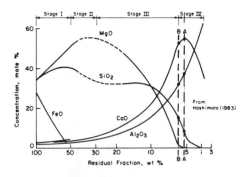

CHEMICAL COMPOSITION, WEIGHT PERCENT			
Compound	Solar Elemental Abundances		Alumina Cement
	@ A-A	@ B-B	
CaO	42.7	40.0	36–42
Al_2O_3	52.3	48.8	36–51
SiO_2	5.0	11.0	4–9

Fig. 3 Change of composition of residual molten oxide material as a charge is vaporized away into a vacuum at 2073°K.

mineralogy. The simplest and most useful classification
is based on specific gravity. Lunar soils and rocks all
have specific gravities higher than 2.6 and are believed
to be quality material for aggregate use. To produce
concrete on the Moon, lunar rocks can be crushed to
suitable coarse aggregates size, and the abundant
lunar soils can be sieved to good gradation of fine
aggregates.

Glassy soils used as aggregate may develop
alkali-aggregate reaction that could cause the concrete to
crack or spall. The lunar materials have never been
exposed to oxygen and water since the creation, and the
chemical and physical stability of these materials when
exposed to water are not yet fully known. Research on
lunar soil for possible aggregate application is indeed
important.

Water Production

There have been studies on oxygen and metal production
using lunar materials. Proposed methods include an alkali
hydroxide based scheme,[9] hydrogen reduction of ilmenite,[10]
and others. The ilmenite reduction reaction yields iron
and water.

$$FeTiO_3 + H_2 \rightarrow TiO_2 + Fe + H_2O$$
$$\text{ilmenite} \qquad \text{iron} \quad \text{water}$$

Hydrogen is not readily available on the Moon and may
have to be imported from Earth. The terrestrial hydrogen
can be transported in the form of liquid hydrogen, (H_2),
or methane (CH_4) or ammonia (NH_3).[11] In considering the
need of carbon and nitrogen for life support and the
higher boiling points of methane -322F (-161C) and ammonia
-91F (-33C) than liquid hydrogen -486F (-252C), it may be
more advantageous to import methane and ammonia to the
Moon rather than liquid hydrogen.

Reinforced Concrete

Concrete is basically a mixture of two components:
aggregate and cement paste. The paste, comprised of
cement and water, binds the aggregate into a rocklike mass
as it hardens.

The flexural strength of plain concrete is generally
low, about one tenth of its compressive strength.
However, concrete reinforced with either steel or glass
fibers has increased flexural strength, strain energy

capacity and ductility.　Test data reveal that concrete reinforced with four percent, by weight, of steel fibers possesses nearly twice the flexural strength of plain concrete.[12]　These fibers act as crack arresters, that is, the fibers restrict the growth of microcracks in concrete.

Structural Design

Design of structures for a lunar base differs from design of structures on Earth:　First, there are no wind and earthquake loads on the Moon.　Second, the lower lunar gravity, 1/6 that of Earth, could permit an increase in the span length of a flexural member to 2.4 times, based on the flexural theory of a simply supported beam.

Figure 4 shows a proposed 3-story, 210-ft (64m) diameter concrete structure.　The structure is assumed to be subjected to 1 atmosphere pressure inside and vacuum outside.　The cylindrical tank at the center of the system serves as safety shelter for inhabitants in case the system suffers damage or air-leak.　It could also serve as a "storm cellar" during solar flares.　The roof will be covered with lunar soil of suitable thickness 6 to 18 ft (2 to 6m)[13] to protect personnel and facilities from harmful effects of cosmic radiation.

Plain concrete is normally weak in tension but strong in compression.　Conceivably, the major demand on the structural system will be the high tensile stresses in the wall resulting from the internal pressure.　To solve the

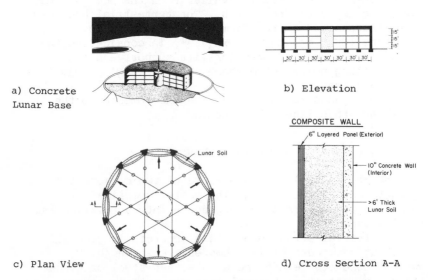

a) Concrete Lunar Base

b) Elevation

COMPOSITE WALL
6" Layered Panel (Exterior)
10" Concrete Wall (Interior)
>6' Thick Lunar Soil

Lunar Soil

c) Plan View

d) Cross Section A-A

Fig. 4　Proposed three-level concrete lunar base

problem, use of circular panels facing outward and supported by columns, will change the tension into compression (Fig. 4c). Steel tendons can then be used to secure the columns in position. For effective use, these tendons could be placed around the cylindrical tank, and stressed to provide hoop forces on the tank, and then anchored to columns at the opposite side. The 6-in. (15cm) thick layered panels at external faces of the wall (Fig 4d) are nonload bearing units. They are used to contain the soil in between the internal and external panels. A layered system that is free to expand can minimize the thermal stresses due to extreme temperature changes on the Moon.

The proposed concrete lunar base structure has 90,000 ft^2 (8,360 m^2) of usable area. Approximately 250 tons of steel and 12,200 tons of concrete would be needed for the construction. That much concrete requires approximately 1,500 tons of cement and 490 tons of water. All these materials can be obtained on the Moon except hydrogen. The needed hydrogen from Earth is about 55 tons.

Advantages of Concrete Lunar Base

Concrete lunar bases offer the following advantages:
1. Economic. Table 3 compares energy requirements for 4 major construction materials:[1] To produce 1 cubic meter of aluminum alloy requires 360 GJ energy; 1 m^3 of mild steel requires 300 GJ; 1 m^3 of glass requires 50 GJ; and 1 m^3 of concrete requires 4 GJ. The energy ratio between aluminum alloy and concrete is 90:1. Less energy requirement in the production can be translated into lower cost.
2. Compartmentalization. One major advantage of concrete is that it can be cast into any monolithic configuration. A lunar structure could

Table 3. Typical Properties of Construction Materials

Materials	α ($10^{-6}/°C$)	k (W/m K)	E_{rq}' (GJ/m^3)
Aluminum Alloy	23	125	360
Mild Steel	12	50	300
Glass	6	3	50
Concrete	10	3	3.4
			(4.0)*

*H$_2$O is made from ilmenite.

be compartmentalized to prevent catastrophic destruction in case of any local damage.

3. Concrete Strength. Lunar surface temperature may vary from -250F (-150C) in the dark to +250F (120C) facing the sun. Figure 5 shows that strength of heated concrete is practically unaffected at 250F (120C).[14] Concrete maintained at 75 percent relative humidity and temperature of -150 F(-100C) increases in strength two and one-half times that of room temperature, and two times at -250F (-156C). Concrete that has 0% relative humidity neither gains nor loses the strength in the course of cooling period, down to -250F (-150C).[15]

4. Heat Resistance. Concrete is thermally stable up to 1100F (600C). The low thermal conductivity as shown in Table 3 and high specific heat make concrete an excellent heat resistant construction material.

5. Radiation Shielding. Most radiation energy will be converted into heat energy in the course of attenuation in an exposed body. In general, a hardened concrete consists of, by weight, 95% aggregate and cement and 5% water. Both aggregate and cement are nonmetallic and inorganic, and are

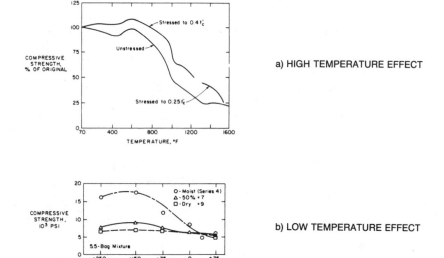

a) HIGH TEMPERATURE EFFECT

b) LOW TEMPERATURE EFFECT

Fig. 5 Effect of temperature on compressive strength of siliceous aggregate concrete

excellent materials for absorbing γ-ray energy.
Water is the best substance for absorbing neutron
energy.[16]

6. Abrasion Resistance. Micrometeorites can strike
the Moon with relative speeds up to 25 miles/s
(40 Km/s). These microparticles may abrade the
surface of the lunar structures. Concrete
possesses high abrasion resistance which increases
proportionally with concrete strength.

7. Effect of Vacuum. Exposed to the lunar surface
the free moisture in concrete may eventually
evaporate, but the chemically bonded water will
not. Again, Fig. 5 shows that the loss of free
moisture, which generally takes place around 212F
(100C) has no adverse effect on concrete strength.

A pressurized concrete structure may not be completely
airtight. To solve this problem, an epoxy coating, or
another sealant that hardens without oxidation, can be
applied on the internal surface.

Conclusion

Reinforced concrete has many material and structural
merits for the proposed lunar base construction. The
attractiveness of this proposal lies in the fact that most
of the components of the concrete can be produced simply
from lunar materials. The scenario for the self-growth
lunar base is as follows.

1. Materials: Cement could be obtained by high
temperature processing of lunar rocks. Aggregates
would be obtained by physical processing of lunar
rocks and soils. Lunar ilmenite would be heated
with terrestrial hydrogen to form water, while the
residual iron could be processed into fibers, wire
and bars for reinforcement.

2. Concrete: Casting and curing chambers for concrete
could be developed from empty shuttle fuel tanks.
Temperature and humidity control, as well as
controlled drying and recycling of excess water,
are vital parameters because water is an expensive
commodity in space.

3. Construction: The evaluation of the most suitable
structural design must include considerations of
constructibility. It is possible to optimize the
concrete properties in order to achieve the most

suitable design, both for ease of construction and
for maintenance free service.

Conceivably, concrete lunar bases will be essential
facilities for the scientific and industrial developments
on the moon. Perhaps concrete will provide the ultimate
solution to the outer space colonization. The task of
constructing lunar bases is a great challenge to
scientists and engineers in this fascinating space age.

References

[1]Mindess, S. and Young, J. F., Concrete, 1981, Prentice-Hall,
New York.

[2]Takashima, S. and Amano, F. "Some Studies on Lower Calcium
Silicates in Portland Cement" Reviews of 74th General Meeting of
Cement Association of Japan, 1960.

[3]Morris, R. V., "Handbook of Lunar Soils," 1983 Part 1:Apollo
11-15, Part 2:Apollo 16-17 Planetary Materials Branch of NASA,
Publication 67.

[4]Ryder, G. and Norman, M. D., "Catalog of Apollo 16 Rocks," 1980,
Parts 1, 2, and 3. Lunar Curatorial Branch of NASA, Publication
52.

[5]Fruland, R. M., "Introduction to the Core Samples from the
Apollo 16 Landing Site," 1981 Lunar Curatorial Branch of NASA,
Publication Nos. 58 and 61.

[6]Wood, J. A., "The Moon" Scientific American September, 1975.

[7]Hashimoto, A., "Evaporation Metamorphism in the Early Solar
Nebula - Evaporation Experiments on the Melt
$FeO-MgO-SiO_2-CaO-Al_2O_3$ and Chemical Fractionations of
Primitive Materials," Geochemical Journal Vol. 17, pp. 111, 1983.

[8]Lea, F. M., Book, "The Chemistry of Cement and Concrete,"
pp 497, Published by Chemical Publishing Company, Inc., 1971

[9]Culter, A. H., "An Alkali Hydroxide Based Scheme for Lunar
Oxygen Production," Lunar Bases Symposium, of NASA, 1984.

[10]Agosto, W. N., "Electrostatic Concentration of Lunar Soil
Elmenite on Vacuum Ambient," Lunar Bases Symposium, NASA, 1984.

[11]Friedlander, H. N., "An Analysis of Alternate Hydrogen Sources
for Lunar Manufacture" Lunar Bases Symposium, NASA, 1984.

[12]Amir N. Hanna, "Steel Fiber Reinforced Concrete Properties and
Resurfacing Applications," Research and Development RD049.01P,
1977, Construction Technology Laboratories of the Portland Cement
Association.

[13]Summer Workshop on Near-Earth Resources, NASA Conference Publication 2031, August 1977.

[14]Abrams, M. S., "Compressive Strength of Concrete at Temperatures to 1600F," Research and Development Publication RD016.01T, 1973, the Portland Cement Association.

[15]Monfore, G. E. and Lentz, A. E., "Physical Properties of Concrete at very Low Temperatures," Research Department Bulletin 145, 1962, the Portland Cement Association.

[16]The Effects of Atomic Weapons, Book, published by the Combat Forces Press, Washington, D. C., 1950.

Physical Properties of Concrete Made with Apollo 16 Lunar Soil Sample

T. D. Lin,* H. Love,† and D. Stark‡

Construction Technology Laboratories, Inc., Skokie, Illinois

Abstract

On March 6, 1986, the National Aeronautics and Space Administration awarded Construction Technology Laboratories with 40 grams of lunar soil. These samples were to undergo special testings to evaluate the feasibility of using lunar soil as an ingredient for concrete on a proposed lunar-based construction program. Two types of investigations were performed. One was to evaluate the performance of lunar soil as an aggregate for concrete and the other was to determine the physical properties of concrete made from this soil.

Prior to use, portions of the lunar sample were microscopically examined to determine their petrologic characteristics. The samples were also analyzed with the scanning electron microscope to determine their morphology and elemental composition. The lunar soil appeared to have good characteristics as a fine aggregate. High alumina cement and distilled water were then mixed with the lunar sample to make a 25 mm cube, a 13 mm cube and three 3x15x80 mm slab specimens. Tests were performed on these specimens to determine the compressive strength, the modulus of rupture, the modulus of elasticity and the thermal expansion coefficient. The results are as follows:

Compressive strength: 75.7 MPa;
modulus of rupture: 8.3 MPa;
modulus of elasticity: 21,400 MPa
thermal expansion coefficient: 5.4×10^6 cm/cm/c

*Principal Research Engineer.
†Petrographer.
‡Manager.

It was concluded that lunar material can be used as an aggregate for concrete construction on the moon.

Materials

Materials used in the fabrication of the test specimens are described below:

Fine Aggregate: Lunar soil sample, graded Ottawa sand and glassy rhyolite sand were used.

Analysis of Lunar Soil Sample

The 40 gram, graded lunar soil sample, as shown in Fig. 1 had a similar particle size distribution as the graded Ottawa sand used in the preliminary test program.[1] Table 1 shows the particle size distribution.

Fig. 1 40 gram graded lunar soil

Table 1. Particle Size Distribution of
the 40 Gram Lunar Soil Sample

Sieve	Retained Wt., grams
No. 100 (150 μm)	8.0
No. 50 (300 μm)	18.0
No. 40 (425 μm)	12.4
No. 30 (600 μm)	1.6
No. 16 (1.18 mm)	0.0
TOTAL =	40.0

The particles were subangular to subrounded in shape, somewhat friable, and appeared to have a relatively high porosity. It was estimated that 40 to 50% of the sample consisted of white grains, while 50 to 60% consisted of grayish black, glassy-appearing grains. The white particles were more friable than the grayish black particles.

The dark gray to black particles were crystalline, somewhat friable, relatively fine-grained, and subrounded in shape. In polarized light, they were found to consist of about 60 to 75% plagioclase feldspar in the compositional range of bytownite bordering on anorthite. Twenty-five to 40% of the particles consisted of minerals in the pyroxene family. Traces of opaque minerals, possibly iron-rich, were also present. Optical characteristics of the feldspar were similar to those in the white particles except that individual crystals were of much smaller size.

Glassy clear particles were present in trace amounts. These particles consisted of individual crystals of plagioclase feldspar in the compositional range of bytownite to anorthite.

A portion of the lunar sample was analyzed under an ISI-SX40 Scanning Electron Microscope (SEM) equipped with a Tracer Northern Energy Dispensive X-Ray (EDX) spectrometer system to determine the morphology and elemental composition.

Electron Micro Probe Analysis (EMPA) Spectra indicated that the major elements in the sample were Ca, Al, and Si. Minor elements, listed in the order of abundance, were: Mg, Fe, (Ka-Kß), Ti, Na and K. Many of the lunar particles examined had two good cleavage directions oriented at approximately 90° to one another. This indicated a crystalline structure.

In summary, the analyzed lunar material consisted primarily of particles of anorthite, a triclinic mineral ($Ca\ Al_2\ Si_2\ O_8$) with two directions of cleavage 86° to one another. The measured volume, of the 40 gram lunar soil was 29.2 cm^3, the bulk unit weight was 1.37 gram/cm^3, the void percentage was 45% and the specific gravity was 2.5.

Lunar Soil Simulant

Following procedures cited in ASTM, Designation: C 136[2], natural Ottawa sand and crushed glassy rhyolite were sieved and mixed to produce fine aggregate with a particle size distribution as shown in Table 1. The glassy rhyolite was an acid volcanic material that consisted of low calcium, high sodium and high potassium. The material was a high absorptive, good quality aggregate.

The bulk unit weights of the graded Ottawa sand and the lightweight sand were 1.6 and 1.2 gram/cm^3 respectively, while their specific gravities were 2.67 and 2.35, respectively.

Cement: Calcium aluminate cement was used for the mortar mixes. The specific gravity of the calcium aluminate cement was 3.08.

Water: Distilled water was used for all mixes.

Fabrication and Curing of Specimens

The following describes the fabrication and curing of the specimens:

Molds: Three cube molds with 13, 25, and 50 mm sides were used for casting the cube specimens, while a plastic rectangular mold was used to cast the slab specimens.

Mix Proportions

ASTM C 109[3] recommends a water-cement ratio of 0.485 and a sand-cement ratio of 2.75:1 for portland cement used in mortar mixes. No ratios are specified for calcium-aluminate cement. However, to maintain the recommended 0.485 water-cement ratio for calcium-aluminate cement and to achieve a suitable workability for the cement, the recommended 2.75:1 sand to cement proportion must be altered.

The sand:cement ratio was selected to produce a flow of 110+5% as described in Section 8.3 of ASTM Designation C 109. Six trial mixes using high absorptive glassy rhyolite of the same particle size distribution as that of the lunar soil sample were prepared and a 1.75:1.00:0.485 (sand:cement:water) was selected for the test program.

Fabrication: The mixing, casting and curing procedures are presented in Reference 4.

Determination of Specimen Age for Tests

During the hydration process, calcium aluminate cement behaves in a substantially different manner than portland cement. The process of hydration is described in Reference 5. In general, the strength decreases as the duration of water exposure increases. For this reason, twenty-four 1-in. cubes made with graded Ottawa sand were tested to determine their compressive strengths in

relation to specimen ages. The test results show that
cubes between 3 and 4 days old would give the optimum
strength. The 3 1/2 day age was thus, selected for
testing the specimens.

Test Specimens

A 25 mm cube, a 13 mm cube, and three 3x15x80 mm slab
specimens were fabricated from the lunar soil and calcium
aluminate cement mortar. In addition to these specimens,
18 companion specimens (6 of each kind) were fabricated
using the simulated lunar material made with glassy
rhyolite and calcium aluminate cement.

Compression Tests

Figure 2 shows the loading portion of the compression
machine with a 25 mm cube specimen positioned below a
76x76x38 mm steel plate. The steel plate is secured to
the center of the upper bearing block of the closed-loop
system.

During the compression test, an input signal is fed
into the function generator. The feedback signal is
compared with this input signal (in the function
generator) by the command module. The difference between
the two signals generates a variational signal and the
servo-valve controls the movement of the platen such that
this variational signal is minimized.

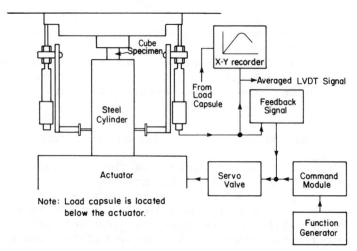

Fig. 2 Closed-loop system of compression test machine

Prior to the load application, the spherically seated upper bearing block was checked for freedom to tilt. A slight load was applied to ensure full contacts between the steel plate, the bearing block and the test cube. Subsequently the bearing block was fixed and the load was applied slowly to follow the programmed deflection control curve. The rate of loading was controlled to produce a cube contraction of approximately 0.5µm per second. In general, the test load reached a maximum in 20 minutes and then slowly decreased. Each compression test was completed in about 45 minutes.

Figure 3 shows the stress-strain curve for a cube made with the lunar soil sample and for a representative companion cube made with highly absorptive rhyolite obtained during the compression tests. The average compressive strength for the lunar cube was 75.7 MPa while the average strength of the companion cubes was 54.9 MPa.

Static moduli of elasticity were estimated as 12,400 MPa for the lunar cube and 7,600 MPa psi for companion cubes. These values were calculated by taking the steepest slope of the rising portion of the stress-strain curve shown in Fig. 3.

The longitudinal and lateral deformations of test cubes under the compressive load were manually recorded. From the data, Poisson's ratios were calculated. Calculations were made up to the peak load of each test. No lateral deformations were observed in the lunar cube for stress below 45.5 MPa The companion cube did not show any appreciable lateral deformation until a stress of 33.1

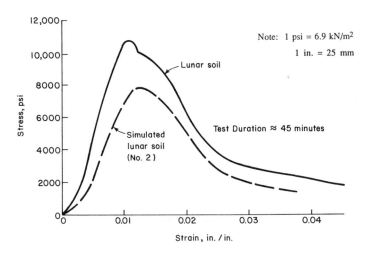

Fig. 3 Stress versus strain curves for 1 in. cubes

MPa was reached.   Figure 4 shows the calculated Poisson's
ratios for both types of cubes in graphical form.   At peak
load, Poisson's ratio for the lunar soil was 0.39, while
for the rhyolite the ratio was 0.27.

Modulus of Rupture

One minislab made with the lunar sample and three
slabs made with rhyolite material were tested using an
Instron testing machine to determine flexural strengths.
Figure 5 shows a diagrammatic view of the flexural test
setup.   Each slab specimen was subjected to two concentric
loads at the third points of the span.   The section
modulus, Z, of the slab was calculated from the geometry
property of the slab.   The maximum bending moment was
computed using the equation, M = PL/6 where P is the the
maximum load at rupture and L is the span length between
supports.   In this case, L was 78 mm.   The modulus of
rupture was obtained by dividing M by Z.   The average
modulus of rupture for slabs made with rhyolite material
was 8.6 MPa while that of the slab made with the lunar
material was 8.3 MPa.

Dynamic Modulus of Elasticity

The dynamic modulus of elasticity of the hardened
mortar was determined with a sonometer.   Measurements of
the fundamental flexural resonance frequencies of the
thin-slab specimen were made in a glove box to ensure a

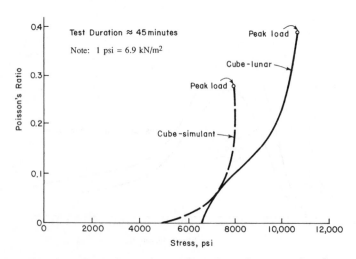

Fig. 4 Poisson's ratio as function of stress level

CO_2-free atmosphere, a controlled relative humidity and a temperature of 24°C.

The slab was supported near a node and vibrated by a wire clip driven by a crystal phonograph cartridge. The other end of the slab was supported at a node by a soft foam plastic pad. The vibration was detected by a wire probe cemented to a Sonotone 3P-1S ceramic cartridge. The cartridge was mounted in a modified Rek-o-Cut S-320 tone arm. The stylus pressure was adjusted to the minimum required to maintain contact between the specimen and probe. The arrangement is shown in Figure 6. The modulus of elasticity for the slab made with lunar soil was

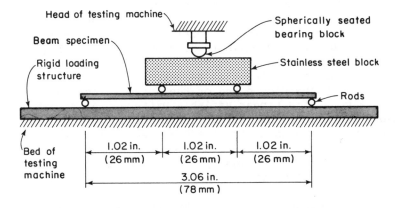

Fig. 5 Diagrammatic view of flexural test machine

Fig. 6 Measurement of fundamental flexural resonance frequencies

21,400 MPa while the average of modulus of elasticity for the slabs made with rhyolite material was 19,400 MPa. It is not uncommon that the modulus of elasticity obtained by the resonant method is higher than that obtained by the static method. This is due to the fact that the specimen under the sustained load experiences creep which adds to the total elastic-plastic deformation.

Thermal Expansion Coefficients

Figure 7 shows the test setup of a commercially manufactured dilatometer and a heating/cooling unit for measuring thermal expansion of slab specimens.

Thermally induced deformations of the specimens were transferred through a fused silica rod attached to an Invar bar that rode on ball-bearing pulleys. The end of the Invar bar rested against the plunger of a dial gage with a calibrated sensitivity of 0.002 mm. Pressure from the light spring of the dial gage kept the specimen in contact with the closed end of the fused silica tube.

The core of a Linear Variable Differential Transformer (LVDT) was mounted axially on the outer end of the dial gage plunger. Housing for the dial gage and LVDT were mounted on an adjustable assembly fixed to the slate base. In addition to lightly loading the specimen, the dial gage was also used to calibrate the response of the LVDT.

Length changes measured by the LVDT included a component caused by thermal expansion of the fused

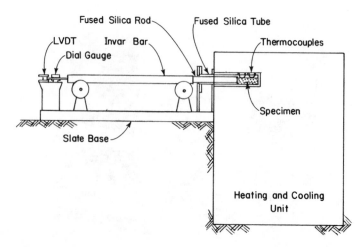

Fig. 7   Test setup for coefficient of thermal expansion

silicate tube and rod used to contain the specimen. Expansion of the fused silica components was approximately 2-7% of the length changes of the specimen. Data was corrected for this effect using results of calibration tests on fused silica specimens.[6]

Thermal expansion of the specimen measured by the LVDT, and specimen temperature measured by thermocouples, were continuously recorded on an X-Y plotter. Each slab specimen was subjected to heating and cooling for two complete cycles to determine its response to temperature changes, that varied from -38 to 177°C.

Measured data was used in calculating coefficients of thermal expansion at elevated temperatures as well as low temperatures. A five percent correction was made for each computed value to justify the thermal expansion of the fused silica rod. The average thermal expansion coefficient for slabs made with lunar sample was 5.4×10^{-6} cm/cm/c.

Examination of Lunar Cube After Test

Following the determination of the compressive strength, the tested 25 mm cube was examined microscopically to characterize the microstructure of the mortar. A stereomicroscope with a magnification range from 7 to 35 was used for this work.

The examination revealed that the calcium aluminate cement paste matrix was light brown in color and contained a uniform distribution of entrapped air voids. The estimated amount of air voids was 8% by volume of mortar. The paste matrix displayed a uniform and intimate bond with the aggregate particles. There were essentially no entrapped air voids in direct contact with the particles.

The examination also revealed that fractures that had developed during compression testing almost invariably passed through aggregate particles. In only a few cases did cracks pass around the particles at paste-aggregate interfaces. In addition, virtually, no aggregate sockets were visible on the fractured surfaces. This further attests to the tendency of compression cracks to pass through the aggregate particles.

Discussion

a) The mechanical properties of the lunar soil are governed by the distribution of grain sizes, the angularity of the grains and porosity. The lunar soil sample had particle sizes ranging from about 0.1 to 0.8 mm. The particles were subangular to subrounded

in shape. It should be noted that properties of fresh
concrete are more significantly effected by particle
shape and surface texture than the properties of
hardened concrete. Rough-textured, angular, elongated
particles require more water to produce a workable mix
than do smooth, rounded, compacted aggregates. This
suggests that the use of lunar soil as aggregate for
concrete mix will require more water than well-rounded
terrestrial sands.

b) The bond between cement paste and a given aggregate
    generally increases as particles change from smooth
    and rounded to rough and angular. This increase in
    bond is a consideration in selecting aggregates for
    concrete where flexural strength is important or where
    high-compressive strength is needed. The lunar soil
    sample revealed that the lunar soil has suitable
    physical properties for aggregate use.

c) For aggregate of the same grading, the water
    requirement for mixing tends to increase as aggregate
    void content increases. The lunar soil is dry and has
    45% of void content, about 5% higher than that of
    graded Ottawa sand (40%). Again, the lunar soil tends
    to require more water to produce workable concrete
    than the smooth, rounded Ottawa sand.

d) An examination of the lunar soil sample using the SEM
    revealed that lunar soil includes breccia lithick
    grains, mineral grains, glass fragments and lunar
    agglutinates. Some grains developed microfractures
    due to impact of micrometeorites. The immediate
    concern for use of lunar soil in making concrete was
    the effect of agglutination and microfractures on
    concrete strength.

    A careful examination of expanded shale light-
    weight aggregates commonly used in construction today
    shows that a great percentage of the material is of
    agglutinates formed during the sintering process at
    temperatures ranging from 980 to 1,200°C. It appears
    that the agglutinated joints often develop strength
    higher than the strength of the expanded shale itself.

    Test results of the cube specimen made with the
    lunar soil sample and examination of the tested cube
    provide convincing evidences that the agglutination
    and microfractures constitute no negative effect on
    the cube strength.

e) Mature lunar soil consists of about 110 ppm solar wind
    hydrogen and noble gasses such as argon and
    helium.[7] It is believed that the gas rich surface
    of soil particles has no drawback on the cube

strength. On the contrary, it may improve the quality of concrete.

The use of admixtures in mixing concrete will help explain this phenomenon. In the process of making concrete, air-entrained agents are often used to create air-bubbles and thus to increase the durability of concrete. This demonstrates that nitrogen, about 78% in air, has no negative effect on the strength of concrete. Argon and helium, like nitrogen, are inert gases and are believed to have no effect on cement hydration.

Conclusion

Forty grams of lunar soil provides a relatively small amount of aggregate for making mortar specimens. However, considering the value and scarcity of the lunar soil, 40 grams is a great quantity. A great deal of effort and time has been devoted to the development of the testing methods and apparatus. The obtained data provide scientific evidence that lunar soil can be used to produce quality concrete for construction on the moon or in space.

Acknowledgment

The authors would like to express their gratitude to NASA for awarding the 40 grams of graded lunar soil collected during the Apollo 16 mission.

References

[1]Lin, T. D., "Testing Concrete Specimens Made with Lunar Soil," Reported to National Aeronautics and Space Administration, November 1985.

[2]"ASTM Designation: C-136, Standard Method for Sieve Analysis of Fine and Coarse Aggregates," 1984 Annual Book of ASTM Standards, Philadelphia, PA.

[3]"ASTM Designation: C-109, Standard Method of Test for Compressive Strength of Hydraulic Cement Mortars," 1984 Annual Book of ASTM Standards, Philadelphia, PA.

[4]Lin. T. D., Love, H., and Stark, D., "Testing Concrete Specimens Made with Lunar Soil," Report to NASA, September 1986.

[5]Neville, A., "High Alumina Cement Concrete," p. 35 John Wiley & Sons, New York, 1975.

[6]Cruz, C. R. and Gillen, M., "Thermal expansion of Portland Cement Paste, Mortar, and Concrete at High Temperatures." PCA Publication, p. 2 RD074.)1T, 1980.

[7]Williams, R. J. and Jadwick, J. J., "Handbook of Lunar Soils," p. 93, NASA Publication 1057 Scientific and Technical Information Office, 1980.

Author Index

PROGRESS IN ASTRONAUTICS AND AERONAUTICS SERIES VOLUMES

VOLUME TITLE/EDITORS

*1. **Solid Propellant Rocket Research** (1960)
Martin Summerfield
Princeton University

*2. **Liquid Rockets and Propellants** (1960)
Loren E. Bollinger
The Ohio State University
Martin Goldsmith
The Rand Corporation
Alexis W. Lemmon Jr.
Battelle Memorial Institute

*3. **Energy Conversion for Space Power** (1961)
Nathan W. Snyder
Institute for Defense Analyses

*4. **Space Power Systems** (1961)
Nathan W. Snyder
Institute for Defense Analyses

*5. **Electrostatic Propulsion** (1961)
David B. Langmuir
Space Technology Laboratories, Inc.
Ernst Stuhlinger
NASA George C. Marshall Space Flight Center
J.M. Sellen Jr.
Space Technology Laboratories, Inc.

*6. **Detonation and Two-Phase Flow** (1962)
S.S. Penner
California Institute of Technology
F.A. Williams
Harvard University

*7. **Hypersonic Flow Research** (1962)
Frederick R. Riddell
AVCO Corporation

*8. **Guidance and Control** (1962)
Robert E. Roberson
Consultant
James S. Farrior
Lockheed Missiles and Space Company

*9. **Electric Propulsion Development** (1963)
Ernst Stuhlinger
NASA George C. Marshall Space Flight Center

*10. **Technology of Lunar Exploration** (1963)
Clifford I. Cummings and
Harold R. Lawrence
Jet Propulsion Laboratory

*11. **Power Systems for Space Flight** (1963)
Morris A. Zipkin and
Russell N. Edwards
General Electric Company

*12. **Ionization in High-Temperature Gases** (1963)
Kurt E. Shuler, Editor
National Bureau of Standards
John B. Fenn, Associate Editor
Princeton University

*13. **Guidance and Control—II** (1964)
Robert C. Langford
General Precision Inc.
Charles J. Mundo
Institute of Naval Studies

*14. **Celestial Mechanics and Astrodynamics** (1964)
Victor G. Szebehely
Yale University Observatory

*15. **Heterogeneous Combustion** (1964)
Hans G. Wolfhard
Institute for Defense Analyses
Irvin Glassman
Princeton University
Leon Green Jr.
Air Force Systems Command

*16. **Space Power Systems Engineering** (1966)
George C. Szego
Institute for Defense Analyses
J. Edward Taylor
TRW Inc.

*17. **Methods in Astrodynamics and Celestial Mechanics** (1966)
Raynor L. Duncombe
U.S. Naval Observatory
Victor G. Szebehely
Yale University Observatory

*18. **Thermophysics and Temperature Control of Spacecraft and Entry Vehicles** (1966)
Gerhard B. Heller
NASA George C. Marshall Space Flight Center

*19. **Communication Satellite Systems Technology** (1966)
Richard B. Marsten
Radio Corporation of America

*Out of print.

(Other Volumes are planned.)